# The Advent of PhyloCode

Biological nomenclature is an essential tool for storing and retrieving biological information. Yet traditional nomenclature poorly reflects evolutionary theory. Current biological nomenclature is one of the few fields promoting deliberately vague usage of technical terms. A new code based on evolutionary studies and phylogenetic results (the PhyloCode) will be a major milestone in biological nomenclature. Unfortunately, The PhyloCode and the companion volume are highly technical publications intended for practicing systematists. This book will reach a broader readership of those using nomenclature but who remain unaware of its theoretical foundations.

Key Features

- Responds to the biodiversity crisis and the recent implementation of the PhyloCode.
- Summarizes the spectacular progress of phylogenetics, which makes it both increasingly easy and crucially important to define precisely taxon names.
- Provides a 300-year historical perspective featuring high-profile characters, such as Linnaeus and Darwin.
- Summarizes for a broad readership a widely scattered, highly technical and underappreciated scientific literature.
- Documents the activities of the International Society for Phylogenetic Nomenclature, a scholarly society in which the author has played a prominent role.

# The Advent of PhyloCode

## The Continuing Evolution of Biological Nomenclature

Michel Laurin

**CRC Press**
Taylor & Francis Group
Boca Raton  London  New York

CRC Press is an imprint of the
Taylor & Francis Group, an **informa** business

Designed cover image: Shutterstock contributor, lessysebastian

First edition published 2024
by CRC Press
6000 Broken Sound Parkway NW, Suite 300, Boca Raton, FL 33487–2742

and by CRC Press
4 Park Square, Milton Park, Abingdon, Oxon, OX14 4RN

*CRC Press is an imprint of Taylor & Francis Group, LLC*

© 2024 Michel Laurin

ISBN: 978-0-367-55288-6 (hbk)
ISBN: 978-0-367-55210-7 (pbk)
ISBN: 978-1-003-09282-7 (ebk)

DOI: 10.1201/9781003092827

# Contents

# Preface

When I started this book, my original purpose was to write a semi-popular essay that would show how the history of biology explained the development of biological nomenclature through time. **Nomenclature** (terms in **bold type** outside citations are defined in the glossary, whereas bold font in citations is used for emphasis and is mine, unless otherwise noted) designates the specialized vocabulary used by scientists, as well as the rules that regulate how the technical terms are coined and applied. Biological nomenclature thus encompasses the terminology used to designate taxa (the core subject of this book) as well as terminology used to discuss other biological entities, such as structures (from the molecular to the morphological level), behaviors and so on. Each field has its own methods and traditions, and even though all of them are worthy of interest, this book only covers biological nomenclature of taxa: how biologists name taxonomic groups of organisms and the codes that regulate how to form and use these names. Along with taxonomy, it is part of systematics, the science that studies biodiversity and its classification. Thus, while focusing on the biological nomenclature of taxa, this book covers also key notions of evolutionary biology, phylogenetics, systematics, paleontology and philosophy of science to provide the context that determines, to an extent, the development of biological nomenclature.

It soon became obvious that the level of coverage of these various fields required to understand how and why they developed and how they interacted with each other necessitated use of a language much more technical than what is expected in a semi-popular book. Thus, this book is aimed at university students who have had at least one year of biology instruction, graduate students and practicing biologists who are interested in the fields previously listed.

The growth of biology since the beginning of the 20th century is such that nobody can claim to be an expert in all its branches, and I am certainly no exception. For this reason, and to avoid producing an encyclopedia rather than a book, the coverage of various fields is very uneven. While I tried to cover biological nomenclature fairly thoroughly, the coverage of phylogenetics, systematics and (when relevant) paleontology is much less thorough because only a basic knowledge of these fields is required to understand how they conditioned the development of biological nomenclature.

It is well-established that the explosion of our knowledge of biodiversity, linked to a large extent with the great explorers who revealed new life forms from other continents to European scientists, created a strong need for a better nomenclatural system, and that the development of biological nomenclature by Linnaeus in the mid-18th century must be viewed in this context. For instance, Henry Walter Bates (1825–1892) collected specimens of 14,712 species of animals, more than 8,000 of which were new to science (Egerton 2012: 35).

The early nomenclature developed in this context was necessarily based on phenotypic overall similarities between organisms, and it referred to **absolute ranks**. The latter are subjective categories of groups of biological organisms that form a series, from **subspecies** and **species** at the least inclusive end to **kingdoms** at the most inclusive end. Rank-based nomenclature is still used by most practicing systematists. However, progress in phylogenetics allowed for the development of a classification of organisms reflecting their evolutionary history. This shift occurred in the second half of the 20th century (but it was already considered the ultimate goal by Darwin and several of his predecessors) and continues to this day. This allowed the development of *phylogenetic nomenclature*, which is the core subject of this book. A good classification is important for any knowledge management, notably because it allows inductive generalizations (for instance, if two species of a genus share a property, a third species of the same genus in which this property has not been assessed may plausibly be inferred to possess it too), and a good nomenclature is required to name the classified entities (Minelli 2022: 203).

Evolutionary biology, genetics, folk taxonomy and philosophy of science are covered more superficially in this book. Thus, acceptance of the idea of evolution among systematists in the 19th and early 20th centuries was critical for the development of biological nomenclature (but with a lag of a half-century, as we will see in Chapter 4). Likewise, the acceptance of a branching (divergent) rather than linear or web-like evolutionary pattern for biological organisms, as well as a better understanding of the basic mechanisms of inheritance (which prevent inheritance of most characters acquired by organisms in their life to their progeny, with some exceptions, notably in bacteria), proved critical for nomenclature. However, it seems unnecessary to cover in depth the tremendous progress made by genetics since the early 20th century. In any case, a much larger book than this one would be required to consider genetics alone in a depth similar to what I have done here for biological nomenclature, and numerous excellent genetics textbooks are available.

Rather than conducting a detailed, technical review of all the aspects of phylogenetic nomenclature treated in the *PhyloCode*, this book focuses on explaining the basic concepts and why they make sense in the context of 21st-century systematics. It also embraces the history of the field, presents current debates and identifies the main challenges. Among the latter, the weight of tradition and practical difficulties in training entire generations of systematists to a new, revolutionary nomenclature (phylogenetic nomenclature) have been widely debated. Less frequently discussed among nomenclature specialists is the debate (see Chapter 6) among bacteriologists about the extent of horizontal gene transfer (HGT). This is relevant for biological nomenclature because a massive amount of HGT could result, according to some authors, in a Web of Life, rather than a Tree of Life, at least among bacteria and archaeans, whereas phylogenetic nomenclature rests on the assumption that biodiversity was generated mostly through divergent (rather than reticulate) evolution.

This book includes many quotes. Just like paleontologists reconstruct the evolutionary history of life through the fossil record, historians must rely on ancient written texts. Just like paleontology textbooks should provide some pictures of fossils, I found it important to document my interpretation of others' works through quotes. This is crucial for authors who worked and lived long ago because their culture was very different from ours, but this is also useful for recent authors. Throughout my career, I have been surprised by the variety of interpretations that others expressed of what I had written, not only in reviews of first drafts of my papers that might have been ambiguous, but also in final, published versions, which I had tried to clarify as best I could. Now, considering the difficulty that two contemporary specialists of the same field (say, vertebrate paleontology of Permo-Carboniferous tetrapods, which may seem an incredibly narrow field, with just a dozen established specialists worldwide) face to understand each other, imagine the difficulty of understanding fully an early 19th-century systematist or evolutionary biologist (like Lamarck). Thus, I have provided many quotes, and sometimes lengthy ones, in an attempt to distort as little as possible other authors' thoughts.[1] For texts that were originally in French, I provide both the unedited French original (mostly for the fairly numerous European readers who can read French) along with my own translation.[2] While this approach still leaves many possibilities of misinterpreting another author's works (and considerable uncertainty will always remain concerning some points), I hope to have provided the readers with sufficient evidence to make their own opinion about what earlier authors thought. However, ideas change, not only through centuries, but also within the career of a single author. Thus, when summarizing the statements of an author, even a contemporary one, I use the past tense and refer to a specific publication because the author may have changed his mind and may no longer view his earlier statement as correct. In some cases, this is documented in papers that contradict some statements found in earlier publications by the same author. I noticed this phenomenon years ago in my field (vertebrate paleontology) and I encountered such

cases in my few readings on folk taxonomy and document a few of them in Section 1.1.1.

I felt that documenting my interpretation of the history of biological nomenclature, systematics and evolutionary biology through quotes was particularly important because of the controversies that inevitably arise during scientific revolutions, and I argue in this book that biological nomenclature is in the midst of a scientific revolution associated with the advent of phylogenetic nomenclature. One of my goals in telling this fascinating history is to show that many authors, such as Charles Darwin, Alphonse de Candolle and Willy Hennig, expressed views that called for changes in biological nomenclature very much like those introduced by phylogenetic nomenclature. Thus, despite what some of the most conservative proponents of rank-based nomenclature claim, the current revolution brought about by phylogenetic nomenclature was long overdue and was (vaguely) predicted by the greatest minds since the mid-19th century.

## NOTES

1 To avoid the risk that taking quotes out of their context represents, I tried to read the entire paper or book from which they originate. For long books, this has not always been possible. But, in these few cases, I have read at least several pages around each quote.

2 The translations were typically produced with the help of DeepL Translate (www.deepl.com/translator) or Google Translate (https://translate.google.com/) to obtain an approximate first draft, but were always carefully edited before being inserted into the draft.

## REFERENCES

Egerton, F. N. 2012. History of ecological sciences, part 41: Victorian naturalists in Amazonia-Wallace, Bates, Spruce. Bulletin of the Ecological Society of America 93:35–60.

Minelli, A. 2022. The species before and after Linnaeus: Tension between disciplinary nomadism and conservative nomenclature; pp. 191–226 in J. Wilkins, I. Pavlinov, and F. Zachos (eds.), Species problems and beyond: Contemporary issues in philosophy and practice. CRC Press, Boca Raton, FL.

# *Abstract*

Everybody knows many names of animals and plants, but very few have the slightest idea about how scientists go about naming groups of biological organisms (called "**taxa**") and how these names are applied to these groups. Yet, with the explosion in our knowledge of biodiversity over the last three centuries (about 1.5 million species have been named, so about as many groups of species could potentially be named), an efficient nomenclatural system is of critical importance. A good nomenclatural practice is also of great societal importance given that it is required to fight the rapid erosion of biodiversity linked with the explosion of human populations in the last centuries. The system currently used by most practicing systematists (the biologists who describe and classify the biodiversity), known as **rank-based nomenclature**, harks back to the works of Linnaeus in the mid-18th century and the Strickland code, which was inaugurated in 1843. When Linnaeus proposed his system, most scientists were creationists and fixists, whereas modern biology has provided ample proof that Life has been evolving on Earth for more than 3 billion years. It is thus not unexpected that a growing number of scientists find rank-based nomenclature inadequate. This problem is linked to the fact that rank-based nomenclature aims at not delimiting taxa precisely, a goal that is arguably opposite to that of most other sciences, such as geology, chemistry and geography, and which hampers our ability to communicate efficiently about taxa. Consequently, a group of scientists has developed a new code of biological nomenclature based on new principles, called **phylogenetic nomenclature**. This new code, called the **PhyloCode**, took effect in April 2020.

This book seeks to describe the history of how groups of animals and plants have been named, starting with the prehistory but focusing on the last three centuries. More importantly, it describes the underlying events and issues that have shaped this history, such as developments in systematics, evolutionary biology and phylogenetics. It outlines the current controversies and challenges facing biological nomenclature in the 21st century.

# *Acknowledgements*

This book benefited from feedback from many colleagues who read earlier drafts. These include, first and foremost, Alessandro Minelli (formerly at University of Padova), who commented on the whole book and provided information on early zoological literature. Philippe Lherminier (Fontenil Castle, Orne, France) commented on the Preface, Introduction and Chapters 1 and 3. Chris Healey (Australian National University, Canberra) commented on the Preface and Introduction, in addition to providing some literature on folk taxonomies. Vivian de Buffrénil (National of Museum of Natural History, Paris) commented on the Preface and Introduction. Manon Thomazo (Paris Nanterre University) made typographic and stylistic corrections on an earlier draft of the Preface, the Introduction and of Chapters 1 and 3. Mary Pickard ("Polly") Winsor (University of Toronto) commented on Chapter 1. The first part of Chapter 1 (from prehistory to the Middle Ages) was also reviewed by Marcel Humar (Humboldt-Universität zu Berlin), and the section of Chapter 1 on folk taxonomy was improved by comments by David Ludwig (Wageningen University). Pascal Tassy (CR2P, Paris) commented on a draft on Aristotle that is extensively cited in Chapter 1 and kindly provided a good scan of one of Gaudry's first evolutionary trees. Chapter 2 was improved by comments from various colleagues: Aharon Oren (Hebrew University) provided highly relevant literature on prokaryote nomenclature and provided many constructive comments on this topic; and Philip D. Cantino (Ohio University, Athens, OH) commented on that chapter and provided information on rank-based codes (especially the *Botanical Code*) and on early botanical literature. Thomas Servais (Université de Lille) commented on much of Chapter 5, which also benefitted from comments by Robert W. Hook (U. of Texas at Austin) on the stratigraphic section and from comments from Cantaura La Cruz (École Pratique des Hautes Études, Paris) for the geopolitical section. Unfortunately, I lacked time to properly integrate many comments about the geopolitical section because this chapter was completed last, shortly before I submitted the draft; any resulting shortcomings are my sole responsibility. Arnaud Horellou (Muséum National d'histoire Naturelle, Paris) commented on the section on subspecies and lists of threatened species (in Chapter 6).

Many colleagues provided information that helped me write the book. I discussed folk taxonomic ranks with Paul Sillitoe (Durham University), who provided very useful explanations on this complex and controversial topic. Andrew Pawley (Australian National University) sent me a large package of printed papers on folk taxonomy and Andrea Guasparri (E Campus University, Como, Italy) provided additional literature on the same topic. John S. Wilkins (University of Sydney) answered various questions relating to his book on species, which I quoted abundantly in various chapters. Robert W. Hook (U. of Texas at Austin), Gerilyn S. Soreghan (University of Oklahoma, Norman) and Kathleen C. Benison (West Virginia University, Morgantown) provided information on various geological topics covered in the stratigraphic section of Chapter 5. Richard Pyle (Architect and Manager of *Zoobank* as well as Commissioner and Councilor for ICZN) provided statistics about *Zoobank* (covered in Chapter 2) and explained how it works. Alain Dubois kindly sent me several of his papers that were most relevant to this book. André Nel (MNHN, Paris) provided information about the nomenclature of beetles in Linnaeus' *Systema Naturae*. Scott Redhead (Ottawa Research and Development Centre) similarly provided information and papers about the nomenclatural debates on the mushroom *Coprinus* and checked the accuracy of my brief review of this topic. My Editor, C. R. Crumly, and his editorial assistants K. Roberts and L. Eberhart (all three at CRC Press) were very supportive throughout this project and always answered my queries promptly. I thank all these people warmly for their help. Even though the final draft was improved substantially through these many reviews, the text reflects first and foremost my view of this field, and the aforementioned colleagues did not necessarily endorse all the statements made in this book. As the common saying goes, the remaining errors are mine alone.

To finish on a personal note, I must thank my daughters, Pénélope and Thaïs, and my spouse, Alexandra, with whom I could not spend as much time as I would have liked during the nearly three years of intensive reading and writing that resulted in this book. This project would not have been possible without the moral support that my parents have always provided, especially when I decided to study biology and paleontology rather than a more lucrative and conventional field, such as medicine.

# *Author Biography*

**Michel Laurin** has been a tenured CNRS (Centre National de la Recherche Scientifique) research scientist since 1998. His specialty is the evolution of vertebrates from the Devonian to the Triassic. His current interests include dating the tree of life, as well as other problems such as the invasion of land by vertebrates, the appearance of the tympanum (ear drum) and the origin of extant amphibians. This work has introduced new methods for paleontology and evo-devo, some of which have been implemented in computer programs. He has supervised seven doctoral students. He has served on the governing council of various scientific societies, including the French Paleontological Association (APF), where he completed a term as president. More importantly, after adopting phylogenetic nomenclature early in his career, he served the ISPN (International Society for Phylogenetic Nomenclature), which develops and promotes a new code of biological nomenclature (the *PhyloCode*) as both secretary (several terms, including a current one) and president. He currently serves on several editorial boards of scientific journals, including *Zoologica Scripta*. In January 2011, he became chief editor of the *Comptes Rendus Palevol*, which is published by the French Academy of Sciences (in collaboration with the Muséum national d'Histoire Naturelle since 2020). He has recently received the Charles Bocquet prize, which is awarded every three years by the Zoological Society of France to honor a career dedicated to evolutionary zoology. Since 2014, he has been the director of the "Phylogeny and Diversification of Metazoans" team of the "Centre de recherche en paléontologie – Paris" (CR2P), located in the National Museum of Natural History and Sorbonne Université (Paris). This team includes 14 tenured or emeritus scientists, in addition to postdoctoral and doctoral students.

# Introduction

## Scope of This Book

All sciences require a precise terminology composed mostly of technical words unknown to the laymen and, for this reason, a glossary is included in this book! Nomenclature is especially important in biology because we have already named about 1.3 to 1.8 million biological **species** (Lücking 2019: 205), all-inclusive (archaea, bacteria, eukaryotes, extant and extinct). This means that we could also recognize nearly as many higher-level **taxa** (though many remain unnamed for now). Most of these taxa have no vernacular names, but even when such names exist, they do not often refer to taxa recognized by scientists, or a given vernacular name as used in different countries refers to several such taxa, which are not necessarily closely related to each other (Dubois et al. 2021: 82). Obviously, this prevents use of these names by scientists, who need a universal language.

The extant lineages that could represent species (according to some definitions of the term "species") that remain to be described are probably much more numerous than those that have already been named and described. Mora et al. (2011) suggested that there might be 8–9 million eukaryotic extant species, and Winston (2018: 1128) reported that about 18,000 new species are erected each year. The fact that so much biodiversity remains to be described reflects the fact that some highly diverse taxa, such as fungi, arthropods and nematodes, have been insufficiently studied. Thus, Hibbett et al. (2005: 661) reported that:

> Analyses of environmental sequences also suggest that there are major groups of homobasidiomycetes [a paraphyletic group of fungi] that have not yet been described (Vandenkoornhuyse et al. 2002; Schadt et al. 2003). This is not surprising, given that only about 70,000 species of fungi have been described (Kirk et al. 2001), whereas it is estimated that there are as many as 1.5 million extant species (Hawksworth 2001).

Similarly, Blaxter et al. (2005: 1935) stated: "the number of described species of nematodes is quoted as between 26 000 and 40 000, but the real total is estimated to be above one million." However, even among the taxa previously considered to be well-known, thorough molecular studies have revealed much cryptic (non-morphological) biodiversity (Minelli 2017: 658). And even in fairly well-known taxa where much of the biodiversity is not cryptic, most of it remains to be described. For instance, among the cichlid teleosts, which produced species flocks in the East Africa Great Lakes, about 1,700 extant species have been described, but there might be between 3,000 and 4,000 (Salzburger 2018: 706).

Systematists cannot cope with the flood of newly discovered life forms because describing properly new taxa takes time. This results in an increasing use of "grey nomenclature" of basically undescribed taxa (Minelli 2017), many of which are identified only though DNA barcoding, the so-called MOTUs (Molecular Operational Taxonomic Units), as defined by Blaxter et al. (2005). Obviously, the layman (and even the professional systematist!) cannot know more than a small fraction of even the properly described taxa. Yet, most of our knowledge about Life only applies to some **taxa** (groups of organisms formally recognized by systematists)—sometimes to large (inclusive) ones (such as *Vertebrata* or *Mammalia*), sometimes only to a single nominal species. Thus, a precise biological nomenclature is essential to store and retrieve biological knowledge.

As we will see in Chapter 1, rank-based nomenclature developed over approximately the last three centuries and owes much to the works of Linnaeus (hence the name "Linnaean nomenclature" that is sometimes used). Its principles are explained in detail in Chapter 2, so only very brief explanations are needed here. This nomenclature uses a **type**, which may be a specimen (the remains of an organisms preserved in a museum) or a low-ranking taxon (like a species, to typify a genus name) and a rank (which includes species, genus, family, order and so on) to define a taxon name.

Is the current system (that is, **rank-based nomenclature**) adequate? About 50 years ago, Raven et al. (1971), who used "taxonomy" to designate both genuine taxonomy as well as nomenclature of taxa, stated: "In fact, the current taxonomic [nomenclatural] system is **hopelessly inadequate** as an information retrieval device, and it must be supplanted with one allowing the characteristics of organisms to be handled and retrieved in a much more efficient manner." (Emphasis mine throughout the book unless stated otherwise.) Unfortunately, very little has changed in the meantime. This is not entirely surprising, because despite the importance of this subject, few scientists have actively worked in the field of biological nomenclature, and far fewer still tried to improve our nomenclatural principles. This was nicely summarized by Bock (1994: 6): "Zoological nomenclature affects the work of all zoologists, yet only **a minuscule fraction of one percent** of zoologists deal directly with problems associated with scientific names of animals." Unfortunately, rank-based nomenclature is much more problematic now than it was in 1971. The rapid developments of phylogenetics in the last decades have greatly improved our ability to resolve the Tree of Life. A correct ranking of taxa, to reflect this recently recognized hierarchy, would require many more ranks (which include Kingdom,

DOI: 10.1201/9781003092827-1

Class, Order and so on) than allowed by the rank-based (often called "Linnaean") codes, of which the most frequently discussed are the *Zoological Code* and *Botanical Code*.

This neglect is surprising given the importance of biological nomenclature beyond the academic community. Indeed, decades of studies on folk taxonomies have shown that most pre-literate societies have (or still had recently) a fairly sophisticated biological taxonomy and nomenclature for at least the most prominent elements of the biodiversity that surrounded them (Berlin 2014). Some of our knowledge on biodiversity, which needs to be accessed through our nomenclature, has great societal importance. For instance, knowledge of biodiversity is crucial in conservation biology, because a precise knowledge of rare taxa is required to select areas to preserve. As explained by Winston (2018: 1128), "Recent examples show that species level names raise endangered taxa from invisibility." To illustrate this, Winston (2018) gives, as an example, the Tapanuli orangutan, *Pongo tapanuliensis*, which was erected on the basis of morphological, behavioral and genomic characters. It is the oldest evolutionary lineage in *Pongo* and is among the most endangered of great ape taxa, with fewer than 800 remaining individuals. Saving such taxa is crucially important because extant biodiversity is declining at an alarming and unprecedented rate, and even populations of mammal species deemed common and "of low concern" are declining sharply (Ceballos et al. 2017). Thus, setting ambitious, costly goals in land conservation and enforcing them would be required to stop or reverse these trends within a few decades (Leclère et al. 2020). Saving taxa from extinction is morally right (all life forms, arguably, have a right to exist) and useful in ecotourism, and there are other potential economic benefits. Thus, various substances naturally produced by biological organisms have become pharmaceutically important (like penicillin), and a good nomenclature could possibly help find more such molecules in the future among the species that persist long enough for us to study them.

Similarly, understanding our history requires a precise biological nomenclature. History recorded in written archives extends about 5,000 years into the past. Archeology allows us to study past civilizations through stone tools over 2 or 3 Ma (Semaw et al. 2003; Domínguez-Rodrigo and Alcalá 2016), but biology and paleontology tell us about our deep history, which extends about 3.5 Ga into the past (El Albani et al. 2016; Dodd et al. 2017), and this great portion of our history (more than 99.9%) is intelligible only through a precise biological nomenclature.

A good understanding of nomenclature, taxonomy and evolutionary biology is also useful to avoid confusing Darwinian evolutionary biology, which is basically sound, with the so-called "social Darwinism" that has been used to support racist policies in the past on the basis of an obsolete idea of progress in evolution (Johnson et al. 2012).

But how to classify and name groups of biological organisms? Given that Mankind has faced these problems for millennia, and that a classification allows inductive generalizations to be made (Minelli 2022b), one might think that they were solved long ago. On the contrary, progress on how to classify organisms was fairly slow over a period of about two millennia that ended approximately with the Renaissance. Since then, it has accelerated, especially in the last century (Hennig 1965;

Felsenstein 1979; Pyron 2011; Ronquist et al. 2012; Goloboff et al. 2019). The established consensus in systematics is now that organisms should be classified according to their evolutionary relationships. The **Tree of Life (TOL)** provides a universal, objective basis to classify groups of biological organisms; each branch of that tree is a group (a taxon). For this group to be natural, it must include a whole branch, from its base to all its tips. In other words, such groups include an ancestor and all its descendants; this is what is called a **clade**, or a **monophyletic group**, and many systematists believe that only such groups should be recognized as taxa.

The science that studies the Tree of Life (TOL) is taxonomy; it is not the main topic of this book, but it must be evoked in various sections because the very fact that biodiversity can be modeled as a tree implies that some nomenclatural principles may be more appropriate than others. Taxonomy has long been viewed as a particular kind of classification, which is a more general process of making classes. However, classifications are not necessarily completely hierarchical; some sets of classes could partly overlap other sets of classes. For instance, animals could be divided between terrestrial, amphibious and aquatic forms, but they could also be divided into oviparous and viviparous forms. There is partial overlap between these sets of classes because aquatic animals (for instance) include both oviparous (such as most teleosts) and viviparous (like all aquatic therian mammals) taxa, and the same could be said of amphibious or terrestrial animals. On the contrary, most systematists (at least those working on eukaryotes) think that taxonomies should be strictly hierarchical, with some groups completely included in others, or mutually exclusive, but without partial overlap, with a few exceptions created by hybridization, which is fairly widespread between closely related taxa in eukaryotes (Lherminier and Solignac 2005). Thus, no taxon can be simultaneously a mammal and a reptile. The old (but still occasionally used) expression "mammal-like reptiles" (for instance, Kemp 2006), which might at first sight seem to provide a counter-example, is a holdover from a period in which **paraphyletic taxa** (groups that do not include all of the descendants of a given ancestor) were accepted and in which some early ancestors of mammals were considered to be reptiles, but this no longer reflects the established consensus (Laurin and Reisz 2020). In fact, Griffiths (1974: 121) stated that "The widespread notion that certain traditional concepts like Reptilia and Pisces can be justified according to the principles of evolutionary classification indicates merely that these principles have not been adequately formulated." The shift from accepting paraphyletic taxa to rejecting them, which was initiated by Hennig (1966), has had profound nomenclatural implications, and this played a critical role in the development of phylogenetic nomenclature, even though the latter could, in principle, be used to define paraphyletic groups.

Most systematists now view taxa as individuals rather than classes (see the section "**Ontology of Taxa,**" which follows); from this perspective, taxonomy is not really classification, in the sense that it is not a hierarchy of classes. Hence, in this book, the term "classification" will be used specifically to designate "ordering into classes," as argued by Griffiths (1974: 85).

Once we have recognized taxa (which are clades, except possibly for species-level taxa, given the vague ontology of species), we need to name them and provide a precise definition of these names. This is the field of biological nomenclature, the central topic of this book. Prehistoric men already had names to designate various taxa, but a scientific nomenclature was developed much later as scholars amassed considerable knowledge about animals and plants that was not general knowledge. A turning point of this field was the mid-18th century, when the Swedish botanist Linnaeus developed binominal nomenclature (for instance, *Homo sapiens*), which is still with us. This early history of biological taxonomy nomenclature is covered in the first chapter.

Biological nomenclature is now regulated by codes that are developed and updated by international organizations. Such regulation is required to prevent the proliferation of synonymous names (different names for the same taxa) and of homonyms (several meanings attached to a single name). Most codes of biological nomenclature are based on rank-based nomenclature, which is covered in Chapter 2. Most laymen have heard of species, but most are much less familiar with more inclusive taxa, which under rank-based nomenclature are grouped into categories ranked as genera, families and orders, among others.

The third chapter covers the growth of phylogenetics since the 19th century and its implications for biological nomenclature. Most importantly, it shows how the Linnaean categories are incompatible with our quickly improving knowledge of evolutionary processes and patterns. Yet, as mentioned in the *PhyloCode* (Cantino and de Queiroz 2020: xii), "For many researchers, naming clades is just as important as naming species." It also explains why the first detailed phylogenies that are required to apply phylogenetic nomenclature only started becoming available in the 1970s, which partly explains the continued use of rank-based nomenclature.

Chapter 4 covers the history and principles of phylogenetic nomenclature, from Hennig, who stressed the importance of divergences between extant taxa, through a period of theoretical developments and early applications between the late 1980s and 2020, to the current edition of the *PhyloCode*. Some controversies among proponents of phylogenetic nomenclature are discussed, such as the controversy between converting long-established, well-known names and promoting an integrated, intuitive nomenclature that entails forming new names to replace legacy names.

Comparisons between biological nomenclature and other fields show that precisely delimiting named entities is useful, as explained in Chapter 5. Geological nomenclature is especially enlightening that way because its recent history has featured a transition from **undelimited** geochronological units to strictly delimited ones, which is reminiscent of the ongoing drive to replace rank-based nomenclature, which does not delimit taxa, by phylogenetic nomenclature, which strives at delimiting them precisely.

The sixth chapter discusses controversies in biological nomenclature that concern basic issues, such as the ontology of taxa, including (but not limited to) species concepts, and more practical debates, such as whether or not codes should include rules to delimit taxa.

The seventh and concluding chapter evokes the future of biological nomenclature and the importance of innovation in this field.

As this outline shows, to fully understand how and why biological nomenclature changed over time and what the next logical steps should be in its future development, some basic knowledge of the evolutionary theory and phylogenetics are required, as well as their history. This book attempts at providing this broad perspective, but given the diversity of topics covered and the organization of the text into chapters, the length of the various chapters is uneven. Thus, Chapter 1, about the early (pre-20th century) history of systematics and evolutionary biology, is long, whereas Chapter 7, about the future of biological nomenclature, is much shorter. The diversity of topics covered in the book also prevents an in-depth treatment of such a wide diversity of topics, about which many books and thousands of scientific papers have been published. However, a reasonably extensive bibliography provides an entry into that large body of literature.

## Controversies in Nomenclature

Laymen often imagine scientists as extremely objective, logical people. This is indeed an objective toward which many (hopefully most) systematists strive, but with a variable degree of success, as in any scientific field. The highly technical field of biological nomenclature has the undeserved reputation of being boring, as illustrated by this quote from Berlin (2014: 26): "In Western scientific biology, nomenclatural concerns have become essentially legalistic, pedantic, and tedious. Many systematists treat nomenclatural problems as a necessary evil." Yet, biological nomenclature has generated several heated debates, which refute the opinion that it is tedious. These debates may have occurred partly because, rather than a science, biological nomenclature is a form of highly codified technical language. Thus, subjective choices must be made, and personal preferences play a part in this activity, even though these choices should be guided by our understanding of biological evolution. The reader who ventures into the primary literature must be warned that he will occasionally stumble upon surprisingly harsh criticism of other points of view, even for apparently trivial details. This can be illustrated by a quote from the "Saint-Louis" *Botanical Code* (Greuter et al. 2000: xvii–xviii):

> We have, however, been saddened by the context in which these decisions took place. Passion in nomenclatural discussions is fine and (which is perhaps surprising with as dry a subject) has a solid tradition of long standing; but **hatred** has not. The **Jacobine frenzy** with which the Section was induced to eradicate all traces of registration from the Tokyo Code is we believe unprecedented. The refusal to listen to others, to let contradictory arguments be exposed and explained, has worried us deeply. With such a large and largely novel audience, nomenclature had a unique chance to prove itself a rational discipline. In this it has failed.

**(Bold font used for emphasis is mine, unless noted otherwise.)**

Yet, this hot debate concerned a fairly trivial, common-sense issue in biological nomenclature regarding the requirement to register new names in online databases. The benefits of such a registration to have an exhaustive database are pretty obvious. The reader can easily imagine the even more passionate (and perhaps not entirely rational) reactions (for instance, Brower 2020) that have been triggered by the much more substantial, ongoing revolution that is the main topic of this book, namely the development of phylogenetic nomenclature and of a code based on these principles, the *PhyloCode*. As this book will show, this amounts to a paradigm shift in nomenclature that is at least as important (and arguably more so) as the introduction of rank-based nomenclature through Linnaeus' works. After all, absolute ranks (the so-called "Linnaean categories") had been used before Linnaeus, notably by Magnol (1689), who named several families.

This book will try to cover the main controversies in biological nomenclature and show how our choices may be guided by our growing knowledge of the evolution of biological organisms.

## Rank-Based vs Phylogenetic Nomenclature

The shift from rank-based to phylogenetic nomenclature is much more radical than deciding to require registration of some or all nomenclatural acts into an online database, notably because it entails a shift from **undelimited** to **delimited** taxa (de Queiroz and Gauthier 1990), which is a basic difference between both systems. Rank-based nomenclature indeed rests on absolute ranks, which are arbitrary concepts (they have no objective existence, as explained in Chapter 3) that cannot delimit taxa since no systematist has ever observed a family, an order or any other absolute rank. Even if we had a perfect, exhaustive knowledge of all extant and extinct biodiversity, the definitions provided by rank-based nomenclature would be unable to provide precise, stable delimitation of taxa, as explained in Chapter 3. By contrast, phylogenetic nomenclature ignores such arbitrary ranks; it relies on definitions and the phylogenetic context to precisely delimit taxa. For instance, the taxon *Archosauria* (which includes birds and crocodiles, but not other extant reptiles) could be defined under phylogenetic nomenclature as the smallest clade (monophyletic group) that includes *Crocodylus niloticus* (the Nile crocodile) and *Passer domesticus* (the house sparrow), which under all recently published phylogenies would include all known extant and extinct crocodilians and birds, as well as additional extinct taxa (such as all Mesozoic dinosaurs), but would exclude turtles, snakes and other squamates ("lizards," a paraphyletic group).

Other important differences between both kinds of nomenclature are that under rank-based nomenclature, taxa need not be monophyletic (for instance, the rank-based codes used by most zoologists and botanists do not require, or even recommend, that taxa be monophyletic; see Chapter 2). Thus, many taxa named under that system were long known to be paraphyletic (not to include all descendants of their last common ancestor), but this was considered unproblematic. Well-known examples include *Protista* (unicellular eukaryotes), *Invertebrata* (animals without backbones), *Pisces* (fishes), *Reptilia* (which traditionally includes ectothermic amniotes, such as turtles and crocodilians, but not birds, which are closely related to crocodilians). By contrast, under phylogenetic nomenclature as implemented under the *PhyloCode*, monophyly is always required. Thus, the taxa previously mentioned are either not recognized under that system (such as *Protista*), or re-delimited to become monophyletic (to include all descendants of their last common ancestor), such as *Reptilia*, which includes birds under phylogenetic nomenclature, because it is well-established that crocodilians are more closely related to birds than to other extant reptiles (Brusatte et al. 2010 and references cited therein).

Because of all these differences between both systems, the correspondence between nomenclature and our knowledge of phylogeny is relatively loose and imperfect under rank-based nomenclature, whereas it is by definition perfect under phylogenetic nomenclature. Of course, our knowledge of phylogeny requires improvement; therefore, the contents of taxa will continue to vary under both systems. But, under phylogenetic nomenclature, such changes will only be caused by changes in our objective knowledge of nature (about the phylogeny and biodiversity). By contrast, under rank-based nomenclature, in addition to these sources of changes in taxon composition, personal preferences of systematists add another, subjective source of instability that no amount of knowledge could possibly eliminate. Thus, as our knowledge of biodiversity and phylogeny improves, the contents of taxa should stabilize under phylogenetic nomenclature, but there is no reason to expect that it would stabilize under rank-based nomenclature. Paradoxically, it has even been argued that progress in our knowledge of biodiversity and phylogeny creates nomenclatural instability under rank-based nomenclature (see Section 3.5). Conversely, the development and adoption of phylogenetic nomenclature could not have occurred before phylogenetics had become a mature science because it requires reasonably well-resolved, reliable trees to be applied to any taxon. Given that an evolutionary point of view is increasingly present in current biology, and given the spectacular progress in phylogenetics since the 1950s, there are substantial advantages (to all biologists; not only for experts in biological nomenclature) in adopting phylogenetic nomenclature. Demonstrating this is a major goal of this book.

Putting the profound differences between rank-based and phylogenetic nomenclature into a proper perspective requires key notions of phylogenetics and of evolutionary biology. Thus, even though the central topic of this book is biological nomenclature, substantial space has been allocated to these topics (Chapter 3). A brief summary of the developments of phylogenetics since the mid-20th century (Section 3.1) is important to show how we are now able to obtain fairly robust, reliable, well-resolved phylogenies of many taxa. Without this, phylogenetic nomenclature would be a theoretically sound, but inapplicable, method. Notions of evolutionary biology (Section 3.2) are required to show how rank-based nomenclature is artificial and cannot be reconciled with our factual knowledge of evolution, despite attempts using recent methods, such as time banding (Section 3.3.3.1).

## Ontology of Taxa

What are taxa? This topic is of central importance to biological nomenclature because nomenclatural principles may not be equally adapted to all kinds of entities. For instance, members of a class can be defined by their intrinsic, essential properties, whereas individuals may be better defined by their history. This topic will be covered in greater detail in Chapter 6, but some basic notions need to be presented here.

A controversy in this field centers on the place of essentialism in early systematists. Essentialism has been defined in several ways (Winsor 2006: 151), but in this book, it is considered to imply that essence is prior to existence; applied to taxa, this implies that taxa have ideal types, in the sense of Platonic ideas, and that these exist independently of actual organisms. According to an established view that came under strong criticism in the last two decades (Winsor 2006; Wilkins 2018), early systematists were essentialists in a narrow sense of the word, which implied that they considered that taxa possessed definitional essences "in terms of necessary and sufficient, intrinsic, unchanging, ahistorical properties" (Winsor 2003: 388). If this view existed at all among early systematists, Wilkins (2018: 122) suggested that it may have been a reaction to Darwin's theory of evolution through natural selection. In the extreme form portrayed in some of Ernst Mayr's writings, essentialism involved "a transfer of these features of Schindeworlf's [one of the main proponents of essentialism in systematics] types down to the species level, where no one believed they belonged" (Winsor 2006: 159). Essentialism in systematics was thus linked to typology (Winsor 2006: 167), which created confusion because this word can refer to the abstract types (Platonic ideas) or to museum types, which are specimens that represent a taxon. Current systematics is undoubtedly typological in the latter sense, but this is not a problem because museum types help to stabilize nomenclature; the debate centers on whether pre-Darwinian systematists were typological in the former sense.

Under this essentialist view (which may never have been very widespread among systematists), taxa are considered universals (classes), in the philosophical sense of the word. Classes have defining properties and, under this concept, an organism belongs to a taxon if it shares these defining property. Classes are often considered to be defined by eternal (hence unchanging and ahistorical) intrinsic necessary and sufficient properties, even though this point is disputed (Boyd 1999: 151). Under this concept, a class has no precise spatiotemporal delimitation. Atoms of a given atomic number and of a given isotope fulfill this condition. A hydrogen atom found on Earth today differs in no respect from one that existed 10 Ga in the past at a very great distance from here. However, if classes are defined this strictly, it is difficult to view taxa as forming classes (Boyd 1999). For instance, *Tetrapoda*, which means "four feet," refers to extant limbed vertebrates and a number of related extinct taxa. Thus, it includes, among others, frogs, salamanders, mammals, turtles and even birds because the wings are modified arms. But the birds show that for this concept of *Tetrapoda* to work, the property of having "four feet" has to be interpreted fairly liberally. In fact, the situation is even more complex than this because gymnophionans (limbless amphibians) and numerous squamates lack feet or structures derived thereof, but they are universally considered tetrapods by systematists. Thus, it is not really the presence of four feet that makes a tetrapod; obviously, any descendant of a four-footed vertebrate qualifies as a tetrapod. Such historical concepts have no status in rank-based (Linnaean) nomenclature and depart strongly from the classical view of classes as ahistorical entities, but they can be captured by phylogenetic definitions associated with names that are regulated under the *PhyloCode*.

Given the difficulties in considering that taxa can be defined by intrinsic properties (Griffiths 1973), many systematists, following Ghiselin (1966a, 1974) and Griffiths (1974), now view taxa as individuals in the philosophical sense (for instance, O'Hara 1996). Actually, the idea that species are individuals apparently "goes back at least to Buffon" (Ghiselin 1974: 536) and may even hark back to Thomas Aquinas (Wilkins 2018: 44), but recent works provided many new arguments showing why this perspective makes sense and how it is advantageous.

Individuals are unique, and do not have intrinsic defining properties, but are spatiotemporally delimited. They have an origin (like egg fertilization by a sperm cell, for an individual organism) and end (death for an organism, or extinction for a taxon), and they are not found everywhere in the universe (cats and humans are only found on Earth, as far as we know). This was expressed by the English botanist George Bentham (1800–1884), who distinguished between **definitions**, which applied to "genera," here taken in the philosophical, logical sense of universal kinds ("common names") defined by (intrinsic) properties, and **individuation**, which he defines thus:

> the corresponding operation performed on the name of an **individual**, viz. the exhibition of its genus and characteristic properties. This mode differs from definition; because, in this case, **the only characteristic properties are those of time and place**, which must both be exhibited.
>
> **(Bentham 1827: 79)**

Later works suggest that individuals are the product of a history (for instance, Wiley 1980). A corollary is that taxa are defined by relationships. Species, under the so-called "biological species concept," can be defined by the nexus of reproduction and gene flux, a concept that can be traced back to Buffon (Stemerding 1993; Hoquet 2007: 416). Higher taxa can be defined by the phylogeny, which is precisely the basic idea of phylogenetic nomenclature, and this idea was clearly expressed by Hennig (1969, 1981) and even, as a wish for the future, by Darwin (1859: 486), who stated (to take one of several similar quotes from the same book): "Our classifications will come to be, as far as they can be so made, genealogies." Ghiselin (1966b: 128) colorfully illustrated the futility of trying to define taxa by their intrinsic properties by comparing this with an attempt at defining the word "aunt" by age, behavior (someone who sends gifts) and resemblance to parents. Obviously, the only correct definition is relational: it is the sister of someone's parent. Individuals "can undergo an indefinite amount of change over time, and yet remain the same thing"

(Ghiselin 2002: 157). This is obvious for organisms; between the zygote, through birth (for viviparous forms) and till death, they change in many fundamental respects. Likewise for taxa; the first synapsid, which lived over 320 Ma ago (Didier and Laurin 2020), may have looked more like a squamate than like an extant mammal, and *Synapsida* diversified from a single lineage to more than 5,400 species today (www.gbif.org/, consulted on October 20, 2020), but it is still the same taxon. To sum this up, "Individuals such as species and clades owe their properties to history, not laws of nature" (Ghiselin 2002: 151).

By contrast, classes or universals, like atoms of a certain kind (such as H or He) originated soon after the Big Bang and could exist till the end of Times, and they can be found anywhere, even in distant galaxy clusters. About 100 seconds after the Big Bang, there were already nuclei of these elements (Hawking 2011). Even larger, more complex nuclei of atoms such as those of C (carbon), O (oxygen), Si (silicium) and Fe (iron) already existed in stars that formed less than 1 Ga (one billion years) after the Big Bang (Becker et al. 2011). On the contrary, most present-day species, as they are currently conceptualized, date from less than 1 Ma (million years) ago (e.g., Geraads et al. 2020), and many taxa that originated early in Earth's history are long-extinct (Carroll 1988). These facts are consistent with such taxa being individuals rather than universals. Note that the parts, if any, of individuals need not be physically contiguous (an important point, as taxa can have disjunct distributions); as pointed out by Ghiselin (1974), "The United States of America is an individual in the class of national states, and this is true in spite of the interposition of Canadian territory and international waters between Alaska and the rest." The fact that taxa are individuals may explain why there are very few evolutionary laws. Nevertheless, the history of individuals (hence, taxa) can be used to test laws because history cannot violate natural laws, as Ghiselin (1984: 107) stated succinctly: "What can happen to individuals is a matter of law; what has happened to them [is] a matter of history." This clarifies a bit how systematics and other historical sciences can contribute to human knowledge.

Higher taxa (leaving aside the problematic "species problem" for now) can be objectively temporally delimited if they are monophyletic. Indeed, the "extinction" of a paraphyletic taxon is only a pseudoextinction (some of its descendants survive) and recognizing this event depends on the arbitrary decision to exclude some descendants (Smith and Patterson 1988). On the contrary, there is nothing subjective about the extinction of clades. Thus, the requirement of monophyly, first emphasized by Hennig (1965), is important. It long proved controversial; Boyd (1999: 179) once called it "extreme cladism," and, at the time, many systematists probably shared his opinion, but the consensus has shifted distinctly in favor of monophyly in the following decades. It is likely that a majority of practicing systematists now consider that higher taxa should be monophyletic. This is a basic requirement of the *PhyloCode*.

A somewhat intermediate view on the ontology of taxa (between that of typical classes and of individuals) is to consider taxa as homeostatic property clusters (for instance, Boyd 1999). Homeostatic property clusters (HPCs) are classes or universals, but in a broader sense; there is no singly, necessary and sufficient defining property. Instead, each is characterized by a combination of features, most of which are present in each element of the HPC. Boyd (1999) argues that this is a perfectly valid concept of universal, and that HPCs are far more common than the narrowly defined classes or universals described previously. HPCs are relatively stable because of cohesion mechanisms, such as gene flow and stabilizing selection, and they need not be monophyletic (Boyd 1999: 182). This concept was formalized in the late 20th century, but Winsor (2003: 388) argued that pre-Darwinian systematists conceptualized taxa as HPCs, and she considered that this is a form of essentialism—one that may hark back to Aristotle. In any case, this quote of Aristotle's *On Interpretation*, taken from Wilkins (2018: 13), does suggest a concept of classes much broader than the much more recent, now-classical view (though not necessarily matching HPCs):

> Some things are universal, others individual. By the term "universal" I mean that which is of such a nature as to be predicated of many subjects, by "individual" that which is not thus predicated. Thus "man" is a universal, "Callias" is an individual.

We will return to the advantages and drawbacks of viewing taxa as HPCs in Chapter 6, but note that in this book, taxa will be considered as individuals unless otherwise stated.

It could be argued that if taxa are individuals rather than classes, the best approach is not to classify them; rather, they should be integrated into a **system** or **arrangement** (Griffiths 1973, 1974), and these differ from classifications by including more information. For instance, systems typically include more than the binary inclusion/exclusion information about the entities that it contains. O'Hara (1996) illustrates the difference between a classification and a system through a geographic example. A classification of Europe would include countries and cities (among various entities). It would indicate that its countries include, among others, Belgium, France, Germany, Italy and Greece. And it would indicate that Italy includes cities like Milan, Rome and Napoli. But this does not give an accurate idea of the distance between cities. A system could be a map, which shows the countries and cities (information also found in the classification), but in addition, the map shows the distances between the various cities and the relative position; the map would show, for instance, that Milan is located North of Rome. Given the now well-accepted paradigm of the evolutionary tree (O'Hara 1996), the most appropriate system for taxa is the evolutionary tree, in which taxa represent branches. The tree shows not only the inclusion/exclusion information conveyed by a classification or a taxonomy (for instance, insects are not vertebrates, and vice versa) but it also shows distance; insects and vertebrates share a last common ancestor that probably lived at least 550 Ma ago (in the Ediacaran; Ogg et al. 2016) according to the fossil record (Benton et al. 2015), and possibly about 700 Ma, according to molecular dating studies (Hedges et al. 2015). Thus, in systematics, the system is a tree, and branch length information (which typically represents evolutionary time but may also represent amount of change, for example, nucleotide substitutions or number of changes in a phenotypic character matrix) represents the additional information included in this system that cannot be included in a classification (or a taxonomy). At a more basic level, Griffiths (1974) argued that species are systems and that individual organisms are its parts. In fact, organisms are also systems, with organs or organelles being their parts (Griffiths 1974: table 1).

There are other viewpoints on the ontology of species. For instance, Casetta (2010) suggested that species are **conventional objects**, which means that they exist independently of us, but their boundaries are decided by convention. This applies to individual species, but not to the species category. Indeed, given the multiple species concepts, Ereshefsky (2002) has claimed that Linnaean ranks, including the species, are "ontologically empty designations." Casetta (2010) focused on species because "species seem to enjoy a sort of ontologically privileged status; they are the 'currency' of every biologist, and whereas the reality of [taxa belonging to] higher classes is often denied, the reality of species is rarely in debate." Similar claims (and confusions between taxa and their rank) are common in the literature (for instance, Henry 2011: 214). But the contrary could be argued because there is a broad (though not universal) agreement that higher taxa should be clades (Hennig 1965; Ax 1987; de Queiroz and Gauthier 1990; Brochu and Sumrall 2001; Dubois et al. 2021: 10), whereas there is no agreement on what a species is. On the contrary, dozens of species concepts have been proposed and used in biology (Lhermin[i]er and Solignac 2000). Cowan (1968), quoted in Stackebrandt and Goebel (1994: 846), even suggested that "there are as many ideas on species as there are biologists, and many a biologist has changed his idea during the course of his working life," and he "regarded the form genera as natural entities but considered species largely artificial." Even reaching a consensus on the definition of species might not solve the problem because Rieppel (1986) has shown that under some of the most widely accepted species concepts, temporal delimitation of species is problematic, which raises doubts about the thesis of species as individuals. Last, but not least, Minelli (2022a) pointed out that the word "species" is currently applied, often without explicit distinction:

> (1) to named taxa such as *Homo sapiens* and *Panthera tigris*, (2) to a rank (usually but not necessarily the lowest and/or the most fundamental one) of the biological classification, and (3) as a variegated set of notions, the most important among them being the morphospecies, the biospecies, the ecospecies, the evolutionary species, the agamospecies and the taxonomic species.

And the fact that the word "species" in English is spelled the same in the singular and plural (this is not the case in some other languages) contributes further to the semantic confusion. Such considerations (all those mentioned in this paragraph) led Pleijel and Rouse (2000) to suggest dropping species altogether and simply referring to the smallest subdivisions of biodiversity as **LITUs**, for **Least Inclusive Taxonomic Units**. Under this system, LITUs are simply the smallest recognizable taxa, given the available methods and data. As research progresses, some of these LITUs may well be subdivided into smaller LITUs, thus losing their LITU status.

These considerations led the founders of phylogenetic nomenclature to devise nomenclatural principles to name and delimit clades (de Queiroz and Gauthier 1990; Cantino et al. 1999; Lee 1999; Sereno 1999). In due time, these developments were incorporated into the *PhyloCode* (Cantino and de Queiroz 2020). It could be argued that this is the most recent (though perhaps not the last) phase of the Darwinian revolution. When discussing "tree thinking," an expression inspired by Mayr's expression "population thinking," O'Hara (1988: 153) wrote:

> In what is surely one of the most remarkable understatements in English literature Darwin observes near the end of his great work of 1859 that "when the views entertained in this volume on the origin of species, or when analogous views are generally admitted, we can dimly foresee that there will be a considerable revolution in natural history." In the hundred or more years since Darwin's revolution began it has extended far beyond natural history, touching nearly every division of science and art. But the muse of history says here that when the future looks back through the lens of narrative, not only on Darwin's age but also on our own, it will see that **the revolution of 1859** did not come to a close even within natural history until it **came to a close with us at the end of the Twentieth Century**. It was not until the end of the Twentieth Century that the inertia of pre-evolutionary thought—of state questions and group thinking—was at last overcome by force of history—of change questions and tree thinking—and Clio came down from the rafters of our museums, shook off the dust, and took her rightful place in the director's chair.

This book will show that, contrary to what O'Hara (1988) suggested, the Darwinian revolution was not complete by the end of the 20th century. This is the core subject of this book, but to put it into proper perspective, it will be necessary to give some background information on the development of biological nomenclature and taxonomy since Antiquity, and to follow their developments through the Renaissance and more recent times.

---

## Typographic Conventions throughout This Book

The various codes differ somewhat in typographic conventions, as will be explained in Chapters 2 and 3. Throughout this book, I will follow the recommendations of most codes (the obvious exception being the *Zoological Code*) in italicizing all taxon names (not only genus and species names), except in quotes and titles of cited works, in which I do not change the original typographic conventions adopted by the various authors.

## REFERENCES

Ax, P. 1987. The phylogenetic system: The systematization of organisms on the basis of their phylogenesis. John Wiley & Sons, Toronto, 340 pp.

Becker, G. D., W. L. Sargent, M. Rauch, and R. F. Carswell. 2011. Iron and α-element production in the first one billion years after the big bang. The Astrophysical Journal 744:91.

Bentham, G. 1827. Outline of a new system of logic: With a critical examination of Dr. Whately's "elements of logic". Hunt and Clarke, London, xii + 287 pp.

Benton, M. J., P. C. Donoghue, R. J. Asher, M. Friedman, T. J. Near, and J. Vinther. 2015. Constraints on the timescale of animal evolutionary history. Palaeontologia Electronica 18:1–107.

Berlin, B. 2014. Ethnobiological classification: Principles of categorization of plants and animals in traditional societies. Princeton University Press, Princeton, NJ, 354 pp.

Blaxter, M., J. Mann, T. Chapman, F. Thomas, C. Whitton, R. Floyd, and E. Abebe. 2005. Defining operational taxonomic units using DNA barcode data. Philosophical Transactions of the Royal Society of London, Series B 360:1935–1943.

Bock, W. J. 1994. History and nomenclature of avian family-group names. Bulletin of the American Museum of Natural History (USA) 222:1–281.

Boyd, R. 1999. Homeostasis, species, and higher taxa; pp. 141–185 in R. A. Wilson (ed.), Species: New interdisciplinary essays. MIT Press, Cambridge, MA.

Brochu, C. A. and C. D. Sumrall. 2001. Phylogenetic nomenclature and paleontology. Journal of Paleontology 75:754–757.

Brower, A. V. 2020. Dead on arrival: A postmortem assessment of "phylogenetic nomenclature", 20+ years on. Cladistics 36:627–637. DOI: 10.1111/cla.12432

Brusatte, S. L., M. J. Benton, J. B. Desojo, and M. C. Langer. 2010. The higher-level phylogeny of Archosauria (Tetrapoda: Diapsida). Journal of Systematic Palaeontology 8:3–47.

Cantino, P. D., H. N. Bryant, K. de Queiroz, M. J. Donoghue, T. Eriksson, D. M. Hillis, and M. S. Y. Lee. 1999. Species names in phylogenetic nomenclature. Systematic Biology 48:790–807.

Cantino, P. D. and K. de Queiroz. 2020. International code of phylogenetic nomenclature (PhyloCode): A phylogenetic code of biological nomenclature. CRC Press, Boca Raton, Florida, xl + 149 pp.

Carroll, R. L. 1988. Vertebrate paleontology and evolution. W. H. Freeman, New York, 698 pp.

Casetta, E. 2010. Categories, taxa, and chimeras; pp. 264–278 in M. D'Agostino, G. Giorello, F. Laudisa, T. Pievani, and C. Sinigaglia (eds.), New essays in logic and philosophy of science. College Publications, London.

Ceballos, G., P. R. Ehrlich, and R. Dirzo. 2017. Biological annihilation via the ongoing sixth mass extinction signaled by vertebrate population losses and declines. Proceedings of the National Academy of Sciences 114:E6089–E6096.

Cowan, S. T. 1968. A dictionary of microbial taxonomic usage. Oliver & Boyd, Ltd., Edinburgh, x + 118 pp.

Darwin, C. 1859. On the origin of species by means of natural selection or the preservation of favoured races in the struggle for life. John Murray, London, 502 pp.

de Queiroz, K. and J. Gauthier. 1990. Phylogeny as a central principle in taxonomy: Phylogenetic definitions of taxon names. Systematic Zoology 39:307–322.

Didier, G. and M. Laurin. 2020. Exact distribution of divergence times from fossil ages and tree topologies. Systematic Biology 69:1068–1087.

Dodd, M. S., D. Papineau, T. Grenne, J. F. Slack, M. Rittner, F. Pirajno, J. O'Neil, and C. T. Little. 2017. Evidence for early life in Earth's oldest hydrothermal vent precipitates. Nature 543:60–64.

Domínguez-Rodrigo, M. and L. Alcalá. 2016. 3.3-million-year-old stone tools and butchery traces? More evidence needed. PaleoAnthropology 2016:46–53.

Dubois, A., A. Ohler, and R. Pyron. 2021. New concepts and methods for phylogenetic taxonomy and nomenclature in zoology, exemplified by a new ranked cladonomy of recent amphibians (Lissamphibia). Megataxa 5:1–738.

El Albani, A., R. Macchiarelli, and A. R. Meunier. 2016. Aux origines de la vie: Une nouvelle histoire de l'évolution. Dunod, Paris, 221 pp.

Ereshefsky, M. 2002. Linnaean ranks: Vestiges of a bygone era. Philosophy of Science 69:S305–S315.

Felsenstein, J. 1979. Alternative methods of phylogenetic inference and their interrelationship. Systematic Zoology 28:49–62.

Geraads, D., G. Didier, A. Barr, D. Reed, and M. Laurin. 2020. The fossil record of camelids demonstrates a late divergence between Bactrian camel and dromedary. Acta Palaeontologica Polonica 65:251–260.

Ghiselin, M. T. 1966a. On psychologism in the logic of taxonomic controversies. Systematic Zoology 15:207–215.

Ghiselin, M. T. 1966b. An application of the theory of definitions to systematic principles. Systematic Zoology 15:127–130.

Ghiselin, M. T. 1974. A radical solution to the species problem. Systematic Zoology 23:536–544.

Ghiselin, M. T. 1984. "Definition," "character," and other equivocal terms. Systematic Zoology 33:104–110.

Ghiselin, M. T. 2002. Species concepts: The basis for controversy and reconciliation. Fish and Fisheries 3:151–160.

Goloboff, P. A., M. Pittman, D. Pol, and X. Xu. 2019. Morphological data sets fit a common mechanism much more poorly than DNA sequences and call into question the Mkv model. Systematic Biology 68:494–504.

Greuter, W., J. McNeill, F. R. Barrie, H. M. Burdet, V. Demoulin, T. S. Filgueiras, D. H. Nicolson, P. C. Silva, J. E. Skog, P. Trehane, N. J. Turland, and D. L. Hawksworth. 2000. International code of botanical nomenclature. Koeltz Scientific Books, Königstein, Germany, xviii + 474 pp.

Griffiths, G. C. D. 1973. Some fundamental problems in biological classification. Systematic Zoology 22:338–343.

Griffiths, G. C. D. 1974. On the foundations of biological systematics. Acta Biotheoretica 23:85–131.

Hawking, S. 2011. A brief history of time. Bantam, London, 272 pp.

Hedges, S. B., J. Marin, M. Suleski, M. Paymer, and S. Kumar. 2015. Tree of life reveals clock-like speciation and diversification. Molecular Biology and Evolution 32:835–845.

Hennig, W. 1965. Phylogenetic systematics. Annual Review of Entomology 10:97–116.

Hennig, W. 1966. Phylogenetic systematics. University of Illinois Press, Urbana, Chicago, London, 263 pp.

Hennig, W. 1969. Die Stammesgeschichte der Insekten. Kramer, Frankfurt am Main, 436 pp.

Hennig, W. 1981. Insect phylogeny. John Wiley & Sons, Chichester, xi + 514 pp.

Henry, D. 2011. Aristotle's pluralistic realism. The Monist 94:197–220.

Hibbett, D. S., H. R. Nilsson, M. Snyder, M. Fonseca, J. Costanzo, and M. Shonfeld. 2005. Automated phylogenetic taxonomy:

An example in the homobasidiomycetes (mushroom-forming Fungi). Systematic Biology 54:660–667.

Hoquet, T. 2007. Buffon: From natural history to the history of nature? Biological Theory 2:413–419.

Johnson, N. A., D. C. Lahti, and D. T. Blumstein. 2012. Combating the assumption of evolutionary progress: Lessons from the decay and loss of traits. Evolution: Education and Outreach 5:128–138.

Kemp, T. S. 2006. The origin and early radiation of the therapsid mammal-like reptiles: A palaeobiological hypothesis. Journal of Evolutionary Biology 19:1231–1247.

Laurin, M. and R. R. Reisz. 2020. Reptilia; pp. 1027–1031 in K. de Queiroz, P. D. Cantino, and J. A. Gauthier (eds.), Phylonyms: An implementation of phyloCode. CRC Press, Boca Raton, Florida.

Leclère, D., M. Obersteiner, M. Barrett, S. H. Butchart, A. Chaudhary, A. De Palma, F. A. DeClerck, M. Di Marco, J. C. Doelman, and M. Dürauer. 2020. Bending the curve of terrestrial biodiversity needs an integrated strategy. Nature 585:551–556.

Lee, M. S. Y. 1999. Reference taxa and phylogenetic nomenclature. Taxon 48:31–34.

Lhermin[i]er, P. and M. Solignac. 2000. L'espèce: définitions d'auteurs. Comptes Rendus de l'Académie des Sciences—Series III—Sciences de la Vie 323:153–165.

Lherminier, P. and M. Solignac. 2005. De l'espèce. Syllepse, Paris, XI + 694 pp.

Lücking, R. 2019. Stop the abuse of time! Strict temporal banding is not the future of rank-based classifications in fungi (including lichens) and other organisms. Critical Reviews in Plant Sciences 38:199–253.

Magnol, P. 1689. Prodromus historiae generalis plantarum in quo familiae plantarum per tabulas disponuntur. Pech, Montpellier, 79 pp.

Minelli, A. 2017. Grey nomenclature needs rules. Ecologica Montenegrina 7:654–666.

Minelli, A. 2022a. Species; in B. Hjørland and C. Gnoli (ed.), ISKO Encyclopedia of Knowledge Organization (IEKO). www.isko.org/cyclo/species

Minelli, A. 2022b. The species before and after Linnaeus: Tension between disciplinary nomadism and conservative nomenclature; pp. 191–226 in J. Wilkins, I. Pavlinov, and F. Zachos (eds.), Species problems and beyond: Contemporary issues in philosophy and practice. CRC Press, Boca Raton, FL.

Mora, C., D. P. Tittensor, S. Adl, A. G. B. Simpson, and B. Worm. 2011. How many species are there on Earth and in the ocean? PLoS Biology 9:1–8.

Ogg, J. G., G. Ogg, and F. M. Gradstein. 2016. A concise geologic time scale: 2016. Elsevier, Amsterdam, 240 pp.

O'Hara, R. J. 1988. Homage to Clio, or, toward an historical philosophy for evolutionary biology. Systematic Zoology 37:142–155.

O'Hara, R. J. 1996. Trees of history in systematics and philology. Memorie della Società Italiana di Scienze Naturali e del Museo Civico di Storia Naturale di Milano 27:81–88.

Pleijel, F. and G. W. Rouse. 2000. Least-inclusive taxonomic unit: A new taxonomic concept for biology. Proceedings of the Royal Society of London, Series B 267:627–630.

Pyron, R. A. 2011. Divergence-time estimation using fossils as terminal taxa and the origins of Lissamphibia. Systematic Biology 60:466–481.

Raven, P. H., B. Berlin, and D. E. Breedlove. 1971. The origins of taxonomy. Science 174:1210–1213.

Rieppel, O. 1986. Species are individuals: A review and critique of the Argument; pp. 283–317 in M. K. Hecht, B. Wallace, and G. T. Prance (eds.), Evolutionary biology, vol. 20. Plenum Publishing Corporation, New York.

Ronquist, F., M. Teslenko, P. van der Mark, D. L. Ayres, A. Darling, S. Höhna, B. Larget, L. Liu, M. A. Suchard, and J. P. Huelsenbeck. 2012. MrBayes 3.2: Efficient Bayesian phylogenetic inference and model choice across a large model space. Systematic Biology 61:539–542.

Salzburger, W. 2018. Understanding explosive diversification through cichlid fish genomics. Nature Reviews Genetics 19:705–717.

Semaw, S., M. J. Rogers, J. Quade, P. R. Renne, R. F. Butler, M. Dominguez-Rodrigo, D. Stout, W. S. Hart, T. Pickering, and S. W. Simpson. 2003. 2.6-million-year-old stone tools and associated bones from OGS-6 and OGS-7, Gona, Afar, Ethiopia. Journal of Human Evolution 45:169–177.

Sereno, P. C. 1999. Definitions in phylogenetic taxonomy: Critique and rationale. Systematic Biology 48:329–351.

Smith, A. B. and C. Patterson. 1988. The influence of taxonomic method on the perception of patterns of evolution; pp. 127–216 in M. K. Hecht and B. Wallace (eds.), Evolutionary biology. Plenum Press, New York.

Stackebrandt, E. and B. M. Goebel. 1994. A place for DNA-DNA reassorciation and 16S rRNA sequence analysis in the present species definition in bacteriology. International Journal of Systematic Bacteriology 44:846–849.

Stemerding, D. 1993. How to make oneself nature's spokesman? A Latourian account of classification in eighteenth-and early nineteenth-century natural history. Biology and Philosophy 8:193–223.

Wiley, E. O. 1980. Is the evolutionary species fiction? A consideration of classes, individuals and historical entities. Systematic Zoology 29:76–80.

Wilkins, J. S. 2018. Species: The evolution of the idea. CRC Press, Boca Raton, xxxviii + 389 pp.

Winsor, M. P. 2003. Non-essentialist methods in pre-Darwinian taxonomy. Biology and Philosophy 18:387–400.

Winsor, M. P. 2006. The creation of the essentialism story: An exercise in metahistory. History and Philosophy of the Life Sciences 28:149–174.

Winston, J. E. 2018. Twenty-first century biological nomenclature: The enduring power of names. Integrative and Comparative Biology 58:1122–1131.

# 1

## The Roots of Biological Nomenclature – A Short History of Systematics through the 19th Century

## 1.1 From the Prehistory to the 15th Century

### 1.1.1 Preliterate Classifications

#### 1.1.1.1 Ancient Roots, Debates and Relevance to Biological Nomenclature

Humans must have felt a need to name animals and plants a very long time ago, in the prehistory, but this can be assessed only through ethnobiological studies of the spoken language of various people, particularly those that have (or had until very recently) a stone-age culture, and more globally, those who still rely on hunting, fishing and gathering food in natural environments, even though many of these supplement this with farming. These peoples live in environments with a much greater biodiversity than today's urban or even countryside environments currently inhabited by most of humanity. Given their strong dependence on many species that they harvest, they are very knowledgeable about many taxa (much more so than the average urban dweller), and Western science is barely starting to recognize the value of some of these cultures. For instance, indigenous peoples in northern Canada long recognized the importance of leaders in caribou migrations, an importance that was validated by recent ecological research (Pierotti 2020: 3). Thus, it is worth looking into their biological classification and nomenclature.

Generally, no written records attest the antiquity of indigenous taxonomy and nomenclature, but the shared properties of most folk taxonomies should hark back to a distant past that predates writing. This is important because archaeology, zooarchaeology and archaeobotany yield interesting clues about habitat use by early humans, such as which species were eaten or used to make clothes or tools, or even about the environment (which plant communities were locally present, for instance) inhabited by our ancestors (for instance, Nagaoka and Wolverton 2016; Wolverton et al. 2016), but this yields no data about how these early human populations classified biodiversity and named taxa.

This section (1.1.1) on folk taxonomy needs not cover all aspects of ethnobiology. For instance, topics such as how contemporary indigenous communities struggle with issues such as biodiversity management and climate change adaptation are just as interesting as those covered here, but this section focuses on two key questions that are directly relevant to understand the current debates in biological nomenclature, namely: to what extent is folk taxonomy hierarchical, and does folk taxonomy and nomenclature incorporate cryptic absolute ranks? The first question determines to an extent the parallels that

can be drawn between folk and "scientific" taxonomies, which are based on a hierarchy of taxa that are either completely nested in each other (like *Mammalia* is part of *Vertebrata*) or mutually exclusive (no partial overlap occurs, as between *Vertebrata* and *Mollusca*), with some notable exceptions (concerning hybrids, notably), and at least among eukaryotes (see Section 6.4). The second question determines to what extent folk taxonomies and nomenclatures resemble either of the two current competing scientific nomenclatures. These are respectively rank-based nomenclature (abbreviated RN from here on), which has explicit absolute ranks called "Linnaean categories" (see Chapter 2), and phylogenetic nomenclature (abbreviated PN from here on), which lacks them (see Chapter 4). Whether or not cryptic categories (absolute ranks) exist in folk taxonomies can also yield clues about how humans spontaneously classify living beings, a topic that is also relevant to biological nomenclature. Finally, folk taxonomy is interesting in its own right simply because it attests to how most of Mankind has classified biodiversity throughout most of its (pre)history.

Ethnobiologists pointed out long ago that folk taxonomies strongly differ from scientific taxonomies, both in purpose and in many characteristics. Thus, Atran (1985: 306) suggested that science and common sense (which he interprets as underlying folk taxonomies) are "logically independent approaches" and they may rely on different ontologies and conceptual terminologies. For instance, folk taxonomies rely heavily on gross morphology, ecological niche and societal use, as we will see in the following, whereas scientific taxonomies have emphasized evolutionary relationships for many decades (see Chapter 3). However, comparisons between both can be enlightening, partly because folk taxonomies may reveal how humans tend to classify living beings naturally, using common sense. We will also see that some debates among experts in folk taxonomies parallel those among systematists. For instance, folk taxonomies, like the rank-based nomenclature used by most practicing systematists, may rely on absolute ranks (see Section 1.1.1.2), but what characterizes these ranks, and even their prevalence among folk taxa, is debated (see Sections 1.1.1.3 to 1.1.1.5). An obvious question raised by this similarity is the origin of folk taxonomic ranks: does it represent a shared feature of folk taxonomies generated by general processes operating in human minds independently of the spoken language and the culture, or is it simply an artifact generated by ethnobiologists when trying to make sense of their data? And does it represent an influence of the Linnaean categories used in rank-based nomenclature on ethnobiologists? This is reminiscent of the current debate about the merits of rank-based nomenclature vs phylogenetic nomenclature

DOI: 10.1201/9781003092827-2

(which does not require such ranks) among systematists (see Chapters 3 and 4). Similarly, there are debates about the extent to which folk taxonomies are hierarchical (see Sections 1.1.1.7 and 1.1.1.8). A less compelling parallel (than for ranks) can be made with the controversy among systematists about the extent to which evolution has been divergent (which leads to neat hierarchies), or whether horizontal gene transfer has blurred this divergent signal to such extant that it might be very difficult to detect and nearly meaningless in parts of the living world (see Section 6.4). This parallel is less convincing because this debate among systematists focuses mostly on prokaryotes (bacteria and archaea), which are absent from folk taxonomies because they are microscopic, but the adequacy of a hierarchic representation of biodiversity is a shared similarity between these debates among experts on folk taxonomies and systematists.

### 1.1.1.2 Folk Taxonomic Ranks: Berlin's Hypothesis

Studies of folk biology from the 1960s have shown that most indigenous peoples have fairly complex, mostly hierarchical biological classifications. Berlin (2014: 26) explained it thus:

> However, the empirical comparative data between Western scientific and folk scientific systems of biological classification, as well as among the folk systems themselves, point to a single, preferred ordering that is primary and fundamental in humans' appreciation of nature's plan, lending credence to Gilmour and Walters's assertion that in the classification of living things "one way is more natural than any other" (Gilmour and Walters 1964: 4–5). This plan is, in the main, so striking in its presentation that human observers are highly constrained in the ways that they may choose to deal with it. Biological reality allows for few options.

The extensive knowledge of biodiversity of indigenous peoples reflects their close ties to their natural environment, on which they depend for their survival. However, according to Berlin (2014: 31), this knowledge reflects affinities between the taxa themselves rather than the cultural significance of these taxa; these are not primarily utilitarian classifications, even though taxa exploited by these societies may be better represented than other taxa in such classifications. In an arguably overly optimistic tone, Berlin (1973: 260) even suggested that "The primitive natural systematist is apparently as much concerned with bringing classificatory order to his biological universe as is his western counterpart." Similarly, Berlin et al. (1981: fig. 5) illustrated a rather striking similarity between the scientific systematics and the folk systematics of woodpeckers known to the Aguaruna of the upper Maranon river valley of northern Peru. Both taxonomies were represented as a tree of eight taxa, only one of which was positioned differently according to these two taxonomies. Berlin (2014: 6) even reported that in a monograph, Diamond (who himself cited a draft by Bulmer) indicated that the Fore of New Guinea have names for sibling species of birds (for two *Sericornis* warblers and two *Macropygia* cuckoo-doves)! Ragupathy et al. (2009)

documented a case in which a new cryptic taxon of grass was discovered through comparison between folk taxonomy of two peoples (the Irulas and Malasars) inhabiting the Western Ghats of southern India and scientific taxonomy. DNA barcoding validated the additional (cryptic) taxon *Tripogon cope* that the Irulas and Malasars have long called *Sunai pul*.

As briefly summarized by Hunn (1982: 832), "Folk biological classification has been approached as if information about plants and animals were stored in people's heads in taxonomically organized domains." This is indeed more or less what Berlin (1973: 260) had stated:

> The fundamental organizing principle of folk biological classification—the result partially, perhaps, of the large numbers of classes of organisms involved—is taxonomic, whereby recognized groupings (hereafter called taxa) of greater and lesser inclusiveness are arranged hierarchically (9, 24). It should be noted that the taxa which occur as members of the same folk ethnobiological category are always mutually exclusive.

This quote does evoke a taxonomy as conceptualized by practicing systematists, but as we will see in the following, the claim that it applies to most folk taxonomies is contested by some ethnobiologists.

A common view since the 1970s is that the taxa in these classifications belong to cryptic absolute ranks, even though their exact number varies a little between studies (Berlin et al. 1973; Atran 1998; Berlin 2014). Thus, Berlin et al. (1973) recognized the following ranks (Table 1.1): unique beginner (plant or animal), which is often unnamed but can be recognized otherwise; life forms; intermediate taxa (which may nor may not be present in a given taxon and are often unnamed); generic; specific; and varietal taxa. Atran (1998: 549) recognized the same ranks, but with a different nomenclature: folk-kingdom (equivalent to Berlin et al.'s 1973 unique beginner), life-form (such as bird, mammal, tree, bush), intermediate (not always present), generic or generic-species (such as shark, dog, oak), folk-specific (such as poodle, white oak), and folk-varietal (such as toy poodle, swamp white oak). This hierarchy is apparently not developed in all societies; thus, Berlin (2014: 24) stated that "There is some evidence that foraging societies have poorly developed or lack entirely taxa of specific rank. No foraging society will exhibit taxa of varietal rank." Indeed, varietal taxa mostly represent types of domesticated animals and plants (Pawley 2011: 425). These ranks were mostly defined from linguistic factors (Alves et al. 2016: 118), although Berlin (1973) also mentioned that life forms were recognized based on gross morphological differences. These ranks are often called "folk taxonomic ranks," but they could perhaps be more appropriately be called "folk nomenclatural ranks."

Atran (1987a, 1998) and Wierzbicka (1992: 22) pointed out that hierarchy is common in cognitive domains and that it applies, among other things, to artifacts, such as "chair" (which belongs to the higher-order category "furniture") or "car" (which belongs to the higher category "vehicle"), but that in most cognitive domains, there are no absolute ranks. In this respect, folk biological classifications would be an exception, if the reality of these ranks were established beyond doubt. Given the importance that absolute ranks took in RN (in which they are represented by the so-called Linnaean categories), this topic deserves a brief discussion.

These ranks in folk taxonomies were recognized and named by some ethnobiologists; they were neither formally recognized nor named as such by the users of these classifications, but indirect clues suggest their existence. Berlin et al. (1973) and Atran (1998) pointed out various linguistic clues that arguably reflect this phenomenon. Thus, life forms and generics are designated by primary lexemes (a basic abstract unit of meaning that cannot be reduced to smaller constituents). These primary lexemes may be unanalyzable linguistically (such as "oak" and "robin"), but others are analyzable (such as "tuliptree" and "beggartick"). Specific and varietal taxa are designated by secondary lexemes distinguished by one or only a few semantic dimensions (such as "red rose" vs "white rose"), and the names of varietals are formed by adding an attribute to a species name (for instance, the species "common bean" includes the varieties "red common bean" and "black common bean"). A few exceptions to these rules occur, but they are easily explainable. As Berlin et al. (1973: 224) explained, some specific taxa are designated by a primary lexeme when they are the most common specific taxon of their generic (then, the name of the generic is used as a shorthand for the specific name), or when the specific taxon appears to be in the process of assuming generic status. In this case, the name is usually polysemous (it refers to at least two taxa) and is also used for a more inclusive taxon, typically a generic one (Berlin 1973: 265).

Note that the folk-specific taxa are often composed of two words, which evokes binominal nomenclature, as popularized by Linnaeus. Berlin (1973: 264–265) suggested that this was more than a mere superficial resemblance that arose by coincidence:

> It is perhaps an unintentional bit of western systematic **ethnocentrism** to attribute the "invention" of our current binomial system of nomenclature to Linnaeus (or to Bauhin) if in so doing one is suggesting a radical break with folk tradition. It is more close to the facts to observe that Linnaeus and his predecessors formally codified a system of nomenclature **present in the folk systematics of earliest prescientific man and still recognized in the natural folk biological systems** of classification found in the languages of preliterate peoples today (25).

> **(Bold font used for emphasis is mine, unless noted otherwise.)**

Berlin et al. (1973) also mentioned other properties that characterize various levels. Thus, life forms are always polytypic (include more than one taxon of the rank immediately below) and are not numerous; there are typically between five and ten (Berlin et al. 1973: 261). Generics are much more numerous than life forms (they are arguably the building blocks of these taxonomies). The Tzeltal (a Mayan language spoken in the Mexican state of Chiapas and the people who speak that language) thus named 471 generic taxa, of which

398 are monotypic (include a single species) and 73 are polytypic (include more than one species). Taxa of intermediate rank may correspond vaguely to taxa of family rank recognized by Western science. Varietal taxa are rare; Berlin et al. (1973: 221) reported that they found that only four specific taxa are subdivided into varietals by the Tzeltal Maya and, for three of them, these are cultigens.

Note that even though Tzeltal Maya have been emphasized here and in Berlin et al. (1973), that study also found support for its conclusions in previous works on various other indigenous peoples. These include the Hanunóo (from the island of Mindoro, Philippines), who possess an extensive knowledge of biodiversity, considering that they have named 1,625 terminal plant taxa; the Karam (called Kalam in more recent studies) from New Guinea, who recognize very few specific (but many generic) taxa; the Fore, also from New Guinea, who apparently do not recognize intermediate taxa; the Cantonese boat people, who apparently recognize neither varietals nor taxa of the intermediate category (Berlin et al. 1973: 231); the Navajo (from the United States); the Guaraní from South America; and classical Nahuatl as spoken by the Aztecs (who lived in the Central Mexican Plateau).

Atran (1998) argued that the existence of these ranks is supported by the shared properties of folk taxa of a given level, although the evidence for this is less conclusive than the linguistic properties of folk taxon names. For instance, life form taxa are characterized by "general adaptations to broad sets of ecological conditions." More convincingly, Atran (1998) showed that inductive inferences about taxa were preferentially performed at the generic-specific level. Thus, both Itzaj (now called Itza') Maya Amerindians from Guatemala and urbanized Americans from Michigan frequently declared that if a folk-specific or varietal taxon had a property, all members of its folk-genus shared this property. By contrast, far fewer individuals made the same inductive inference about folk kingdoms or life forms if told that one of its generic-specific possessed a given attribute. This finding lends some support to the claimed existence of some ranks in folk taxonomies.

Folk taxonomies are usually better correlated with phenetic than with cladistic taxonomies (Atran 1998: 568–569). This is unsurprising because humans initially classified living beings according to salient (overall) similarity. The distinction between primitive and derived characters was introduced by Hennig (1950, 1966), the founder of cladistics, only in the 20th century, and adoption of cladistics by professional systematists took a few decades (see Chapter 3). Thus, it would be very surprising indeed if any indigenous people had long ago reasoned according to these principles. Nevertheless, for some taxa, the correspondence between folk taxa and taxa recognized by systematists is good (e.g., Ragupathy et al. 2009). Thus, Berlin et al. (1973: 219) wrote that "With the exception of all fungi, lichens, algae, and the like, the boundaries of the domain of plants as conceived by the Tzeltal corresponds almost perfectly with the standard plant division of Western systematic botany."

### 1.1.1.3 Folk Taxonomic Ranks: Very Cryptic

The existence of such absolute ranks in folk taxonomies was accepted by the vast majority of practicing experts of folk taxonomies in the 1970s and 1980s (probably less so now), but it has been contested by various experts in that field (for instance, Ellen 1998; Hunn 1982, 1998), by some cognitive scientists (such as Hatano 1998) and by some systematists (e.g., Mishler and Wilkins 2018). Thus, Mishler and Wilkins (2018: 2) stated: "In most early classifications, taxa were not ranked. While folk kinds are generally nested in a hierarchy, there are no fixed and specified levels or grades of kinds in most folk taxonomies." According to this perspective, the ranks evoked here are concepts developed by Western scientists only; the indigenous peoples had no such notions. Ellen (1998: 572) argued likewise:

> Nuaulu, like Itzaj Maya, **do not "essentialize ranks,"** which would violate their primary concern with "ecological and morpho-behavioural relationships" in favour of abstract properties. The development of worldwide scientific systematics has explicitly required rejecting such relationships (sect. 2.1.2.3) with their cross-cutting classifications.

Ellen (1998: 573) further stated that:

> There are also difficulties with Atran's generalisations regarding the concept of rank (as indicating "fundamentally different levels of reality") in organising the domain of living things. Ranks remain **very difficult to establish** cross-culturally as commensurate entities. Beyond the principle of successive inclusivity and the basic level, the concept is tricky to defend as a universal and hard-wired cognitive tendency. Ontological categories such as "folk kingdom" exist, in the sense that plant and animal are recognised (if not named) in virtually all cultures. But although certain life-forms are remarkably constant (e.g., "tree") many categories that contrast with them are very inconstant, and distinguishing between "lifeforms" and "unaffiliated generics" is not always easy. . . . Successive inclusive division **does not necessarily provide "levels."**

A potential but dubious exception may be found in the Fore folk taxonomy (the Fore inhabit the Papua New Guinea Highlands). Diamond (1966: 1103) reported that the Fore recognized nine high-ranking taxa that they called "tábé aké" (which means "big names") and that each of these is subdivided into lower-ranking taxa that they call "ámana aké" (which means "small names"). However, it is unclear if these two ranks are absolute or relative, and a comparison with the Wola highlanders, who also inhabit the Papua New Guinea Highlands, suggests the latter. The Wola designate various taxa by the word "sem," which translates into "family." It applies to high-ranking taxa, such as life forms, as well as to lower-ranking taxa (Sillitoe 2003: 13) and to kin groups among people, down to single families (Sillitoe 2002: 1163). In an earlier paper, Sillitoe (1995: 207) indicated that various suffixes were added to the word "sem" to denote various ranks of taxa, but these imply only relative ranks, as Sillitoe kindly clarified in a personal communication (June 6, 2022). Thus, "sem" may designate taxon of any rank. If lower-ranking taxa

contained in the first taxon are named in the same conversation, the Wola may use the word "semonda" to indicate this relative ranking. Finally, "semgᵉnk" would refer to still lower-ranking taxa, which would be subdivisions of a taxon deemed a "semonda." All these terms are context-dependent; a given taxon may be called "sem" in a conversation, "semonda" in another and "semgᵉnk" in yet another context. Thus, no absolute ranks are implied, even though the Wola have words to designate relative ranks, and I am unaware of a single convincing report documenting the use of absolute folk taxonomic ranks by indigenous people. Evidence of absolute ranks in most folk taxonomies thus remains elusive.

Not all ranks are equally easy to characterize. Indeed, even some enthusiastic proponents of the existence of ranks in folk taxonomy (Wierzbicka 1992: 5) admit that "the concept of life form—despite its widely recognized importance—is surrounded by a great confusion." This is not just a matter of not having access to much data about exotic languages spoken by a dwindling number of distant tribes; even in English folk taxonomy, the life form concept remains vague, as emphasized by Wierzbicka (1992: 6):

> In my view, however, the situation in English with respect to the concept life form is far from clear. This lack of clarity, prevailing even with respect to a shared language to whose data all the writers on the subject have intuitive access, highlights the conceptual confusion that continues to surround the concept.

Similarly, Wierzbicka (1992: 12) disagreed with the previous claim, stated also by Atran (1985), that life forms encompass all of a given local flora, notably because there are unaffiliated folk generics (which belong to no life form). These disagreements do not necessarily mean that life forms are fictitious; Wierzbicka (1992) simply argued that the criteria that had been suggested in previous studies to characterize them were insufficient. Similarly, Wierzbicka (1992: 10) presented a strong case against the intuitively satisfying but apparently false idea that "animal" (in English) is a unique beginner but argued that it is a life form, instead, and its meaning is closer to that of "mammal," or perhaps "amniote." The unique beginner in English folk taxonomy closer to the scientific meaning of "animal" is "creature" (Wierzbicka 1992: 11; Goddard 2018).

The number of folk taxonomic ranks is controversial. Even though most authors seem to recognize about six such ranks (Table 1.1), others have argued that there are as few as three or as many as eight. Thus, Atran (1987b) recognized only three ranks, which are, in decreasing order of inclusiveness: unique beginner, life-form and basic taxa, also called "generic-specieme," to emphasize that most folk taxonomies do not distinguish between genera and species. This is perhaps best explained in this quote (Atran 1987b: 207–208):

> the distinction between species and genus is largely irrelevant to a basic appreciation of the flora and fauna of a local environment; historically, the distinction only assumes a modern character in

connection with Europe's Age of Exploration, that is, in the context of a search of worldwide order—an order transcending local concerns.

Conversely, Guasparri (2022) has shown that antique Roman folk classification suggests the presence of eight taxonomic ranks (see Section 1.1.4). These debates about the nature and number of folk taxonomic ranks raise doubts about the objective basis of these ranks.

### 1.1.1.4 Folk Taxonomic Ranks vs Linnaean Categories

The variable correspondence between folk taxa of a given rank and the equally ontologically dubious rank that has been attributed to taxa under RN (Table 1.1; also see Chapter 4) raises further doubts about folk taxonomic ranks. For instance, Atran (1998: 549) stated that:

> Generic species often correspond to scientific genera or species, at least for those organisms that humans most readily perceive, such as large vertebrates and flowering plants. On occasion, generic species correspond to local fragments of biological families (e.g., vulture), orders (e.g., bat), and, especially with invertebrates, even higher-order taxa.

Pawley (2011: 424) provided similar arguments illustrated through different examples. For instance, many folk taxonomies include a folk generic name for "pig," which more or less matches the zoological genus *Sus* (in this case, nomenclatural levels match), but many folk taxonomies include a folk generic name for "frog," which matches the zoological taxon *Anura*, which is typically ranked as an order (thus, much higher than genera). Thus, generic species, which are supposed to form the core of folk taxonomies, appear to be composed of quite heterogeneous entities, ranging from species or genera to orders (note that bats, which form the order evoked in the quote, include about 1,400 currently recognized species, according to https://en.wikipedia.org/wiki/Bat, consulted on March 8, 2021) or even higher-level taxa, according to the RN used by most contemporary systematists.

Higher-ranking folk taxa differ even more from biological high-ranking taxa, as explained by Pawley (2011: 425):

> At higher levels the methods and motives underpinning folk and scientific classifications tend to diverge and therefore so do the categories. The higher you go, either in a folk taxonomy or a scientific taxonomy, the less the taxa are likely to conform to "natural kinds." Western biologists aim to capture the evolutionary relationships of organisms, a concern peculiar to science. Ordinary people, on the other hand, impose higher order categories on cultural grounds, be they pragmatic or cosmological. Sometimes the grounds for grouping different kinds of organisms are broad likenesses in form and behaviour. For example, it is fairly common to find a taxon that subsumes both birds and bats, or both fish and whales. In other cases, social factors (e.g. ritual restrictions and taboos) and technological or economic factors (e.g. techniques used to obtain or process foods) peculiar to a society influence the grouping.

As explained in the introduction of this book (and in more detail in Section 6.2), systematists and philosophers of science have long recognized the lack of objective basis for Linnaean categories (e.g., Minelli 2000; Ereshefsky 2002; Laurin 2008); they persist only because they are required by RN. Indeed, a recent multi-authored paper by proponents of the RN stated clearly that:

> nomenclatural ranks do not have biological definitions or meanings and that they should never be used in an "absolute" way (e.g., to express degrees of genetic or phenetic divergence between taxa or hypothesised ages of cladogeneses) but in a "relative" way.

> **(Dubois et al. 2021: 5)**

This fact, along with the heterogeneity of correspondence between generic species and the category that has been attributed to taxa by systematists, suggests that the generic species rank is defined by linguistic and perceptual considerations for their names and delimitation, respectively, but that generic species are not united ontologically. Just like other (higher) Linnaean categories, they are "ontologically empty designations" (Ereshefsky 2002: S309).

### 1.1.1.5 Folk Taxonomic Ranks vs Hierarchy

Incoherence or uncertainty in the ranking of folk taxa (without regard to how equivalent taxa are ranked by systematists) also raises doubts about the validity of the concept of folk

**TABLE 1.1**

Schematic summary of commonly recognized folk taxonomic ranks, with a few selected examples of taxa for each rank (using vernacular names when possible, or the closest scientific name otherwise), and a highly tentative, admittedly imperfect, proposed correspondence between the folk taxonomic ranks and Linnaean categories (many exceptions show that these correspondences are imperfect).

| Folk taxonomic rank name | Example of taxa thus ranked | Potential equivalent rank in biological nomenclature |
|---|---|---|
| unique beginner | plant, animal | Kingdom |
| life form | bird, mammal, tree, bush | Class |
| Intermediate | *Curimatidae* (*Teleostei*)[1] | Class to genus |
| generic or generic-species | shark, dog, oak | Order to genus |
| folk-specific | poodle, white oak | Species, subspecies, race |
| folk-varietal | toy poodle, swamp white oak | Species, subspecies, race |

[1] Berlin (2014: 96, fig. 2.7). From the Huambisa (also known as the Wampis, from Peru and Ecuador) classification.

taxonomic ranks. Thus, Berlin et al. (1973: 230) reported incoherence between inferred folk taxon rank and hierarchy. Quoting a personal communication from Bulmer, they wrote of the Karam folk taxa:

> Thus if one regards *jejeg, lk* and *gwnm* [three Karam folk taxa] as polytypic genera, one is faced with the awkward situation that the subdivisions of one are conceptually **varietals**, of another are conceptually **specifics** and of the third include one **specific** and one which is itself at least **covertly generic**.

This ranking difficulty was also discussed by Bulmer (1974: 23):

> If one only tabulated those Kalam uninomials which can **not** form part of binomials except as the first segment of these, then their correspondence would be overwhelmingly with what Berlin regards as "life-forms", and as "unaffiliated generics". This raises two further problems, if one attempts to follow Berlin's schema. One is how to distinguish clearly between unaffiliated generics and life-forms.

Bulmer (1970) and Pawley (2011: 424) interpreted Kalam folk taxonomy as including many folk specific taxa with uninomial names (composed of a single primary lexeme), which is at odds with Berlin's hypotheses. Bulmer (1974: 20) seemed reasonably confident in this interpretation because he stated that "While I have not been able to make anything like an exhaustive count, my strong impression is that the **majority of uninomials** is normally applied at a level which corresponds logically to that of 'species.'" Note that, in this quote, "logically" indicates that the word "species" is used in its philosophical, rather than biological sense, but this does not mean that these entities are higher-ranking than biological species. This is clarified in the following quote (Bulmer 1974: 21):

> **Uninomially named cultivars** or groups of cultivars [which include cultivated forms of a single species of plant, or of a taxon of lower rank] are in most cases seen by Kalam as contrasting in multiple characters. For example, sweet potato taxa are distinguished by shape, colour and quality of leaf and of vine, and hairs on leaf and vine, as well as by size, shape, skin-colour, flesh-colour, texture and flavour of the tuber. Thus **logically they are "species" rather than "varieties."** And we may again say, as with vertebrate animals and wild plants, that **most uninomial taxa are logical "species."**

If correct, this is a major departure from Berlin's conclusions of how folk taxa are named (with a generic name and a modifier). Bulmer (1974: 21) was fully aware of this because he mentioned that "Berlin suggests that the great majority of what I have here called uninomial taxa should be termed 'generics' or 'folk-genera.'" After developing his arguments against Berlin's views on this point, Bulmer (1974: 22) concluded that:

> there are, empirically, too many cases on record where uninomials in fact apply to logical (and often also biological) species.

Thus in three areas out of four in Kalam folk-biology which I have described above, uninomials are applied in the majority of cases to taxa Kalam see as "species," not "genera."

To document this further, Bulmer (1974: 22) provided rather convincing basic statistics:

> In the case of vertebrate animals, 66 % of Kalam uninomials appear to correspond to biological species, 21 % to groups of species, 11 % to divisions of species (polymorphic or sexually dimorphic forms, or life-stages), and 2 % to taxa which cross-cut species, classing a division of one species with either a division, or the totality, of another (see Table C). If Kalam conceptualisations are taken into account, 77% of uninomials apply to categories within which they do not recognise complex internal differentiation, 16 % to groups within which such internal differentiation is recognised, and 7 % to divisions (e.g. sexually dimorphic forms, or life-stages) within or cross-cutting "species."

Berlin et al. (1973: 239) reported that another earlier study on Gimi (a language spoken in Papua New Guinea) classification (by Glick, published in 1964) also expressed difficulties in ranking folk taxa:

> There are more than twenty [taxonomically defined first level] botanical categories [taxa], ranging in size from *da* "tree," with at least two hundred members, through *koi* "ginger," with four, and on down to several problematical sets containing only two or three members apiece. At the lower end [of the taxonomic hierarchy] it becomes difficult to decide whether one is justified in calling a pair or trio of closely related plants a category: **does this have the same taxonomic rank** as say, *da* in Gimi thought? My answer is, probably not.

These concerns were echoed by Hunn (1998), who stated that "Atran's own data suggest that rank is not an essential feature of the ethnobiological module" and that:

> In fact, there is substantial evidence gathered by other ethnobiologists that these "life-form" taxa are a **motley crew of categories** grounded in whatever association is handy, which may be morphological similarity, ecological contiguity, common utility, or some other symbolic linkage.

Note that, in this quote, the word "category" is used in the sense of "taxon," rather than for the folk equivalent of Linnaean categories. This use of the word "category" in the literature on folk taxonomy is extremely common and confusing to systematists who distinguish between taxa and the category (absolute rank, such as genus or family) that is typically attributed to them.

Pawley (2006) went a bit further in stating that:

> one can interpret the Wayan data variously to arrive at more or fewer levels, depending on what status is

given to narrow and broad generics in polysemous lexemes and on whether certain taxa that cross-cut other taxa are accepted as representing a level in the taxonomy.

A similar problem had been noticed by Healey (1979: 376) in the taxonomy of raptorial birds used by the Maring-speaking people of the Simbai and Jimi Velleys of the Papua New Guinea highlands. After presenting the most hierarchical version that could be produced of that taxonomy, which included a maximum of five levels, he acknowledged that:

> The most important complication, if it can be called that, is that as far as most Maring are concerned, the classification of birds can be accom[m]odated in a **much more simple arrangement** than that discussed above and summarised in Figure 1.1. In general, **intermediate taxa are rarely employed** by the Maring in everyday discourse. In discussing birds with informants I often asked if certain groups of terminal taxa were grouped together in a named higher-order category. In reply, informants often agreed that the birds in question were similar, but with few exceptions they usually asserted that **there were no intermediate taxa**. There was, they said, an inclusive name kabaq for all birds, and separate names for each different type of bird[4]. For the most part, then, the Maring recognize **only a two level hierarchy**, and even when called upon to detail more complex taxonomic arrangements **will refuse to do so**.

Other authors have pointed out that inferring folk taxonomies is a complicated, difficult task and that the neat picture depicted by Berlin et al. (1973) may not be as well-supported by all the relevant data as it might seem at first glance. Thus, Ellen (1998: 572) indirectly raised doubts about the existence of the folk taxonomies that have been summarized in various publications because "this [taxonomic] knowledge is not carried around in its entirety in the heads of individuals, but is socially distributed. The things we call ethnobiological classifications are an **emergent product** of applying core folk biological knowledge."

The fact that folk taxonomies are so variable and difficult to interpret, with a highly variable number of ranks, raises doubts about the reality of absolute folk taxonomic ranks, or at least their prevalence in such classifications.

### 1.1.1.6 Polysemy and Variability in Delimitation in Folk Taxonomies

Folk taxonomies attest to the antiquity of the phenomenon of polysemy (more than one meaning associated with a given name), which in the context of folk taxonomy typically occurs to two nested taxa (one of which includes the other) of consecutive ranks, most frequently when a given name is applied to a folk species and a folk genus, or a folk genus and a folk life form. In these cases, the lower-level taxon is usually considered the most typical, or most salient (easiest to recognize, based on various criteria, such as size, shape or color)

representative member of the more inclusive taxon (Berlin et al. 1981: 106–107; Berlin 2014).

Pawley (2006, 2011) documented this phenomenon in various Oceanic languages spoken in Pacific archipelagos that are presumed to derive from Proto Oceanic. In these languages, such as Wayan (which is spoken on some western Fiji islands), some folk taxon names are applied to two nested taxa. This can be illustrated by the Proto Oceanic name "*ikan*," which became "*ika*" in the Proto-Central-Pacific language, as shown by this name in many contemporary languages thought to derive from it and documented by Pawley (2011). The Wayan folk taxon name "ika" thus designates two nested taxa that are similar to those designated by the word "fish" in English (Table 1.2). The most inclusive taxon designated by "ika" includes actinopterygians (ray-finned fishes, in vernacular language); actinopterygians plus chondrichthyans (sharks, skates and rays); the former plus some marine mammals (whales and dolphins); and, according to some (but not all) Wayan speakers, all the former plus cephalopods (squid and octopus). English once had such a comprehensive meaning for "fish" as shown by the etymology of various taxon names, such as "cuttlefish," which is a cephalopod mollusk (like the squid and octopus). The least inclusive taxon designated by "ika" includes only actinopterygians, or these and chondrichthyans (variable among Wayan speakers), which coincides fairly closely to the meaning of "fish" since the late 19th century.

### TABLE 1.2

Successive sets of taxa that may be included in the taxon called "*ikan*," which can be translated as "fish" in Proto Oceanic, an ancient language that has been reconstructed on the basis of several contemporary languages spoken in some Pacific islands and archipelagos, such as New Guinea, Vanuatu, the Solomon Islands and Micronesia. The taxon sets are listed in decreasing order of relative frequency of inclusion in "*ika*" (the Proto-Central-Pacific variant), meaning that taxa included in category 1 are considered core taxa (always included in *ika*), whereas those in category 9 are the least frequently included. These frequencies are obtained by counting the languages in which taxa of each category is considered part of *ika*. Slightly modified from Pawley (2011: table 8.1) by adding, in brackets, the scientific names of selected relevant taxa; otherwise, the original wording has been retained. The title of this table in Pawley (2011) is "Implicational scale of types that may be included in reflexes of *ika*."

| Rank | Included taxa |
|---|---|
| (1) | typical fish [actinopterygians] |
| (2) | 1 + sharks and rays [elasmobranchs] |
| (3) | 2 + cetaceans (whales, dolphins) and dugongs [sirenians] |
| (4) | 3 + eels [anguilliforms] |
| (5) | 4 + turtles [testudines] |
| (6) | 5 + crocodiles |
| (7) | 6 + cephalopods (octopus, squid, etc.) |
| (8) | 7 + decapod crustaceans (crabs, crayfish, prawns and their relatives) |
| (9) | 8 + other aquatic invertebrates (mollusks with shells, sea hares, nudibranchs, echinoderms, sea urchins, sea cucumbers, jellyfish, etc.) |

Pawley (2011) documented the meaning of "ika" and variants thereof (called "reflexes" by Pawley) in several other Oceanic languages (such as Tongan, Tahitian and Hawaiian) and found that they correspond to an even greater variety of nested taxa. Each Oceanic language may have a different delimitation of "ika," and some (but not all) recognize two delimitations simultaneously. Pawley (2011) even sorted the component taxa (included lower-ranking taxa) in decreasing order of frequency of inclusiveness (Table 1.2). The core taxon, which is always included in "ika," is *Actinopterygii* ("ray-finned fishes" in vernacular English). Elasomobranchs (sharks, skates and rays) are frequently included too, whereas various marine mammals are less frequently included. Echinoderms (starfish, sea urchins and sea cucumbers, among others), gastropod and bivalve mollusks, and cnidarians (but only jellyfish, typically) are included the least often.

The application of the name "ika" to two nested taxa within a single linguistic community, and even more nested taxa in a set of closely related linguistic communities (Table 1.2), is not completely congruent with Berlin et al.'s (1973) hypotheses. The two nested taxa called "ika" should be life forms, given their hierarchical relationships with lower-ranking taxa, but life forms should be characterized by prominent morphological characters. Yet, the largest of these two nested taxa (according to Wayan use; this differs in other Oceanic languages), "ika" includes both taxa with gills (actinopterygians and chondrichthyans) and some without. In other Oceanic languages, such as Samoan, "ika" can include even mollusks and crustaceans, in addition to the core taxa; it is even more difficult to see how such a taxon could be "recognized on the basis of numerous gross morphological characters" (Berlin 1973: 261).

To an extent, polysemy exists in contemporary systematics. In the context of the taxon names regulated by rank-based codes, this situation is called "homonymy" (a single name associated with two or more taxa). Officially, homonymy is not tolerated by these codes (only the oldest, first-established name is considered valid), except in the case of tautonymy, the practice of using the same name at two levels of a classification for two or more nested taxa. This practice is allowed by the *Zoological Code* (ICZN 1999), which regulates zoological RN, but is forbidden by article 23.4 of the *International Code of Nomenclature for algae, fungi, and plants (Shenzhen Code)*, or "*Botanical Code*" for short (Turland et al. 2018). Tautonymy can be illustrated by various zoological taxon names in which the same word is used both for the genus name and for the specific epithet, as in *Bufo bufo* (the common toad, also known as European toad) and *Anser anser anser* (the western greylag goose; the third "*anser*" designates the subspecies). However, given the fact that RN does not delimit taxa (see Chapter 2), polysemy is actually quite common under RN, simply because a given name is applied to several nested clades but is considered to be the same taxon simply because it is ranked the same (for instance, a genus, family or order).

Variability in delimitation in folk taxonomies does not only occur in cases associated with polysemy. Another source of uncertainty is the exact meaning of folk taxonomic names, or rather, the variability of this meaning. As Pawley (2011: 426) indicated,

> For any generic term, the assumption that there is a single correct definition of its semantic range that can be uncovered by careful research is itself questionable. The fact is that members of a speech community often do not agree completely as to the scope of generics.

In addition to this variability within a fairly small linguistic community, there may be important variations between more distant communities, as shown by the nested taxa referred to by "ika" previously discussed (see Table 1.2).

Hatano (1998), a cognitive scientist, doubted the existence of ranks in folk biology, for various reasons. Notably, he argued that biological classifications may vary more between and within cultures than supposed by some proponents of these ranks, such as Atran (1998). He also disputed the interpretation of Atran's (1998) findings, among other reasons because the properties of taxa that the participants were questioned about (to make inferences) may have influenced the outcome of the test, and because the differentiation of life forms into generic species plausibly varies along with familiarity with various taxa. Thus, he suggested that "we cannot generalize from the results with conspicuous organisms to the entire folk kingdom. It is conceivable that American adults have almost no differentiation among generic species, life forms, and intermediate groups for less conspicuous organisms (e.g., bugs, mushrooms)."

### 1.1.1.7 Hierarchy and Ranks versus Societal Importance of Taxa

A basic debate about some of the principles of folk taxonomies proposed by Berlin et al. (1973) and Berlin (2014; note that this is a reprint of a book published in 1992) is that they may not be entirely or even predominantly hierarchical, and that they may be more utilitarian than Berlin suggested. Thus, Newmaster et al. (2007) suggested that, contrary to Berlin's interpretations, folk taxonomies include "multifarious mechanisms" and, following earlier studies, considered the possibility that folk taxonomies form a "complex web of resemblances."

Ludwig (2018) seriously questioned the universality of some of Berlin et al.'s (1973) conclusions, which were instrumental in the rise of "convergence metaphysics" and insisted on the diversity of folk taxonomies, which he argued are based on several criteria, including utilitarian ones. This was hardly a new claim, because Diamond (1966), in his analysis of the Fore people of the New Guinea Highlands, had already concluded that "The origin of Fore classification is probably utilitarian." This point about the importance of utilitarian criteria in folk classification also echoed claims by Hunn (1982: 831) that can be illustrated by this brief quote:

> Berlin, in a theoretical stock-taking addressed to an audience of biosystematists (1973), felt called upon to stress that "less than half of the named folk generic classes [i.e., basic core folk taxa] of plants in the folk botany of the Tzeltal . . . can be shown to have any cultural significance whatsoever."

Hunn (1982: 831) then explained that "Some [of these culturally insignificant taxa] are poisonous, others invasive weeds, other inedible 'twins' closely resembling edible forms, others useful 'just' as firewood, and so forth." Indeed, such taxa can hardly be called "culturally insignificant"! Cultural influence on folk taxonomy is well-documented. For instance, Hunn (1982: 834) mentioned that "For example, Sahaptin speakers, who depend heavily on fish as a staple food, recognize 60% of the native fish species nomenclaturally, but only 25% of the native bird species." However, note that the utilitarian criteria used by some preliterate societies could still lead, under some conditions, to recognition of a similar set of low-level taxa as Western systematics (genera or species under RN), as Diamond (1966: 1104) suggested might have happened with the Fore of New Guinea.

The difficulty in assessing the practical importance of folk taxa was perhaps best explained by Bulmer (1974: 12–13):

> The recognition of both the objective and subjective importance of ecology to human communities throws light on the problem of the classification and naming of **apparently useless** animals and plants. If one sees individual plant and animal categories [folk taxa] solely in their direct relationships to man, there are many which appear irrelevant, neither utilised nor noxious. However if the relationships between different kinds of plants and animals are recognised as relevant, then a great range of additional forms will very usefully be identifies and classified. If hunting or collecting of wildlife are economically or socially important, or even if these activities are unimportant but accurate observation of wildlife is still culturally relevant for other reasons (e.g. signs of seasonal change, omens) then because animals exist in significant relationships with other animals and plants, there is a **considerable impetus to classify these forms** also.

This passage explains well why various taxa that are not exploited by various societies might still be socially significant and thus, why names were coined for such taxa, even if the motivation was ultimately utilitarian, in the broad sense of the word (also see Sillitoe 1995: 209). Another factor militating for utility as a significant criterion guiding the choice of taxa to name among pre-literate people is that vertebrates are over-represented compared to other metazoans. Thus, Bulmer (1974: 18) stated that: "Unlike vertebrate animals, not all invertebrates are named by Kalam. There are many creatures, especially very small ones, of which Kalam will say that they do not know their names, or they have no names."

Ludwig (2018) also noted that Berlin's earlier work had expressed interpretations that recognized more importance of practical criteria and emphasized less taxonomic hierarchy. For instance, Berlin et al. (1966) reported that out of a sample of 200 native plant names used by the Tzeltal Maya, only 68 corresponded to a botanical species and of these, 40 were introduced by the Spanish and the majority of the Tzeltal names for these plants were derived from the Spanish. They concluded that the Tzeltal had produced a special, rather than general (that is, phylogenetic) classification that maps in a complex way to the scientific taxonomy. This contrasts rather sharply with Mayr's (1949: 371) earlier claim that the

"primitive Papuan of the mountains of New Guinea recognizes as species exactly the same natural units that are called species by the museum ornithologist." According to Ludwig (2018: 417), this claim bolstered the species realism that underlies the ideas proposed by Berlin et al. (1973). However, Ludwig (2018: 417) concluded that "Berlin's metaphysical picture of 'objective discontinuities in nature' that ground cross-cultural convergence in classifications has largely vanished from the [ethnobiological] research literature." Nevertheless, Ludwig (2018) did not reject the existence of cryptic ranks, and neither did most of the eight colleagues who commented on that paper (in the "Comments" section of the paper).

Similarly, after expressing positive comments on Berlin's views but suggesting that Berlin downplayed the importance of utilitarian criteria (such as societal importance for agriculture, medicine and rituals, among others) in folk taxonomies, Alves et al. (2016: 120–121) concluded:

> Furthermore, the information presented highlights naming standards related to a hierarchical classification and to criteria based on morphological characteristics for the classification of living things. However, a more careful analysis does leave aside the idea that perhaps these findings [including the existence and generality of cryptic categories in folk taxonomies] are **merely artifacts** of the collection procedure or the **interpretation** of data, which forces an adaptation of the findings to the principles formulated by Berlin. Much criticism has been directed to the universality of Berlin's principles.

Similar concerns had been expressed decades earlier by Bulmer (1974: 23–24):

> Thus I do not believe that we can yet demonstrate correspondences between the cognitive statuses of folk-taxa and the nomenclature applied to these which are of sufficient intra- and cross-cultural regularity to enable us to arrive at a **simple typology**. There is an obvious danger in advancing, prematurely, a typology of the kind Dr. Berlin proposes, namely that it may lead ethnographers and lexicographers to **distort data by forcing it into inappropriate pigeonholes**, and in particular into failing to appreciate and record the degree of **flexibility and elasticity** which is probably a very general feature of folk taxonomies. Thus for the time-being we should continue to use **two sets of terms**, one to describe the **cognitive status** of taxa, and also of the "covert categories" which are related to these, for which "genera", "species" and "varieties", and perhaps also "life-forms", may be appropriate designations; and the other to describe the **nomenclature** applied[9].

The most forceful, explicit statement that I have found about this possible "pigeonholing" problem was made by Sillitoe (2002: 1168):

> people are quite capable of perceiving mid-groups between formal named ones, [but] to call these "covert categories" (i.e., classes for which people have no

names but that the analyst deduces to exist using his or her taxonomic logic) is, I think, to **misinterpret their significance**. The status of these hidden classes and the ability of researchers to elicit them have been debated for some years (see Atran 1990: 43–46; Berlin 1974, 1992: 176–181; Berlin et al. 1968; Brown 1974, 1984; Descola 1994: 81; Ellen 1993: 119–121; Hays 1976; Taylor 1984). They are ad hoc and variable descriptions that vary from individual to individual and occasion to occasion.

If we interpret taxonomy as facilitating agreed communication, this accounts for some other anomalies. It accounts, for example, for the absence of any kingdom-level terms equivalent to animals or plants. These would be redundant. There is no reason to seek to integrate the primary life-form taxa into a higher-level order, **to complete some nonexistent hierarchy**. Closure at a higher level assumes a hierarchical conception that only makes sense backed up by a cultural construct such as an evolutionary theory, prompting taxonomic completion. For the Wola, the classification of animals is inherently dynamic and subject to negotiation; there can be no closure or final bounded version, no authoritative comprehensive arrangement. The absence of any life-form classes for some animals, such as bats, pythons, pigs, and dogs ("unaffiliated generics" [Berlin 1992: 171–181]) also complies with this interpretation. For a long time (to confess to an embarrassing ethnocentric exercise), **I have worried my friends into inventing "family" classes** for these, in which others have invented "covert classes." They are an **unnecessary fiction**.

This long quote does not deny that there is some hierarchy in folk taxonomies, but it emphasizes that it may not be as extensive as scientific taxonomy. In an earlier paper, Sillitoe stated that the names that some of his Wola friends invented after "repeated questioning" for covert life form taxa represented the result of "a foreign exercise prompted by the author" (Sillitoe 1995: 205). Note the evolution of his perspective between this statement and the conclusion in the more recent quote that these covert life forms are "an unnecessary fiction."

This next quote (Sillitoe 2002: 1169) further explains why folk taxonomies may be less hierarchical (or have fewer nomenclatural levels) than commonly believed:

The Wola have a taxonomy less in the Greek sense of a taxis (arrangement) of natural phenomena for intellectual purposes than an arrangement in the **political sense** of an agreement for settling differences. They employ their taxonomy to communicate everyday with one another about animals, without untoward disputation, not to arrange closely defined classes in a hierarchical scheme for debate about **evolutionary-ecological relations** nor symbolic exegeses. The "solution" to Wola disagreements, taxonomic anomalies, and **indifference to an integrated classification is so straightforward that I find it hard to believe that I have puzzled over**

**them for so many years**. Perhaps it was, I have to confess, that **a scientific education had blinded me to their reality**.

Hunn (1982) was more critical than most other authors about Berlin's view of a strictly hierarchical folk taxonomy because he thought that many folk taxa were artificial and based on utilitarian criteria. According to Hunn (1982: 835–836), the fact that taxa at a given folk taxonomic rank included both natural (i.e., monophyletic) and artificial taxa also raised doubts about the reality of these ranks, which he suggested were:

an attempt to fit the natural, polythetic core of a folk biological domain into the **procrustean bed** of a taxonomic hierarchy by interpreting this core as equivalent to a single taxonomic hierarchic rank, the **generic** partition. The fit is not adequate, as the examples discussed above should make clear. Artificial taxa creep into the generic "partition" as residuals . . . . I have demonstrated that both **natural and artificial** folk biological taxa may occur at Berlin's folk **specific rank** (Hunn 1977: 53). Here I will show that **the same is true** of taxa at Berlin's **life-form rank**, a fact with serious theoretical consequences for the taxonomic hierarchy model of folk biological classification. If there is no necessary correlation between the taxonomic rank of a taxon and its status as natural or artificial, the **notion of taxonomic rank** is shown to be a **purely formal distinction** imposed by the analyst.

While this quote may represent a minority point of view among ethnobiologists, it suggests that the folk taxonomic data that he has analyzed cannot be reduced to a neat hierarchy of near-phylogenetic taxa that can fit into five or six nomenclatural categories, contrary to the suggestions of Berlin et al. (1973) and Berlin (2014).

Hunn (1982) also demonstrated that some taxa that are typically considered life forms do not match the definition of this folk taxonomic level given by Berlin (1973: 261) as "easily recognized on the basis of numerous gross morphological characters." To support this point, Hunn (1982: 838) provided the example of folk taxonomic names that are the closest equivalent of birds (the taxon *Aves*), which frequently includes other flying animals. Thus, the Kalam life form *yakt* includes bats, which are mammals, but excludes the flightless cassowary. Even more telling, the northern Paiute folk taxon *yozidi* includes bats, birds and flying insects. These groups are united by the capacity for flight rather than morphology, obviously, and this evidence is inconsistent with Berlin's (1973: 261) characterization of life form taxa. Other examples of anomalous Kalam life forms were reported by Bulmer (1974: 23); for instance, the taxon *as* includes frogs and some small mammals. Worse still, some life form taxa may partly overlap with each other, although this varies according to the informant (Bulmer 1974: 17).

Hunn (1982: 838) also argued that "Life forms are often residual with respect to practical significance." For instance, in Sahaptin, the word *c'ic'k*, which can be translated as "grass," designates all herbaceous plants that are not "flowers" unless they are named (a name that is typically associated with social significance). Other examples exist, as in Tzeltal, in which the words for "vine" and "herbaceous plant" are also based on utilitarian criteria. Again, the obviously utilitarian criterion that plays a central role in delimiting these folk taxa does not match Berlin's (1973) definition of life forms as characterized by morphology.

This point is further explained, through an example, in the following quote (Hunn 1982: 835):

> The Cha-cha of the Virgin Islands recognize a large, heterogeneous category of fish called *corail* which includes fish on the basis of their "uselessness" for food and their similar patterns of behavior" (Morrill 1967: 408). Such categories are strongly reminiscent of folk English "weed."

According to Hunn (1982: 835), these "residual taxa" lack folk generic taxa, which are typically considered to form the core of folk taxonomies, and this reflects the fact that they are not natural taxa. Instead, they are "special purpose concepts" that represent a "nonresource."

### 1.1.1.8 Strict Hierarchy, or More Overlap than Previously Imagined?

Aside from the reality of folk taxonomic ranks and the role of utilitarian criteria, another important issue is the extent of hierarchy in folk taxonomies. Indeed, folk taxa can partly overlap with each other, as in the case of Zapotec folk taxa referring to birds (Ludwig 2018: 423). Healey (1979: 363) had already warned of the potential pitfalls of assuming that folk taxonomies were structured (with mostly dichotomous hierarchies) as our scientific taxonomies. He also suggested that previous evidence of some taxa being simultaneously included in two higher-ranking taxa of a given rank (that should be mutually exclusive) had been prematurely dismissed as "the result of individual variability or imperfectly shared knowledge" (Healey 1979: 363). Without denying that such explanations might be correct in some cases, he suggested that "the possibility that cross-cutting categories may be unambiguous or generally acceptable features of folk taxonomies does not seem to have received sufficient attention" (Healey 1979: 363). Thus, Healey (1979: 375) described strong evidence of overlap between high-ranking taxa of raptorial birds in Maring folk taxonomy:

> It will be noted that five terminal [low-ranking] taxa of birds can be classed as both tit'p-ronaq and jindak kabaq [two higher-ranking taxa]. In other words, the boundaries of these two secondary [higher] taxa **intersect**. Informants were quite aware of this overlap of categories, but pointed out that there were **good objective grounds** for classing owls with hawks in the tit'p-ronaq category of raptores.

After presenting this evidence, Healey (1979: 378) added that additional examples of overlapping higher taxa (that should not overlap, according to Berlin's suggestions) could be found in Maring folk taxonomy. More importantly, such examples occur in the taxonomy of other peoples, such as the Narak (eastern neighbors of the Maring), which Healey studied, and Healey (1979: 378) cited other studies that documented similar

examples among the Kalam (western neighbors of the Maring) and the Rofaifo of the Eastern Highlands.

These partly overlapping taxa do not match what Healey (1979: 379) called the "principle of taxonomic rigidity," which stipulates that:

> each species is included in a single genus, which is in turn included in a single family, and so on. Or, to restate the principle in terms applicable to folk taxonomies: no taxon may be included simultaneously in two or more higher-order taxa of coordinate [identical] hierarchic rank.

Healey (1979: 379) pointed out that the principle of taxonomic rigidity underlies the proposals of Berlin et al. (1973). However, Healey (1979: 380) concluded that, "Quite simply, there seem to be no good logical grounds for presuming that folk taxonomic systems should be structured on the principle of taxonomic rigidity." Healey (1979: 381) also concluded that the formalized data collection methods that had been used since the 1970s could generate new problems because "they impose the ethnographer's cognitive framework upon our informants, not always explicitly, instead of allowing our informants to speak for themselves."

Similar doubts were subsequently expressed in more general terms by Ellen (1998: 573): "In positing a universal 'abstract taxonomic structure,' the approach all too often seems to be **to delete features of people's classifying behaviour** of living organisms **that do not fit the expected pattern**, until such a pattern is obtained." Ellen (2016: 14) concluded similarly that classification of palm trees by the Nuaulu contains overlapping groups and that Nuaulu linguistic and ethnographic data "do not conform to any pseudo-Linnean local ontology." These findings support Ludwig's (2018: 417) conclusion because, in biological taxonomy, taxa normally do not overlap with each other, except for hybrids, and except for the trivial cases of taxa that are completely contained in higher-ranking taxa (such as birds and vertebrates, respectively).

These statements, without precluding an important hierarchical component in folk taxonomies, seem to support Hunn's (1982: 835) distinction between two models. The first, the **"taxonomic hierarchy model,"** advocated by Berlin, "envisions folk biological domains as sets of taxa at various levels related by set inclusion. This model owes its form to a Linnean analogy and a set theoretic formulation." The second model, which Hunn (1982: 835) calls the **"natural core model,"** includes a general purpose core of natural taxa (which often match those of scientists, but often include morphologically salient, paraphyletic or polyphyletic taxa) surrounded by **special purpose** (artificial) groups (which are not clades).

### 1.1.1.9 Conceptual Differences between Folk Taxonomic Ranks and Linnaean Categories

When summarizing Berlin's ranking scheme, Newmaster et al. (2007: 234) explained that according to that view, "These ethnobiological categories or ranks are similar to taxonomic ranks in Western zoology and botany." This is not entirely accurate, for a few reasons.

First, systematists and philosophers of science have long been aware that Linnaean categories (absolute ranks) are merely artificial constructs required by RN (Ereshefsky 2002; de Queiroz 2005; Laurin 2008), even though a few unsuccessful attempts have been made to render at least some (for instance, Dubois 1982) or all (such as Avise and Johns 1999) of these ranks objective. However, the recent literature suggests that many ethnobiologists and linguists interested in folk taxonomies may be unaware of the artificial, subjective nature of the Linnaean categories (for instance, Wierzbicka 1992). Similarly, many folk biologists (such as Pawley 2011) still seem to conceptualize taxa as natural kinds (classes), even though Wierzbicka (1992: 23) supported the claim that "terms for living things are a kind of **proper name**, while at the same time maintaining my earlier claim (1985) that all **natural kind terms** can, and should, be defined." Wierzbicka (1992) generally seemed to consider that folk taxa are natural kinds, but proper names are usually applied to individuals, which leaves some ambiguity in her position. On the contrary, many systematists now conceptualize them as individuals (see the section entitled "Ontology of Taxa" in the Introduction). Arguably, it makes less sense to attribute absolute ranks to taxa if they are individuals than if they are classes (because classes could be subdivided into smaller classes of a given rank using the same set of attributes, but such a procedure is inappropriate to rank individuals). However, this consideration is only moderately relevant to folk biologists because the conceptualization of taxa among pre-literate peoples must be closer to natural kinds (classes) than to individuals. Nevertheless, the latest developments in systematics, like considering taxa individuals and possibly dropping Linnaean categories, make it increasingly more different from folk taxonomies.

Second, indigenous peoples rarely refer to ranks and, when they do, these are only relative (rather than absolute) ranks. The Fore of New Guinea may constitute the sole exception, if they named absolute ranks, but this seems dubious (see Section 1.1.1.3). On the contrary, practitioners of RN (which includes all systematists in the 20th century and still includes most of them today) routinely explicitly name these ranks in the context of their nomenclatural work. Indeed, this is a requirement of RN because rules of priority to resolve cases of synonymy and homonymy apply within a given rank (see Chapter 2).

A third, more minor difference is that Linnaean categories are much more numerous, up to about 36 (Laurin 2005: 77), than folk taxonomic categories (levels), only five or six of which are present in many folk taxonomies, with a maximum of at least eight in Roman folk classification (Guasparri 2022). This, of course, reflects the great level of technical sophistication of Western science (for which 36 ranks is insufficient; see Section 3.5), with specialized taxonomists that have no true equivalents in any indigenous culture. The high number of ranks in Roman folk classification is consistent with this hypothesis given that antique Romans had an advanced civilization and access to written documents, including a large corpus of older Greek literature. Similarly, for English folk classification, Wierzbicka (1992: 16) suggested recognizing folk subgeneric taxa, for taxa such as siamese cats (which in zoological nomenclature are ranked below the subspecies level) or blue spruce (a species, under the *Botanial code*). But

it looks like as folk taxonomies are scrutinized more closely, the six ranks originally proposed (Berlin et al. 1973) are insufficient for some folk taxonomies.

A fourth difference is that systematists working under RN strive to assign every known biological organism in taxa of all ranks, from kingdom (or even higher, such as domain) down to the species level (sometimes with some uncertainty). Thus, a specimen attributed to a species is also attributed to a genus (as required by binominal nomenclature), family, order, class and so on. In contrast, in folk taxonomies, some organisms are simply assigned to an incomplete set of high-ranking taxa (in which some ranks are missing). As explained by Hunn (1982: 834):

> The fact that **only a fraction** of the potential natural discontinuities may be recognized in a folk biological classification creates theoretical difficulties for the taxonomic model. What is to be done with all those **unclassified entities**? In some cases they are simply left out of the basic level of classification—there will be empty regions in taxonomic "space," regions where many or all tokens are recognized only in very general terms, for example, as **some kind of "bird,"** but are not recognized as some particular kind, that is, in Berlin's terms, as a member of a folk generic taxon.

The best-documented examples of this are the unaffiliated generics, which do not belong to a life form and that can only be attributed to a unique beginner (Berlin et al. 1973; Atran 1985). Unaffiliated plants include corn (cultivated), bamboo and agave (morphologically aberrant) in Tzeltal (a Mayan language spoken in the Mexican state of Chiapas). Among the Cantonese boat people, unaffiliated animals include morphologically aberrant taxa such as the horseshoe crab (a chelicerate, a distant relative of arachnids), starfish (an echinoderm taxon) and jellyfish (a large group of cnidarians), if considered generic taxa, but they could also be considered life form taxa (Berlin et al. 1973: 239). Unaffiliated taxa are not numerous; Berlin et al. (1973) reported that in Tzeltal, the four plant life form taxa (trees, vines, grasses and broad-leafed herbaceous shrubs) include at least 75% of the 471 plant generic taxa. The unaffiliated generics are typically morphologically aberrant or economically important taxa.

### 1.1.1.10 Significance of Folk Taxonomic Ranks

If Berlin et al. (1973) and Atran (1998), among others, are correct in considering that folk taxonomies are tacitly ranked, this teaches us about how the human mind works in classifying and naming biodiversity rather than about the taxa themselves because there is now a consensus among systematists that there is no objective basis to attribute such absolute ranks to taxa themselves (see Chapter 3), with a possible exception of species, for which there is less consensus (see Chapter 6). Berlin et al. (1973: 233) may have alluded to this in a passage that includes a quote from an earlier study of the Cantonese boat people published by Anderson in 1967:

> Finally the author [Anderson] notes that the structural characteristics of English, Cantonese, Tahitian,

and Latin nomenclature "are more similar than coincidence alone [can] explain" [end of quote from Anderson]. As our broader comparative evidence reveals, these similarities must certainly reflect **identical ethnobiological nomenclatural principles** employed by many prescientific peoples in their linguistic treatment of the biological universe.

Ludwig (2018) pointed out the influence of Ernst Mayr (a proponent of rank-based nomenclature, as were all systematists at that time) on Brent Berlin's ideas. This raises the possibility that the cryptic ranks in folk taxonomies were inspired by the biological systematic tradition because, at the time, biological nomenclature of taxa was necessarily rank-based, as all practicing systematists used RN. This may have led Berlin and his followers to think that a scientific nomenclature of taxa required absolute ranks (like Linnaean categories), but the subsequent development of PN, which started in the late 1980s, shows that this is not the case (see Chapter 4). Of course, whether or not Mayr's influence played a major role in subsequent development of folk taxonomic studies, Berlin's ideas were obviously influential and widely accepted in the field (though some components of his proposals are more controversial). This (along with a lack of awareness about the development of PN) can be illustrated by this quote from Wierzbicka (1992: 4):

> although the **importance of absolutely ranked taxonomies** in human thinking and in human representation of knowledge **has not been called into question** (on the contrary, it has been reaffirmed), the universality of this principle (across domains, not across cultures) has been questioned. Indeed, the special link that this principle appears to have with the domain of "living things" has been used as a major argument in favor of "domain specificity" in human cognition in general.

Folk-taxonomic ranks are primarily defined on the basis of linguistic criteria and they are not tightly associated with rank-dependent ontologies, contrary to the futile search for the right definition of "biological species" among biologists (see Section 6.2.2). Berlin (1973) proposed some perceptual criteria to define some folk taxonomic ranks, but as we have seen, these criteria have been criticized far more than linguistic criteria. In other words, most ethnobiologists did not suggest that folk generics share similar intrinsic properties; they are classified as such primarily because of how various indigenous folks perceive and (more importantly) name these taxa. Thus, folk taxonomic ranks reflect human language and cognitive aspects of the human mind rather than any putative property of the folk taxa of a given rank, and it seems that folk taxonomists are well aware of this. This was nicely summarized by Wierzbicka (1992: 13):

> **Psychological salience** may be a characteristic feature of folk genera in many speech communities (especially those living largely in a natural, not human-made, environment), but it **need not be, and should not be, the basis for defining** and

distinguishing these categories (cf. Berlin 1990: 89). If it were, it would lose most of its value as a theoretical construct and as a tool for cross-linguistic comparisons. The **criterion of polytypicity is incomparably more useful and more illuminating**.

This conclusion seems sensible because the sets of taxa recognized by indigenous people presumably reflect many relevant considerations, such as the local biodiversity, perceptual salience and societal relevance (interpreted widely). However, one possible drawback of this position is that this may make distinguishing life forms from societally important folk generics (with several recognized folk-specific taxa) difficult, although the latter may be characterized by several cultivated forms (recognized as either specific or varietal taxa).

To sum up, folk "taxonomic" ranks, if they exist, are essentially characterized by linguistic criteria along with the hierarchy of named taxa and, as such, it would be more correct to call these "nomenclatural ranks." Berlin (1973: 260) had suggested some perceptual criteria (like "gross morphological similarities and differences") for life form taxa, but these have been more contentious and do not seem to hold up to close scrutiny. Given how systematists and philosophers of science now conceptualize taxa (generally as individuals defined by their history; see Section 6.2), it is clear that there can be no link between the folk nomenclatural ranks (or Linnaean categories) and the ontology of the named taxa. The great differences in taxonomic resolution between large animals (especially vertebrates) and smaller ones (like insects) in folk taxonomies also strengthen this conclusion.

Ethnobiologists appear to have used these folk nomenclatural ranks more appropriately than biologists in their comparative studies, presumably because of the primarily linguistic criteria used to define and recognize them. Thus, no ethnobiologist has tried to compare, to my knowledge, the biodiversity of the Neotropics with that of the Paleotropics by comparing the number of generic or folk-specific taxa recognized by indigenous peoples of these two regions. However, this is the procedure that numerous biodiversity studies have followed by counting taxa of a given Linnaean rank to assess the biodiversity of these regions (e.g., Prance 1994). This procedure, called "taxonomic surrogacy," has of course been criticized as ontologically invalid (e.g., Bertrand et al. 2006; Laurin 2010a; also see Section 3.4.2). If such comparisons were ever made using folk taxa, ethnobiologists would interpret the findings as revealing more about how indigenous perceive and use their surrounding biodiversity than about this biodiversity (Pawley 2006; Alves et al. 2016; Ellen 2016; Ludwig 2018), for obvious reasons. Thus, species-level folk taxa may be nearly as numerous as species recognized by biologists among mammals and birds because these organisms are fairly large, easy to recognize with the naked eye and of important societal use (as food, skins for clothes and tents, feathers for ornamentation and so on). However, among insects, zoologists surely recognize far more species than most folk taxonomies, and the discrepancy would increase further for smaller organisms, some of which (many nematodes, some acarians and nearly all unicellular organisms) simply cannot be observed without microscopes or magnifying glasses. Perhaps because of these considerations, ethnobiologists have used folk taxonomies to assess the societal importance of biodiversity rather than biodiversity itself and have not made the same mistakes as many biologists involved in biodiversity studies who have used taxonomic surrogacy (Bertrand et al. 2006; Laurin 2010a). In this respect, folk nomenclatural ranks have played a more positive role in the development of folk taxonomy than Linnaean categories in comparative biology (see Chapter 3).

### 1.1.2 Protohistoric and Early Antique Biological Nomenclature

Closer to the Western civilization, the oldest written documents attesting to biological nomenclature hark back to about 3000 BCE in Mesopotamia (Sundberg and Pleijel 1994: 19). A bit later, Homeric-era Greek literature (8th and 7th centuries BC) records names of 71 animal taxa. These mostly concern domestic animals (Voultsiadou and Tatolas 2005). Plants used for either agriculture or medicine were also known (Klimis 2008). Many of the plants cultivated in the Homeric era had been grown by people at least since the Neolithic, 6,000–8,000 years ago (Thanassoulopoulos 2008).

In one of the oldest and most influential books of Western civilization, we can read:

> And out of the ground the LORD God formed every beast of the field, and every fowl of the air; and brought [them] unto Adam **to see what he would call them**: and whatsoever Adam called every living creature, that [was] the name thereof. [2:20] And Adam **gave names** to all cattle, and to the fowl of the air, and to every beast of the field.

> **(*Genesis*, quoted from the King James version).**

This brief passage attests to a well-developed biological nomenclature. Of course, when *Genesis* was written, there were no specialists of biological nomenclature, but this passage clearly shows that naming animals was important. *Genesis* also attests to a first-species concept, as an elementary unit of biodiversity, vaguely akin to the so-called "biological species concept" to the extent that it implies reproduction between members of the same species. In this regard, *Genesis* 6: 17–20, about Noah's ark, is particularly interesting and deserves to be quoted in full:

> [17] And, behold, I, even I, do bring a flood of waters upon the earth, to destroy all flesh, wherein is the breath of life, from under heaven; and every thing that is in the earth shall die. [18] But with thee will I establish my covenant; and thou shalt come into the ark, thou, and thy sons, and thy wife, and thy sons' wives with thee. [19] And of every living thing of all flesh, **two of every sort** shalt thou bring into the ark, to keep them alive with thee; they shall be **male and female**. [20] Of fowls after their kind, and of cattle after their kind, of every creeping thing of the earth after his kind, two of every sort shall come unto thee, to keep them alive.

This passage, which makes Noah a mythical predecessor of conservation biologists, clearly shows a concept of "sort" of animals that can reproduce with each other, and the fact that two of each "sort" must be brought onto the ark, and that these are male and female, leaves no room for doubt that these designate some form of biological species (Wilkins 2018: 56–62). The writing of *Genesis* is difficult to date (Sommer 2011) and this book may incorporate elements from various periods, especially the 6th century BCE (Davies 2013), but it obviously predates the oldest known zoological works.

### 1.1.3 Aristotle

The great Greek philosopher Aristotle (384–322 BCE) can be considered the founder of zoological classification (Voultsiadou et al. 2017). Mayr (1982: 149) even stated: "The history of taxonomy starts with Aristotle (384–322 BC)," and Charles Darwin (1809–1882) stated, in a letter (Gotthelf 1999), that "Linnaeus and Cuvier have been my two gods, though in very different ways, but they were mere school-boys to old Aristotle." Aristotle apparently had predecessors in Athen's Academy, such as Speusippus (c. 407–339 BCE), among others, who also proposed a classification of animals (Lloyd 1983). Unfortunately, only fragments of Speusippus' works remain. Aristotle's zoological work was not meant to be primarily taxonomic (Pellegrin 1986); instead, his focus was apparently on explaining animals' design (structure) and lifestyle (Pratt 1984: 272) and showing character linkage (Fürst von Lieven and Humar 2008: 244). Interpreting Aristotle's intentions concerning animal classification is hampered by the fact that at least one of his works, the *Anatomai*, is lost. The existence of this lost work is attested by citations in Aristotle's preserved texts (notably, in his *History of Animals*), as well as in citations in the *Apologia*, written by the Roman author Apuleius in the 2nd century BC (Fürst von Lieven et al. 2021). This raises the possibility that some of the taxonomically most significant writings of Aristotle may have been lost as well, even though such works are not documented through citations, contrary to the *Anatomai*. Lennox (2006) discussed the debate about whether Aristotle's aim was more at defining taxa or only their attributes; in any case, both are linked because Aristotle found groups of animals that were produced by dividing according to several correlated characters. Similarly, Stoyles (2013: 5) argued that Aristotle's classification aimed at finding "the widest classes possessing the various animal features," and that this avoided repetition in Aristotle's descriptions. We now know that the best such classification is a taxonomy reflecting the phylogeny. For instance, Aristotle noticed that many animals possessed feathers, wings and beaks; these are birds. Aristotle's taxonomy apparently used both genealogical (but not phylogenetic) criteria, in that he recognized that parents give birth to offspring of the same kind, and functional ones, because Aristotle was interested in structure and function of organs (Pratt 1984: 274).

From this, it should be clear that Aristotle was a **realist**, at least concerning animal classification; in this case, this means that he apparently thought that taxa existed independently of our concepts, and he strived at discovering, through his method, their **essences**, which he thought were eternal. Indeed, in his

*Posterior Analytics*, Aristotle distinguished "nominal definitions," which yield only opinions, from knowledge of essences. Discovering the latter through an act of intellectual intuition was his main goal, and not only in zoology (Sloan 1972: 9). By opposition, a **nominalist** would consider that there is no natural way of dividing biodiversity; nominalism would be justified, for instance, if there were a continuum of forms that blend into each other. More precisely, Henry (2011: 199) argued that Aristotle was a **pluralist** realist, at least when tackling biodiversity, which means that Aristotle denied that "there is only one true set of biological kinds and that a natural classification will divide those kinds into a single set of exhaustive and non-overlapping categories." According to Henry (2011), Aristotle divided animals in several different ways that yielded overlapping sets, which would then be classes in the philosophical sense of the word, not to be confused with classes of the Linnaean hierarchy. For instance, he divided animals according to reproductive modes (live-bearers, egg-layers, larva-producers and animals that are spontaneously generated), locomotor mode (walkers, jumpers, swimmers and flyers), or respiratory organs (lungs, gills, membranes). These division modes yield incompatible results (overlapping sets or classes) because, for instance, among swimmers, there are live-bearers (like dolphins and some sharks) and egg-layers (most teleosts). If Aristotle had been a **monistic realist**, he would have thought that only one way of dividing animals was correct. We will see in the following that the latest works on Aristotle's taxonomy suggest that a case could also be made that Aristotle was a monistic realist when it came to classifying animals. Indeed, Sloan (1972: 6) pointed out that the preface of Aristotle's *Parts of Animals* includes a lengthy attack on the method of dichotomous division and Aristotle argued that each animal kind should be defined by several characters, none of which is sufficient to provide an essential definition.

Opposing views have been expressed on Aristotle's use of the words *genos* and *eidos*, and what this implies about his taxonomy. Some, like Balme (1962), argued that Aristotle often used *genos* and *eidos* to mean "kind" and "form" in general, rather than to designate genera and species as these terms have been employed in biology since Linnaeus. Indeed, according to Balme (1962: 85), in his *History of Animals* (*Historia animalium* in Latin, abbreviated *HA* in the following), Aristotle also used *genos* to designate something else than a kind of animals, and *eidos* designates a kind of animals only in a small minority of cases (24 out of 96). Pellegrin (1986; cited in Romeyer-Dherbey 1986) even argued strongly that Aristotle did not do taxonomy as we intend it, as shown by the fact that the words *genos* and *eidos* are used at various levels in Aristotle's work. Thus, the *eidos* of a given level can become *genos* at a lower level and be subdivided into *eidē*. This suggests that *genos* and *eidos* imply a hierarchy, as is found in modern taxonomy, but that they cannot be equated with the fixed levels in this hierarchy that genus and species represent in rank-based (Linnaean) nomenclature (Laurin and Humar 2022). At the other extreme, some, such as Wiener (2015), argue that Aristotle recognized "atomic species," something that might be vaguely comparable with the concept of species as it has prevailed in systematics since approximately the 18th century. However, given the relative use of *genos* and *eidos* in Aristotle's zoological work

(Balme 1962), one may wonder if Aristotle had such a concept of species, or if he perhaps just subdivided each *genos* into *eidē* as far as his method and data permitted, without necessarily conferring a special ontological status onto his lowermost subdivisions (*atoma eidē*) of animal diversity. This might be vaguely akin to Pleijel and Rouse's (2000) LITU concept that will be presented in greater detail later (Chapter 6). Wiener (2015: 292) admitted that "Not all of Aristotle's classifications divide down to the atomic species," which is compatible with (but does not require) this alternative interpretation. Gotthelf (cited therein) also did not see evidence of a "lowest privileged level" (a special ontological status for atomic species) in Aristotle's zoological work.

Similarly, Henry (2011: 200) pointed out that Aristotle's hierarchical but rank-free classification (in the sense that it does not use absolute ranks like the Linnaean categories) was more similar to PN than RN (though both developed about two millennia later). This is especially surprising given the fairly widespread (but not unanimous) opinion among ethnobiologists that there are cryptic absolute categories in folk taxonomies (see Section 1.1.1). Laurin and Humar (2022) suggested that perhaps Aristotle's animal classification includes cryptic ranks that have not been noticed, perhaps because nobody has looked for them, because they are even more cryptic than in folk taxonomies. Alternatively, Aristotle may have deliberately left such ranks out of his classification, or more simply, folk taxonomic ranks genuinely may not exist.

If the latter interpretations about Aristotle's use of the term *eidos* and the absence of absolute ranks are correct, Aristotle's animal classification looks extremely modern, given that several systematists (such as Pleijel and Rouse 2000; Laurin 2008) and some philosophers of science (Ereshefsky 2002) dispute that the biological species (a concept), as it has come to be conceptualized since the 18th century, confers to the species as a nomenclatural level (see Minelli 2022a for the confusion between these two meanings of "species"), a more valid ontological status than other Linnaean categories. At the other extreme of the taxonomic hierarchy, Stoyles (2013) argues that Aristotle did not consider his expression "greatest kind" (μέγιστα γένη, the plural form, is also translated as "greatest genera") as a technical term, or if he did, that it is of little significance. Again, this evokes a rankless taxonomy, given that current systematic theory attributes no special ontological status to the most inclusive taxa (Cantino and de Queiroz 2020). This apparently rankless, semi-cryptic taxonomy (to the extent that it was never summarized by Aristotle, at least in the works that we know of) in Aristotle's zoological works is a bit surprising given that many ethnobiologists believe that most folk taxonomies feature cryptic ranks (see Section 1.1.1).

Aristotle's criteria for erecting his seven "greatest kinds" (birds, fish, Cetacea, insects, as well as hard-shelled, soft-shelled and soft-bodied animals; see in HA I 6, 490 b7–15) show the tight link between taxonomy and nomenclature. According to Stoyles (2013: 9), Aristotle's usage implies "only the condition that each greatest kind must have in it multiple and **named** forms." Thus, the presence of subdivisions in a class of animals was not sufficient for the class to be considered a "greatest kind."

Pellegrin (1986) argued that Aristotle's biological classification only partly reflected what we would now call phylogeny, but this is unsurprising because the idea of biological evolution came much later, and the idea that evolutionary relationships could be depicted by trees originated later still, with Lamarck, de Barbançois and Darwin in the 19th century (Tassy 2011). Balme (1962: 85) even stated that "there is no classification scheme in the background, and all attempts to construct one for Aristotle have failed."

More recent studies by biologists (rather than by philosophers) have refuted this last claim. By concentrating for the first time on characters and their taxonomic distribution rather than only on the taxa and their names, Fürst von Lieven and Humar (2008) convincingly argued that a parsimony analysis of a data matrix compiled from Aristotle's *HA* produced a hierarchy, which is reminiscent of a taxonomy without absolute ranks (also see Moser 2013: 56). Aristotle indicated that each taxon could only occur once in a classification (Fürst von Lieven and Humar 2008: 243), but some names occur more than once, like *polypoda*, which refers to myriapods and octopods, which are now considered to belong to *Arthropoda* and *Cephalopoda*, respectively (Laurin and Humar 2022). Aristotle obviously realized that these were different taxa, so he apparently occasionally erected **homonyms** (two taxa that bear the same name). A first phylogenetic (parsimony) analysis of 147 terminal taxa included in Aristotle's *HA*, books II—V, scored from these same works for characters attributed to various taxa by Aristotle, produced 58 groups, 29 of which have equivalents (similarly delimited) in Aristotle's work, and a further 12 have equivalents in modern works but not in Aristotle's (Fürst von Lieven and Humar 2008). Of the 47 groups recognized by Aristotle and considered in their study, Fürst von Lieven and Humar (2008: 249) stated that 25 were still valid,

and they also argued that Aristotle distinguished between analogy and homology, which are important evolutionary concepts (Balme 1962: 89; Fürst von Lieven and Humar 2008). The tree resulting from a parsimony analysis of these data only partly matches the currently accepted phylogeny (Fürst von Lieven and Humar 2008: figure 1.4). Thus, insects, crustaceans and teleosts form mutually exclusive clades, as they should, but tunicates are located very far from vertebrates, and echinoderms form a clade with gastropods and bivalves, rather than with chordates.

The interpretation of the findings reported by Fürst von Lieven and Humar (2008) is not straightforward because Hillis and Huelsenbeck (1992: 189) reported that: "Analysis of random data often yields a single most-parsimonious tree, especially if the number of characters examined is large [which is the case in the matrix of Fürst von Lieven and Humar 2008] and the number of taxa examined is small." To better assess the significance of the apparent hierarchical information contained in Aristotle's *HA*, Laurin and Humar (2022) updated the data matrix and performed additional analyses. Their parsimony analysis of the matrix indicates that there are more than 200,000 maximally parsimonious trees; their strict consensus includes several polytomies. One might thus wonder if the similarities between that consensus tree and the currently accepted tree are mere coincidence, and more basically, if the hierarchical structure is spurious. A study of tree length distribution skewness (based on three samples of one million random trees each), which has been shown to yield reliable data about the presence of phylogenetic signal (Huelsenbeck 1991; Hillis 1991; Hillis and Huelsenbeck 1992), refutes the hypothesis of spurious hierarchical structure (**Figure. 1.1**). This reinforces Fürst von Lieven and Humar's (2008) conclusion that Aristotle's work rested on relevant systematic characters.

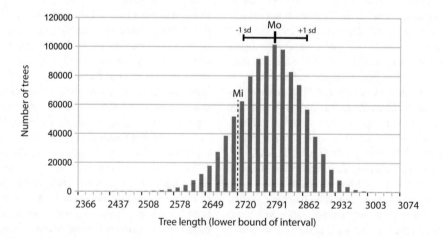

**FIGURE 1.1** Phylogenetic signal in the matrix initially compiled by Fürst von Lieven and Humar (2008) from books II–V of Aristotle's *HA* and updated by Laurin and Humar (2022). This is assessed by tree length distribution skewness on a population of one million random trees obtained from PAUP 4 (Swofford 2002). Skewness (g1 = −0.248; p < 0.01) is smaller than the threshold value (−0.12) for 25 taxa and 100 characters (these values increase with number of taxa and characters; see Hillis and Huelsenbeck 1992). Note that the lowermost and uppermost of the 40 bins of the histograms contains one and two trees each, respectively. To better visualize the skewness, the middle of the range (Mi) is shown. Note the obvious shift to the right of the mode (Mo) compared to the middle of the length range (Mi), which barely fits within 1 sd (standard deviation) from the mode. The shortest of these random trees (2,375 steps) is farther from the mean length (2,788.74 steps) than the longest random tree (3,065 steps); these distances are 413.74 and 276.26 steps, respectively. The shortest tree overall, shown in Figure 1.2, requires only 306 steps.

*Source:* Reproduced from Laurin and Humar (2022: figure 1).

Voultsiadou et al. (2017) similarly argued that Aristotle made tremendous progress in taxonomy and that his attempts at classifying animals according to some criteria, such as reproductive mode and lifestyle, were misinterpreted by subsequent authors. These were not combined into a fuzzy taxonomy because Aristotle distinguished groups of animals that take a certain name, which we call taxa, from groups based on other criteria, such as lifestyle, which he did not intend to be considered as taxa (Voultsiadou et al. 2017: 477). He called many of these groups "anonymous" or "nameless." Some of these groups discussed by Aristotle include the blooded (*enhaima*) and bloodless (*anhaima*) animals (Wiener 2015: 296; Ganias et al. 2017; Voultsiadou et al. 2017), which subsequently became established as taxa as *Vertebrata* and *Invertebrata* (which is no longer considered a valid taxon but the vernacular form "invertebrates" remains widely used), respectively, but this does not imply that Aristotle conferred to these groups the same status as the animal groups that he may have considered to be most natural (those that we now call taxa). Thus, even though Aristotle's primary goal in his biological work does not appear to have been to produce a taxonomy, he may have relied on a taxonomy to organize his zoological knowledge and present it to his readers. Alternatively, this taxonomy may have arisen spontaneously as he described animals and their organs, as the expression of his classification principle based on correlated characters.

The works by Fürst von Lieven and Humar (2008), Voultsiadou et al. (2017) and Laurin and Humar (2022) could thus be used to support the view that in his classification of animals, Aristotle was, to an extent, a **monistic realist**, because some ways of separating animals into sets appeared to be more valid than others to Aristotle, even though the works cited here make no claim about monism in Aristotle. Henry (2011: 206) argued that Aristotle was a **pluralistic realist** because, in his view, Aristotle's "Great Kinds" (major taxa, such as birds and fish) "do not enjoy a privileged status as ontological groupings." This question deserves to be reexamined in light of these recent analyses, although, from a modern perspective, great kinds should indeed not be ontologically different from lower-ranking taxa. Henry (2011: 207) pointed out that "For Aristotle, natural kinds are limited to those groups whose shared similarities are underwritten by common causes." The main common cause of shared similarities between taxa is now known to be the phylogeny, and this, along with the evidence provided by Fürst von Lieven and Humar (2008), Voultsiadou et al. (2017) and Laurin and Humar (2022), logically leads to the conclusion that there is indeed one best way to classify biodiversity (into clades), which is monistic realism and is consistent with modern taxonomic practice, given that we seek to uncover the Tree of Life (TOL). What remains to be determined is whether Aristotle's writings, taken globally and accounting for their inconsistencies, suggest that Aristotle had in mind a classification akin to a taxonomy or was trying to discover one, or if what looks like a taxonomy arose spontaneously as he described biodiversity. Intention is usually difficult to assess, and we may never know what Aristotle's intentions were, which is unfortunate because Hodge (1972: 129) argued that "nothing is more important than intentions" (which may be an overstatement). However,

Aristotle's search for coherent sets of characters that yield appropriate divisions between animal classes simultaneously yields a classification of animals that looks like a taxonomy, as the works cited here show.

A closer look at Aristotle's biological classification shows the extent of the innovations introduced by this great philosopher. In his taxonomy (**Figure. 1.2**), in addition to distinguishing *enhaima* from *anhaima* (see figure), among vertebrates (*enhaima*), he also recognized *Ichthyes*, which is approximately equivalent to the *Pisces* of much more recent authors. This is a paraphyletic group that is unfortunately still used in many scientific papers and books, either in its formal name (*Pisces*) or in the vernacular word "fishes" (e.g., González-Rodriguez et al. 2004; Mottequin et al. 2014). His classification was closer to ours than to Linnaeus' in recognizing a basal dichotomy between *Selachii* (elasmobranchs) and *Lepidotoi* (actinopterygians; Voultsiadou et al. 2017). He probably erected *Selachii* because this name is not known in older works (Laurin and Humar 2022: 3). If correct, this raises doubts about the old claim, recently reasserted by Guasparri (2013: 348) that "Aristotle does not feel the need for a strictly taxonomic nomenclature: he is satisfied enough with the terms used in common parlance by practitioners such as hunters, fishers, and livestock farmers." This may be correct in many cases, when a vernacular term existed, but Aristotle may not necessarily have been limited by this preexisting vocabulary. Among *Ichthyes*, he distinguished 109 taxa among what we now recognize as chondrichthyans and actinopterygians. Among these, the only one that is obviously mis-classified (under the *Selachii*) is the angler (which he called *batrachos*), which is a teleost (Ganias et al. 2017: 1047). But a phylogenetic analysis of the data from Aristotle's *HA* places it among teleosts (**Figure 1.2**), and two millennia later, Linnaeus did far worse in classifying the selachians in his *Amphibia*, along with batrachians, turtles and snakes, among others. Aristotle also recognized diagnostic characters that are still accepted today, such as the fact that *Ichthyes* had no neck or limbs (primitive characters), that they had fins and gills, both unique to vertebrates (Ganias et al. 2017), even though a few other taxa, such as some mollusks and arthropods, have analogous structures also called fins and gills.

Aristotle thought that there should be a name for the group of lunged vertebrates (Pratt 1984: 276). Among the taxa known to Aristotle, this would include tetrapods because Aristotle apparently did not know about lungfishes (Ganias et al. 2017), and he could not know that the swim bladder of teleosts derives from a lung (Laurin 2010b). Aristotle apparently recognized a taxon *Tetrapoda*, subdivided into *ootoka* (oviparous) *tetrapoda* and *zootoka* (viviparous) *tetrapoda*, with a composition fairly close to the currently accepted one (Voultsiadou et al. 2017) except that it excluded some taxa, such as birds (which suggests that Aristotle did not consider the wings "feet") and snakes. However, a phylogenetic analysis of a matrix compiled from Aristotle's *HA* placed cetaceans among tetrapods, as a sister-group of other mammals (Laurin and Humar 2022), which suggests that Aristotle observed fairly well the characters of cetaceans. The confusing position of cetaceans in Aristotle's classification is

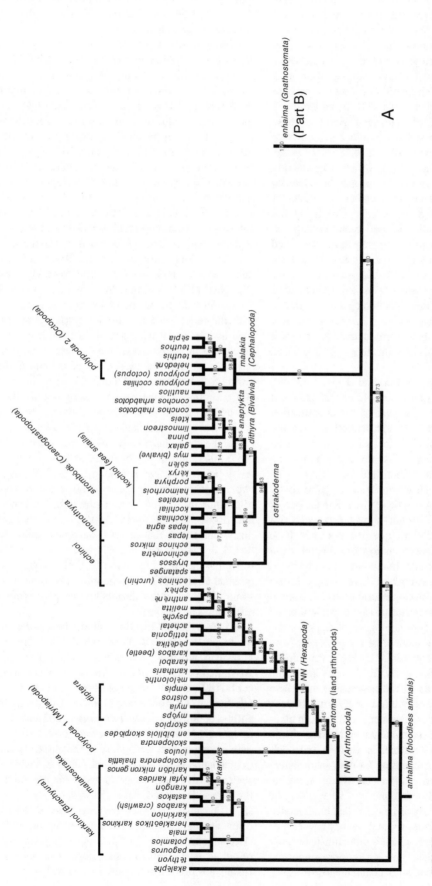

**FIGURE 1.2** Majority-rule consensus of the 200,000 trees produced by a search on the matrix (based on scores extracted from Aristotle's *History of Animals*) produced by Fürst von Lieven and Humar (2008), as updated by Laurin and Humar (2022), with some characters ordered. The numbers on each branch represent the proportion of the source trees that incorporates the various clades. The majority-rule consensus tree with all characters unordered differs only very slightly. The nomenclature follows Aristotle, but the current name of these clades is added in parentheses, when Aristotle's terminology is cryptic to most readers. Thus, *echinoi* is equivalent to *Echinodermata*, *ichthyes* is matches *Pisces*, and *selachē* is reminiscent of *Euselachii*; the modern name is not indicated or such taxa. Conversely, given that the name *dichala* is presumably unknown to most readers, the equivalent name *Artiodactyla* is indicated in parentheses. Part A corresponds to Aristotle's *anhaima*; part B is *enhaima* (gnathostomes).

*Source:* Reproduced from Laurin and Humar (2022: figure 2).

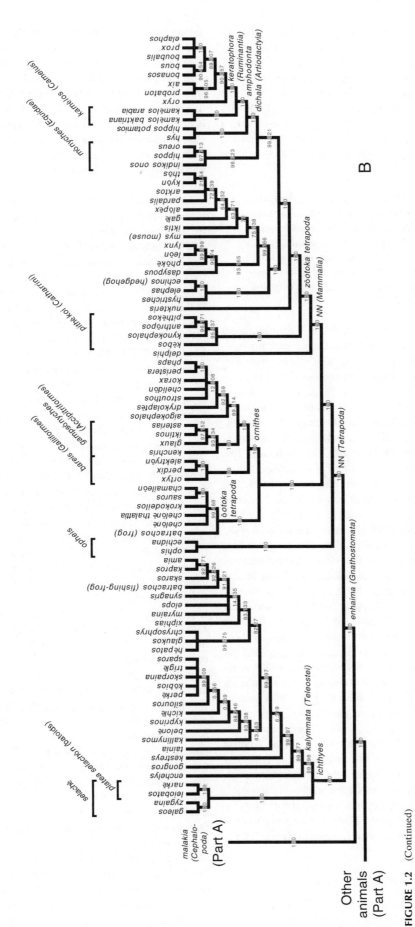

**FIGURE 1.2** (Continued)

hardly surprising given that cetaceans were called "fish" in the English language well into the 19th century, as colorfully illustrated by Melville's (1851: 119) classical novel *Moby Dick*. By contrast, in some languages, the scientific viewpoint that cetaceans were mammals was popularized in the 19th century, as illustrated by Jules Verne's (1871) *20,000 Leagues Under the Sea* (*Vingt Mille Lieues sous les mers*, in the original French). A broad meaning of "fish" has persisted into the 21st century among English speakers, as reported by Pawley (2011: 426):

> A survey of several hundred Australian informants shows that while everyone agrees on including prototypical fish [actinoptegyrians] there is much variation in regard to what other creatures count as fish. Some English speakers exclude eels, sharks, rays, seahorses and certain other 'atypical' fish from this category while others include whales and dolphins and even crayfish and jellyfish.

Laurin and Humar (2022) suggested that Aristotle's *scala naturae* may have acted as a pre-evolutionary polarizing principle, which may partly explain the quality of Aristotle's animal classification, but this suggestion remains to be tested.

Aristotle knew the Mediterranean fauna best, as shown by the fact that among teleosts, the names of taxa commonly fished by Greeks in Antiquity are the most frequent in his writings, and the same can be said about other marine metazoans (Voultsiadou and Vafidis 2007). In fact, he had noticed that scallops had disappeared from Kalloni Bay because fishermen used instruments that scratched the bottom of the sea (Voultsiadou et al. 2010: 38), which may be the first mention of anthropic biodiversity loss recorded in writing. He used the available vernacular names but apparently formulated many new names for groups of animals, and he also proposed new terms for ontogenetic stages, for behavioral patterns (such as migratory), and others for categories used for teleosts in seafood markets (Ganias et al. 2017: 1041). About 30% of the ichthyological names that he used appear for the first time in his work (Ganias et al. 2017: 1045), although it is possible that they were used in earlier works that are now lost. Many of these names continued to be used in the Byzantine empire, and the Greek people still use several of these names with little or no modification. Many of the formal taxon names that Linnaeus and other early systematists established in the 18th century (*Cyprinus, Perca, Muraena*, to mention a few teleost genus names; *Delphinus* and *Phocoena*, among marine mammals) are Latinized versions of Aristotle's taxon names (Ganias et al. 2017: 1045). However, in this process, some Latinized names became attached to different animals than those originally designated by their original Greek names. For instance, Aristotle's *echeneis* (ἐχενηίς) became associated with the remora (*Echeneis remora*) (Linnaeus 1758), but Aristotle's *echeneis* may be the lamprey *Petromyzon marinus* (Humar 2015).

### 1.1.4  Aristotle's Followers in Antiquity

Aristotle's disciple Theophrastus (371 BCE–287 BCE) applied many of Aristotle's ideas and principles to classify plants,

among other topics. He described a great diversity of newly discovered plants (for the Greek) that were brought back from Alexander the Great's conquests. His most famous botanical work, *Enquiry into Plants* (also called *Historia Plantarum*), was extremely influential in the Middle Ages. Because of this and other works (notably "On the Causes of Plants"), he is often viewed as the founder of botany (Wilkins 2018: 17). He emphasized characters of reproductive structures (flowers and fruits), thus establishing a long tradition that persisted well into the 18th century, but he also used other structures to classify plants.

The antique Greeks made the first scientific contributions to biology, but after the rise of the Roman empire, Latin became the language of scholars and would remain so through the Middle Ages and beyond. The famous Roman statesman and scholar Marcus Tullius Cicero (106 BC–43 BC) translated the Greek word *eidos* into the Latin *species* (Lherminier and Solignac 2005: 9). This word now designates one of the lowest-ranking units of biodiversity in biology, although it acquired this status over more than a millennium.

One of the greatest Latin scholars in biology was Pliny the Elder (AD 23/24–79), author of a *Natural History*. Pliny has often been considered a mere compiler, but this is not entirely fair because he was, at times, critical of some beliefs and reports (Steiner 1955; Fögen 2013). Pliny's classification seems to be further from ours than Aristotle's and is less hierarchical. Pliny is among the first to have cited most of the numerous authors (more than 450, in his case, though Aristotle was the most important one) who contributed to his sources (Fögen 2013; Guasparri 2013). Recent studies suggest that he may have contributed original observations (Fögen 2013) and more original thought than previously admitted (Moser 2013). Indeed, Pliny may be the first naturalist who applied the word "fish" ("pisces" in Latin) to many marine metazoans (in addition to vertebrates, as in Aristotle), such as mollusks and crustaceans (Moser 2013: 30), although he was preceded in this by Cicero, who is better known as a statesman, lawyer and philosopher (Guasparri 2022: 56). The group called "fishes" would remain stable, at least in the vernacular language, through the 18th (for instance, French) or at least until the 19th century (for instance, English). This is shown, for English, by the etymology of words such as "starfish" (an echinoderm), "cuttlefish" (a mollusk), "crayfish" (an arthropod) and "jellyfish" (a cnidarian). However, this new delimitation of the taxon Pisces was not an improvement. In the 18th to mid-20th centuries, zoologists returned to a delimitation closer to that of Aristotle before abandoning this name (because it refers to an artificial group). More importantly, Guasparri (2013) argued that Pliny was more concerned than Aristotle about developing an accurate nomenclature for taxa. He did this by combining the Latin name of an animal with its Greek name. In both languages, these terms were often ambiguous (with many synonyms and homonyms), but the combination of the names from both languages allowed a more precise identification of taxa. However, Moser (2013: 48) interpreted Pliny's *Natural History* as implying that there is a genus of dwarf animals, rather than implying (as we now think) that some animals are found in dwarf forms. Clearly, such "genera" have little relationship with taxonomic groups and are simply classes in the philosophical sense of the word. Guasparri (2013: 350) argued that Pliny's animal classification, summarized in an alphabetical list of taxa, suggests ranks as used in folk taxonomies, especially the rank of "generic species."

A recent study of antique Roman classification and nomenclature of aquatic animals (Guasparri 2022) supports the existence of cryptic folk taxonomic ranks, as previously suggested by Berlin et al. (1973). Among the many findings congruent with Berlin et al.'s (1973) proposals is the fact that folk generic taxa were the most numerous (Guasparri 2022: 38). However, in contrast to Berlin et al. (1973) and Berlin (2014), who had suggested that there were six ethnobiological ranks, Guasparri (2022: 36) concluded that there were more Roman ethnobiological ranks because to accommodate the taxonomy of shelled mollusks, he needed three sub-life form ranks, whereas Berlin et al. (1973) recognized no such ranks. Conversely, Guasparri (2022) did not find evidence of folk varietal taxa in Roman ethnobiological classification (but Berlin 2014 did not consider that this rank was encountered in all folk taxonomies). There were several exceptions to the name formation rules proposed by Berlin et al. (1973). For instance, Guasparri (2022: 37) found that 13 folk-specific names out of 57 were composed of uninomials instead of "productive binomials" (consisting in a folk-genus name and a descriptive specific epithet). The folk generic names were less surprising, with 33 "unproductive binomials" (in which the first name is not a generic name, as in "sea hare"), which is unsurprising, but five (out of 282) generic names were composed of "productive binomials." Antique Romans over-differentiated some taxa, especially those that were especially important for their economy, such as purple-dye-producing mollusks (Guasparri 2022: 36). In such cases, a single species recognized by 21st-century zoologists was split by the Romans into two or more species. In any case, the apparent persistence of ethnobiological ranks among antique Romans, who were greatly influenced by Greek culture, raises once more the question of their presence (so far not discussed by other authors) in Aristotle's animal classification, as recently pointed by Laurin and Humar (2022).

## 1.1.5 Systematics in the Middle Ages

Little progress was made in systematics in the Middle Ages, in which most scholars commented on the great classics, especially those of Aristotle. This produced little memorable progress. An exception is Boëtius (480–525 AD, approximately), who was influential through his work on division. The debate between nominalists and realists progressed through the Middle Ages, and this has important implications for the ontology of taxa. Most of this work was not especially biological, although John Scotus Eriugena's (815–c. 877) *De divisione naturae* ("*The division of nature*") features biological examples. Pierre Abélard (1079–1142) is one of the most famous medieval nominalists. It is not necessary to review all of the main medieval works here; those interested in this topic will find a good summary in Wilkins (2018). We can illustrate the kind of work done in the Middle Ages with Michael of Ephesus (probably from the 12th century), who reflected on why we should study biological organisms that most people find repulsive (Arabatzis 2012)—a question that Aristotle had originally raised. A possible answer

is that the world is God's creation and, given that plants and animals are part of it, we should study them. The first mentions of the word "species" or its equivalent in its biological meaning in various European languages appear in the Middle Ages. For instance, the first occurrence of "species" (but in its French equivalent "espèce") for a reproductive lineage is in the *Roman de la Rose* (written in 1260–1280) by Jean de Meung (Lherminier and Solignac 2005: 10).

Most of the medieval works with the greatest systematic content were the herbals and bestiaries. Herbals had a predominantly medicinal purpose (to help use the right plants to prepare medicine), while bestiaries had a moral purpose and used animals to embody virtues and vices. Neither represents systematics at its best. A notable exception is Frederick II of Hohenstaufen (1194–1250), whom Wilkins (2018: 37) nicknamed "the heretic falconer" because he was very interested in falcons (for practical reasons, for hunting) and because he was excommunicated twice for challenging the authority of the Catholic church over secular power. In the preface of his work *De arte venandi cum avibus*, he criticized Aristotle for not checking the relevant facts relating to his quotes from other authors who had not written from first-hand experience. This concern for directly observing nature is strikingly modern. His work (notably Book 1 of *De arte venandi*) also contains one of the earliest mentions of nomenclatural confusion created by lack of regulation (Wilkins 2018: 39):

> the same genera and species are given different names by diverse authors. Sometimes the same bird may have a variety of synonyms; and the same name applied to diverse birds so dissimilar that one cannot establish the true identity of a species simply by its name.

Another notable medieval author is Albert of Lauingen (1193–1280), also known as Albertus Magnus. This Dominican friar taught Thomas Aquinas and was among the first to comment on the new translations of Aristotle's works that were obtained through the Arab tradition. Like Frederick II, he also studied nature directly. He produced some classifications of animals, in which he included bats among birds, presumably because they fly. Wilkins (2018: 42) stated: "That this is a purely arbitrary Aristotelian system based on overall habitat is confirmed by the inclusion of crocodiles, hippopotamuses, seals, dolphins, and whales among the fishes in the book on aquatic animals." This is not entirely accurate. Even though Aristotle used habitat as one of many zoological classification criteria, his *Ichthyes* included only primitively aquatic vertebrates, namely *Lepidotoi* (actinopterygians) and *Selachii* (elasmobranchs). The content of "fishes" listed by Wilkins evokes more Pliny, who included many marine metazoans in his "pisces," and may (possibly) also have included whales (Romero 2012).

Thomas Aquinas (1225–1274), best known for his philosophy and theology, called his *species infimae* (smallest species, something akin to our biological species) "individuals" because they cannot be divided formally. This may well be the earliest mention of species as individuals rather than as simple logical classes (Wilkins 2018: 44), even though Aquinas' argument does not imply the spatio-temporal delimitation inherent in the modern individuality thesis of taxa. Two centuries later, Nicolas of Cusa (1401–1464) also considered that biological species were individuals (Wilkins 2018: 47).

In the Middle Ages, most treatises on animals and plants listed taxa alphabetically, which is not a true taxonomy. The transition to a taxonomic order in monographs only occurred in the Renaissance and can be seen in Conrad Gesner's (1516–1565) work. While he retained an alphabetical order for a work on "quadrupeds" (mammals) published in 1551, he adopted a taxonomic approach consisting of eight orders of birds in his *Icones Avianum*, published in 1560 (Minelli 2022a).

## 1.2 The Renaissance and the Age of Enlightenment

### 1.2.1 Noah's Ark and the Species Concept

We saw in the opening section of this chapter that *Genesis* 6: 17–20, about Noah's ark, evokes a first example of something close to the "biological species concept" to the extent that it implies that individuals reproduce with their "sort." This "sort," vaguely suggestive of the biological species concept, became influential again in the Renaissance when theologians wished to infer how many "sorts" of animals must have taken place on the ark, and thus, how much food had to be brought and what the minimal dimensions of this ark might have been. The first to attempt this calculation was Jean Borrell, also known as Johannes Buteo (1490–1560/1572?). He listed 83 "sorts" (Wilkins 2018: table 3.3), all of which are tetrapods, mostly mammals; birds are not listed. Others attempted similar calculations and came up with different numbers—162 for Walter Raleigh in his *History of the World* (1614), only 50 for Athanasius Kircher in his *Arca Noë* (1675) and 58 for the Bishop John Wilkins (in 1668), who went on to address the problem of species further. Kircher gets a lower number because he considered the others to be hybrids, geographical varieties produced by the influence of local conditions or products of spontaneous generation. Thus, this passage of *Genesis* played an important role in subsequent developments of the species concept (Wilkins 2018: 56–62).

### 1.2.2 Explosion of Our Knowledge of Biodiversity from the 15th to 18th Centuries: Taxonomic and Nomenclatural Consequences

#### 1.2.2.1 Context

Starting with the 16th century through the great explorers, the knowledge of biodiversity among European systematists expanded quickly. This explosion of our knowledge of biodiversity can be illustrated by the fact that Leonhard Fuchs listed 500 plant species in his *De historia stirpium* (1542), whereas John Ray (1627–1705) listed 18,000 in the three volumes of his *Historia Plantarum*, which were published in 1686, 1688 and 1704. This rapid expansion of our knowledge posed new challenges.

First, this new knowledge had to be organized to be useful. Among the first who tried to organize all this new biodiversity into natural groups, as Aristotle and Theophrastus had done 2,000 years earlier, were 16th-century herbalists (Guasparri 2013: 349; Ogilvie 2003). Natural groups can be opposed to artificial ones based on subjective criteria that may have a practical use but that do not attempt to reflect an order existing in nature. This development took time. In the late 15th century, herbalists focused on identifying the plants described in Antiquity by authors such as Pliny the Elder and Dioscorides. Then, herbalists believed that the ancients had described most or all plants. By the 1530s, their successors focused on describing rather than identifying, and slowly realized that there was much more biodiversity than described by the ancients (Ogilvie 2003).

Later, this explosion in our knowledge of biodiversity triggered important progress in taxonomy and nomenclature. In his *Herbarum vivae eicones* (published in three parts from 1530 to 1536), Otto Brunfels (1488–1534) listed "genera" (groups) in a fairly arbitrary order, but each included closely related "species" (kinds). According to Mayr (1982: 157), this can be viewed as an early form of binary nomenclature, which was also used by other herbalists at that time. These names were not necessarily part of a taxonomy, which implies more hierarchical levels (Minelli 2022a), and many 16th century authors of herbals used the term "genus" more or less as an equivalent of what later became the Linnaean species (Minelli 2022b: 196). These names were initially formed by adding an adjective to the name of a previously known relative (Ogilvie 2003: 33). The Swiss botanist Caspar Bauhin (1560–1624) was perhaps the first systematist to use this form of binominal nomenclature, though his use was not nearly as consistent as Linnaeus' would later be (Wilkins 2018: 66). This early form of binominal nomenclature was subsequently obscured as the species names were expanded to remain unambiguous while our knowledge of biodiversity exploded; thus, single-word adjectives were no longer sufficient to distinguish all closely related species, and names became short descriptions that sometimes extended over two or three lines. The concept of genus evolved toward its modern sense at that time through the works of Conrad Gesner and Fabius Columna (1567–1650), also known as Fabio Colonna, who defined plant genera on the basis of flower and seed (Wilkins 2018: 63). We can now follow the impact of this rapid expansion of our knowledge in biodiversity on three fields: the metaphysics of taxa, taxonomy and nomenclature.

#### 1.2.2.2 Metaphysics

On the metaphysical front, Aristotle's works remained influential in systematics well into the 18th century. By then, John Locke's attacks against the Aristotelian theory of classification, published in the 17th century, became influential in systematics and triggered a debate about what we can know about taxa and how to study them (Sloan 1972). Until then, reliance on Aristotelian principles implied that systematists knew, through intuition, which characters were most reliable to yield natural groups (Sloan 1972: 9). This was linked to the belief in immaterial essences, which René Descartes (1596–1650) and later Robert Boyle (1627–1691) rejected. Boyle, like Descartes, also argued that all we know about reality is through subjective experience; hence, essential characters cannot be known (Sloan 1972: 18), and the joint occurrence of several properties defines entities (Sloan 1972: 20). The implications of this new way of thinking about science were not grasped by systematists until much later.

Boyle had worked in what we now call chemistry. John Locke (1632–1704) generalized these ideas into the philosophy of classification in general, and illustrated this through several biological examples (Sloan 1972: 21–22). He logically concluded that real essences cannot be known; only nominal essences, which are human constructs, can be known. Hence, our biological classifications are, to an extent, arbitrary and our knowledge of Nature is probabilistic at best. Consequently, the method consisting of defining by genus and differentia, harking back to Aristotle's *genos* and *eidos*, is not optimal. He considered species as constructs and was, in this respect, a nominalist (Lhermin[i]er and Solignac 2000: 159). Wilkins (2018: 71) refers to his position on this point as species conventionalism. Locke also provided examples of animals, like bats, which seemed to be intermediate between taxa, as well as reports of hybrids and monsters, to point at the limits of current biological classifications. However, this pessimism was tempered by his suggestion that the accumulation of observations will gradually improve our knowledge of Nature, and that this would improve our classifications.

This period also saw significant progress in our concept of species through John Ray's (1627–1705) work, which provided its first exclusively biological definition, according to some authors (Lherminier and Solignac 2005: 24; Wilkins 2018: 73). However, Minelli (2022a: 206) suggested that rather than a species definition, Ray provided a criterion to delimit species, and that he did this only for plants. Before Ray's works, authors viewed species of animals and plants as special kinds of entities, and the word "species" applied to minerals as well as to many other types of entities. In fact, this usage persisted later, as shown by the fact that Linnaeus still recognized mineral species. Thus, the word "species" remained a logical concept, as it had been since the Antiquity. The most commonly accepted concept of species remained that of genera plus differentia until the mid-19th century, but this concept was relegated to specialists of metaphysics and medievalists by new theories and the preeminence of the use of the word in a biological context (Wilkins 2018: 119). Ray argued that the safest way to recognize and delimit species was to look at descent. Thus, individuals that come from the same parents belong to the same species (Wilkins 2018: 73–74). Ray knew about hybridization, but he argued that it occurred between closely related species. He is perhaps best known for having argued that species were fixed since the creation, a view that many later systematists followed, including Linnaeus (Wilkins 2018: 77, 104), but he did not present this conclusion as novel (Minelli 2022b: 205).

Other authors consider that the importance of Ray's work on species was a bit overstated. Thus, Minelli (2022a) wrote that:

> it is hard to construe Ray's actual words [13] as a definition of biological species, rather than as an empirical (experimental) criterion to check the

conspecificity of similar but not identical kinds of plants, i.e., to verify how much variation can be accepted within the limits of genealogical continuity. Moreover, there is no evidence that Ray extended this view to animals too. On the contrary, some passages in his zoological books seem to exclude, on Ray's part, the adoption of a "true breeding" criterion to recognize animal species.

The endnote number 13 cited in brackets in the quote is a translation of a passage from Ray's *Historia plantarum* (published in 1686) that justifies much of Minelli's statement that rather than a definition of species, Ray may have simply provided a criterion to delimit them. This endnote is worth reproducing here:

> Therefore whatever differences arise from a seed of a particular kind of plant either in an individual or in a species, they are accidental and not specific. For they do not propagate their species again from seed . . . if a comparison is made between two kinds of plant, those plants which do not arise from the seed of one or the other, nor when sown from seed are ever changed one into the other, these finally are distinct in species.

Wilkins (2018: 86) similarly argued that Buffon, who worked nearly a century after John Ray, did not propose the first reproductive concept of biological species, contrary to what others have suggested. Instead, he was the first to use reproductive isolation as a tool to delimit species. He even performed crossing experiments to determine if goats and sheep (among others) formed a single species (Stemerding 1993: 206). Minelli (2022a, 2022b) goes further in arguing that Buffon's interest in reproductive isolation was only to ascertain the continuity of lineages, a view supported by some of Buffon's writings (*Histoire Naturelle*, volume XIV) that pointed out that not all interspecific hybrids are sterile (Minelli 2022b: 213). The philosopher Immanuel Kant (1724–1804) accepted Buffon's reproductive view of species over the morphological similarity-based criterion that had previously prevailed. He thus drew a distinction between artificial divisions based on similarities and natural ones based on kinship determined by generation (Wilkins 2018: 97).

### 1.2.2.3 Taxonomy

Because of our expanding knowledge of biodiversity, it was no longer convenient to simply list species or genera in alphabetical order. They had to be organized into larger natural groups, but how to find these groups long remained controversial. This search for natural groups led some systematic botanists, starting with Andrea Cesalpino (1519–1603), through Joseph Pitton de Tournefort (1656–1708) and Carl Linnaeus (1707–1778), among others, to emphasize characters linked with the reproductive system (flower, fruit and seed) to delimit taxa, because they thought that these characters yielded a natural classification (Sloan 1972). But zoologists of that period, such as Jean-Baptiste Lamarck (1744–1829), also searched for natural classifications for similar reasons (the number of known species of animals was also increasing rapidly).

All these developments (in works from Descartes to Locke) are incompatible with Aristotle's thesis that we can know essential properties through intuition. In systematics, this invalidates the premise that justified the a priori weighting of characters that had led, for instance, botanists (starting with Cesalpino) to favor characters linked to reproduction (flower, seed and fruit) over others. John Ray was also involved in this debate. After initially accepting the principle of character weighting (which implies that some characters are more reliable than others in systematics), he rejected this principle in his *Methodus plantarum emendata* (1698). This triggered a debate between the proponents and opponents of character weighting in systematics. In Ray's days, these proponents of character weighting chiefly included Tournefort and August Quirinus Rivinus (1652–1725), but this controversy about weighting still persists today, though in a different form (for instance, Goloboff 1996; Goloboff and Arias 2019). Tournefort emphasized reproductive characters in his taxonomy, and believed that this would yield natural groups. He stated (Tournefort 1694: 20):

> Il ne faut pas écouter ceux qui croyent que **tous les noms** que l'on a donné aux plantes sont également bons, & qu'il n'est pas nécessaire de chercher une méthode si exacte dans la Botanique. Ce serait autoriser un désordre qui a fait beaucoup de tort à cette sience; & tout bien considéré, l'on doit avouer que la **distribution des espèces** sous leurs véritables genres **n'est pas arbitraire**. J'espère que l'on connoîtra dans la suite, que **l'auteur de la nature** qui nous a laissé la liberté de **donner les noms qu'il nous plairoit** aux genres des plantes, a **imprimé un caractère commun** à chacune de leurs espèces, qui doit nous servir de guide pour les ranger à leur **place naturelle**. Nous ne saurions changer ces marques de distinction sans nous écarter trop visiblement de la vérité; mais nous devons bien nous garder de prendre pour ce caractère ce qui ne l'est pas.

> We should not listen to those who believe that **all the names** that have been given to plants are equally good and that it is unnecessary to search for such an accurate method in botany. This would generate a chaos that has seriously hampered this science; and all considered, we must admit that the **distribution of species** in their true genera **is not arbitrary**. I hope that we will know through what follows that the **author of Nature**, who has given us the freedom to **name the plant genera as we wish**, has **impressed a common character** to each of their species, and this must guide us to place them in their **natural place**. We cannot change these distinguishing marks without erring too obviously from the truth; we must not consider just anything as a character. [Translation and emphasis mine; original spelling in French.]

This quote shows that characters are highly relevant for taxonomy, but that they are also important for nomenclature, to which we now turn our attention.

### 1.2.2.4 Nomenclature

The rapid expansion of our knowledge of biodiversity generated a nomenclatural chaos. Each naturalist who discovered a new plant or animal wanted to describe and name it, and many species were thus described under different names by different authors. By the end of the 16th century, many plants were known by a dozen or so names that had been coined by as many authors (Ogilvie 2003: 38). The need for a list of synonyms of the various names of plants was obvious, and Bauhin produced such lists in one of his major works (*Phytopinax seu Enumeratio plantarum ab herbariis nostro seculo descriptarum, cum earum differentiis*) published in 1596. This very problem would much later make scientists realize that biological nomenclature needed to be regulated by codes, but this idea arose only in the 19th century, and the codes came into effect only in the 20th century. This story will be covered in Chapter 2, but it is important to note that these codes were created to solve a problem that arose no later than the 16th century.

To understand how acute the nomenclatural challenges faced by 16th to 18th century systematists were, it is necessary to understand how nomenclature worked in those centuries. Since Aristotle and until Linnaeus (and to an extent, still today), most names were descriptive and typically referred to a character (or a character complex). This is made more explicit in the following quote from Tournefort (1694: 1):

> Connoître les plantes, c'est précisément savoir **les noms qu'on leur a donné** par raport à la **structure** de quelques-unes de leurs parties. Cette structure **fait le caractère** qui distingue essentiellement les plantes les unes des autres.

> To know plants is to know precisely **the names that we have given them** according to the **structure** of a few of their parts. This structure **makes the character** that essentially distinguishes the plants from each other. [Translation and emphasis mine; original spelling in French].

Thus, to Tournefort, structures yield characters which determine taxon names. Of course, this was not a new principle invented by Tournefort; it had obviously been used by Aristotle (for instance, *Tetrapoda* for animals with four feet) and even earlier in folk taxonomies (Berlin 2014: 233) and would still be used by Linnaeus (for instance, *Mammalia*, from the Latin "mamma," which means "breast") and more recent systematists. Names given to taxa today are still often inspired by characters, but not necessarily; they may be attributed to nonessential characters and simply be designed as mnemonic devices. For instance, *Iberospondylus* was erected by Laurin and Soler-Gijón (2001) for a Carboniferous limbed vertebrate found in central Spain. The name was formed from "Ibero," in reference to the Iberian Peninsula, and "spondylos," the Greek word for vertebra that hints at its temnospondyl status. If *Iberospondylus* were later found to have been widespread well beyond the Iberian Peninsula, and if it were taken out of *Temnospondyli*, the name *Iberospondylus* would no longer accurately reflect its geographic distribution and its taxonomic affinities, but this would not invalidate its name (as we will see in the following, current nomenclatural rules require no link between the etymology of a name and the taxon that it designates).

At that time, taxon names were descriptive and can even be viewed as diagnoses (Dubois et al. 2021: 59). Thus, the explosion in our knowledge of biodiversity had resulted in a progressive lengthening of the names, as previously explained. Tournefort (1694: 37–38) thus lamented that some names had become unwieldy:

> Les noms des plantes doivent être les plus courts & les plus clairs qu'il se peut; mais ils doivent renfermer dans leur briéveté ce qu'il y a de plus singulier, & de plus sensible dans chaque espèce. . . . Morison en a donné de si longs qu'on perd haleine en les récitant. . . . Un auteur moderne a donné le nom suivant à une plante d'Afrique.

> *Mesembrianthemum Africanum frutescens minus, erectum, triagularibus foliis viridibus cornuum taurinorum in modum inflexis, fructu turbinato, parvo, pentagono, lignescente, flore albo.*

> Si l'on avait besoin de cette herbe pour traiter une malade, oserait-on charger une ordonnance de ce nom?

> Plant names must be as short and clear as possible; but they must contain, in their brevity, what is most singular and sensitive in each species. . . . Morison has given such long names that you lose your breath reciting them. . . . A modern author has given the following name to an African plant:

> *Mesembrianthemum Africanum frutescens minus, erectum, triagularibus foliis viridibus cornuum taurinorum in modum inflexis, fructu turbinato, parvo, pentagono, lignescente, flore albo.*

> If we needed this herb to treat a patient, would we dare to fill a prescription with that name?

This quote illustrates the drawbacks of linking names and attributes of taxa. As our knowledge of biodiversity expanded, it became increasingly difficult to produce concise, descriptive names that contained all the information required to identify taxa. A solution had to be found and, as we will see, it would be proposed by Linnaeus.

## 1.3 Linnaeus and His Followers before the Advent of the Rank-Based Codes

The context was thus ripe for a nomenclatural revolution. It came from Uppsala, Sweden, where Linnaeus (1707–1778) worked. Linnaeus was primarily a botanist, but two of his major works, *Species Plantarum* (Linnaeus 1753) and the 10th edition of his *Systema Naturae* (Linnaeus 1758), are still considered the starting points of botanical and zoological nomenclature, respectively, to establish the priority of various taxon names. This is because Linnaeus introduced or developed two major nomenclatural innovations that became largely adopted

by systematists and that remain in use today, even though both now create problems. These are absolute categories, which became known as Linnaean categories, and binominal nomenclature. These are now associated exclusively with living beings, but Linnaeus also used them to classify his "mineral kingdom," in which he placed fossils. Indeed, his Petrificata included various genera that refer to fossils, such as *Entomolithus*. More specifically, Linnaeus' *Entomolithus paradoxus* is clearly a trilobite, and other examples are known, such as *Entomolithus succineus*, which Linnaeus applied to various insects preserved in amber (Minelli 2022: 210).

Binominal nomenclature for species name, first used almost consistently in Linnaeus' *Species Plantarum* (Winston 2018: 1123; Witteveen and Müller-Wille 2020: 18), solved the problem of very long species names about which Tournefort (1694: 37–38) had complained. Binominal nomenclature developed gradually, and its complex history can be simplified into four key steps (Choate 1912).

First, Bauhin recommended the use of a form of binominal nomenclature and followed it in his *Pinax Theatri Botanici* (1623), which included an exhaustive synonymy list of all plant species then known. However, this nomenclature is binominal only in the sense that it includes a genus name and a species name. Sometimes, both were composed of a single word, which made them indistinguishable from those that Linnaeus would popularize much later, but this was accidental because, in Bauhin's system, either of the names (generic or specific) could be composed of more than one word. This was increasingly necessary as new species were discovered because adding a single qualifier after a genus name was no longer sufficient to obtain a unique name. Also, the first and second parts of these names do not correspond to genus and species names (respectively) as they came to be established under RN because in some of the taxa that he called "species" he recognized other "species" (Minelli 2022b: 196).

Second, Bachmann (also known as Rivinus) recommended the use of binominal nomenclature in his *Introductio Universalis in Rem Herbariam* (1693). He stressed that a concise method of naming would be advantageous, especially for medicinal plants, and that this could be achieved by adding a second term (which was later named the specific epithet) after the genus name.

Third, Linnaeus developed this system further over time. He initially used a binominal nomenclature in which the genus name was often composed of a single word, but in which the specific name consisted of one word, a few words, or, in a majority of cases, of a descriptive phrase. Thus, in one of his early key publications (*Critica Botanica*, published in 1737), he used mostly long species names (Nicolson 1991). Linnaeus may have developed the simpler, truly binominal (composed of two words only) names to help his students, and the earliest written evidence of them is in the indexes of his travel books *Öländska och Gothländska Resa* (1745) and *Wästgötha Resa* (1747). Thus, Linnaeus may have used these short names for convenience of indexing. He indirectly took the system further in *Pan Suecicus* (1749), a paper written by one of his pupils (Nicolaus L. Hesselgren) under his direction (Minelli 2022a). It includes a list of 866 plants, and about 90% of the names are binominal. Here, contrary to works by previous authors, the second name is not really a species name but a "trivial" or "vulgar" name, which became the specific epithet. The true specific epithet, which no longer aimed at uniquely describing the species, was born. In his *Species Plantarum*, these trivial names were located in the margin opposite the specific names, and Linnaeus emphasized in the preface of that work the novelty of these trivial names. However, Linnaeus did not consistently apply this principle in *Species Plantarum*, in which some trivial names are composed of two words.

In the fourth and last step, binominal nomenclature in its modern form became accepted by systematists worldwide through Linnaeus' use of this system in his subsequent works, through his persistence, his pupils and because of the inherent advantages of this nomenclature over the longer descriptive phrases that had been used previously.

Under this system, species names consist of two words: genus name and the "specific epithet." The early developments (before Linnaeus) of this system may reflect partly Aristotelian nomenclature based on *genos* and *eidos* (later known as "genus and species"), which acquired a different meaning in RN (starting with Linnaeus) to give "genus" and "specific epithet" as the word is now generally understood by contemporary systematists. The genus name is capitalized, but not the specific epithet. The genus name can be used alone to discuss a genus, but on the contrary, the specific epithet cannot be used in isolation because it is not unique. Indeed, each genus name must be unique (within animals and within plants, though the same name can refer to an animal genus and a plant genus, as we will see in Chapter 2), but there is no such requirement for specific epithets. Indeed, some epithets are very common, such as "*californiensis*" (which means "from California"), "*canadensis*" (from Canada), "*japonicus*" (from Japan), "*minimus*" (small), "*grandis*" (large) and so on. The combination of genus name and specific epithet greatly increases the number of taxa that can be named for a given number of possible words used as names. This is how names could be drastically shortened from up to about three lines to only two words.

Linnaeus, like many of his contemporaries, did not pay much attention to priority. Early systematists focused on "replacing the past" (Nicolson 1991: 33)—on building better nomenclatural systems and taxonomies. Priority would only be accepted as important much later, when the nomenclatural codes were developed (see Chapter 2).

While Linnaeus' contribution to species names is major, his species concept was simply that they were lineages that had been created by God, initially from a single individual (for hermaphrodites) or a single couple, for species with classical sexual reproduction (Linné 1751). Late in his career, he seems to have considered genera more real than species and changed his mind during his career about how many species had been initially created and whether or not new ones arose later. He initially thought that all had been created separately, which made species more real than genera, but later suggested that each genus could have diversified from a single species, mostly through hybridization (Wilkins 2018: 107), which increased the importance of genera. Linnaeus even extended this speculation to the genera classified in the same order (Lherminier and Solignac 2005: 29). Similarly, he stated early

on that nature does not make jumps (there is continuous variation), but he removed such statements from his later works. Linnaeus' acceptance of some limited form of evolution does not mean that he had a Tree of Life (TOL) in mind. Indeed, hybridization generates reticulation, but more importantly, Linnaeus' model to depict the affinities between taxa was a geographical map, in which regions represent taxa. In this, he was followed by several contemporary and slightly more recent systematists, such as Antoine-Laurent de Jussieu (1748–1836), Charles-François Brisseau de Mirbel (1776–1854) and Alphonse-Louis-Pierre-Pyramus de Candolle (1806–1893). These authors differed in how much of the map was occupied (how much of the possible range of biodiversity existed); for Linnaeus, it was fully occupied, but it was decreasingly so for his successors (Wilkins 2018: 92).

Linnaeus recognized the following ranks, from least to most inclusive, in the first edition of his *Systema Naturae* (1735): species, genus, order and class, in addition to regnum for the highest-level groups. For instance, our own species in this system is the species *Homo sapiens*, genus *Homo*, order Primates, class Mammalia, regnum Animalia. Other so-called "Linnaean categories" were not used in Linnaeus' work (except variety, used for plants; races are the equivalent for animals, but were not used by Linnaeus) and are thus "Linnaean" only indirectly, in the broad sense of having been integrated into the so-called "Linnaean hierarchy." Some of these ranks had been introduced before Linnaeus. For instance, families (a rank between genus and order) were introduced in botany by Magnol (1689), too early for his taxa to be considered established under the *Botanical Code*, and may have been used for the first time for names that are now considered established in botanical Linnaean nomenclature (that is, after the *Species Plantarum*) by Adanson (1763). Tournefort (1694) was the first to recognized four ranks (Minelli 2022: 198); from most to least inclusive, these included classes, sections, genera and species.

Initially, for plants, Linnaeus used different characters to define taxa of various ranks. For instance, the classes of angiosperms reflected the number and arrangement of stamens, whereas orders were based on the number of pistils (Schmitz et al. 2007). Linnaeus did not propose an equivalent system to provide an objective basis for absolute ranks of zoological taxa. Ranks are such an important component of nomenclature among Linnaeus' followers that we can call this kind of nomenclature "rank-based nomenclature" (RN). We will see in Chapter 3 that binominal nomenclature and Linnaean categories now create problems, but back in the 18th century, nobody had foreseen this.

The initial reception of Linnaeus' proposals was mixed. Some welcomed the convenience of binominal names to replace the clumsy, descriptive names that had become the norm in systematics, but others expressed reservations. For instance, Haller contested the validity of Linnaean categories and disputed the usefulness of binominal nomenclature (Persson 2016: 24). Similar reservations were expressed by contemporary systematists such as Georges-Louis Leclerc, Comte de Buffon (1707–1788), who thought that only taxa considered species (or individuals, in some of his writings) were real or natural (Buffon 1749) and hence, that higher taxa, for instance those ranked as orders and classes, were artificial (Stemerding 1993; Hoquet 2007). Buffon's views were influential, as shown by similar statements by Prichard (1813: 7). Several contemporary botanists, such as Johann Georg Siegesbeck (1686–1755), Lorenz Heister (1683–1758), Michel Adanson (1727–1806), Johann Beckmann (1739–1811), Johann Friedrich Blumenbach (1742–1840) and John Andreas Murray (1740–1791), like the philosopher and physician Julien Jean Offroy de la Mettrie (1709–1751), were especially critical of Linnaeus' taxonomy, both of his method and of the resulting taxa (Schmitz et al. 2007: 183; Persson 2016: 30). Indeed, Linnaeus realized that his system was artificial, and he was apparently very pessimistic about the prospects of discovering a natural system soon (Wilkins 2018: 84, 103). At a more basic level, Louis-Jean-Marie Daubenton (1716–1800), like Buffon, thought that only individuals exist in nature, and he thus argued that many intermediate ranks between those used by Linnaeus (kingdom, class, order, genus, species and variety) could be recognized (Llana 2000: 13; Hoquet 2007). The future would show that he was right (more ranks would be recognized later, and their subjective nature would become better documented), but for the wrong reasons (taxa do exist in nature, at least according to realists). In any case, Linnaeus and his nomenclatural innovations (absolute ranks and binominal nomenclature) had immense success, as shown by the rise of many "Linnaean" learned societies and the implementation of Linnaean nomenclature in various rank-based codes (see Chapter 2).

It has been argued that for Linnaeus and his early (18th century and early 19th century) followers, names of taxa were more closely associated with the taxa (groups of biological organisms) than with ranks, contrary to RN as it came to be implemented in the rank-based codes, starting with the 1840s (de Queiroz 2005, 2012). In the nomenclatural practice of these authors, ranks simply indicate hierarchical position but did not affect the spelling or application of names. Thus, names did not change simply because of a new rank allocation. This can truly be called "Linnaean nomenclature," as opposed to rank-based nomenclature (RN), which was subsequently implemented in the rank-based codes (see Chapter 2) in which rank allocation affected both the spelling and application of names. The examples provided by de Queiroz (2012) to justify this claim include *Amphibia* and *Reptilia*. Thus, *Amphibia* was erected by Linnaeus as a class and recognized in various editions of his *Systema Naturae*. It initially included some taxa that are still considered amphibians (such as anurans and urodeles), but also taxa now considered reptiles (squamates, turtles and crocodylians). Later, Linnaeus added various taxa to this, especially chondrichthyans, but changed neither the name nor the rank of this taxon (*Amphibia*). Merrem (1820) raised the taxon *Amphibia* to a higher (though unspecified) rank, as shown by the fact that he considered that its two primary subgroups were the classes *Pholidota* and *Batrachia*. Another example is provided by the taxon *Reptilia* or *Reptiles*, which Linnaeus (1758) considered to be an order within *Amphibia*. Many subsequent authors used these names for a more inclusive taxon (corresponding in composition to the first version of Linnaeus' *Amphibia*), that they considered a class, while keeping the same name. Of course, this demonstration

would be more conclusive if other examples could be documented among family-group names because only these have mandatory endings under the *Zoological Code* (ICZN 1999). However, such a demonstration would be difficult to produce because Linnaeus did not erect taxa in the family group. Thus, de Queiroz (2012: 137–138) stated that:

> It is important to note that standardized, rank-signifying endings were not used by Linnaeus and other 18th and early 19th Century naturalists. For example, some of the names of taxa that Linnaeus ranked as classes ended in -ia, others in -es, and still others in -a. In addition, these same endings were also used for the names of taxa ranked as orders. In short, particular endings were not used exclusively and universally in association with particular categorical ranks, the way they are today. The standardized, rank-signifying endings were introduced sometime during the early middle of the 19th Century. I have not researched their history thoroughly, but consistent use of the -idae ending for zoological taxa ranked as families can be found as early as 1825 in a paper by the herpetologist J. E. Gray, and this practice was endorsed as a general rule as early as 1835 by W. Swainson. It was also adopted by some of the important precursors of the modern rank-based codes, such as the Stricklandian code in zoology (Strickland et al. 1843). Most importantly, it was adopted by the original international codes of both botanical and zoological nomenclature and all of their subsequent revisions.

To sum up, the lack of change in the names of *Amphibia* and *Reptilia* in the context discussed here could be attributable to the absence of mandatory endings for these names. It would be worth seeking similar examples in the earliest family-group names.

## 1.4 The Rise of Evolutionary Biology and Phylogenetics

### 1.4.1 Biodiversity, Philosophers and Evolution

When most biologists were creationists, as Linnaeus was early in his career, studying biodiversity and its structure was akin to studying the Creator's great design. However, the idea of biological evolution arose and gained in popularity, first among 18th century philosophers, and later, with 19th century biologists. This change in point of view largely resulted from the explosion of our knowledge of biodiversity, which had stimulated systematists to develop new methods and nomenclatural principles, including binominal nomenclature, as we just saw. The idea of evolution influences how we can best classify taxa, and even the type of evolutionary model that we accept (for instance, a largely parallel progress toward greater complexity as suggested by Lamarck or more random, divergent evolution suggested by Darwin and neo-Darwinism) impact our ideas on the ontology of taxa and the optimal nomenclatural systems. Thus, a brief outline of the history of evolutionary biology is relevant in this book. The first part is presented here, and the more recent part of this history (after the 19th century) is continued in Chapter 3.

Improvements in our knowledge of biodiversity were linked with the exploration of other continents, which accelerated greatly starting with the 16th century, and with technological improvements, especially the microscope (notably through Leeuwenhoek's work), which allowed the discovery of minute life forms. Among these, a great diversity of unicellular eukaryotes (including flagellates, such as *Euglena*, and ciliates, such as *Paramecium*) was subsequently described by Christian Gottfried Ehrenberg (1795–1876), who also described fungi and many metazoans (especially cnidarians, insects and vertebrates). Scientists initially imagined a fairly linear organization from simple to complex, which became known as "the Great Chain of Being," and harks back to Aristotle. It became replaced in the works of 19th-century evolutionists by the Tree of Life, but the language used in recent scientific papers suggests that even contemporary systematists have not completely abandoned the concept of "the Great Chain of Being" (Rigato and Minelli 2013).

The concept of biological evolution, in the modern sense of the word (it initially referred to ontogenetic development well through the 18th and early 19th century), was first discussed by philosophers such as Bonnet, Diderot, Kircher, Leibniz, Maupertuis and Robinet in the late 17th and 18th centuries (Mayr 1972; Lherminier and Solignac 2005: 40). For instance, Kircher wrote in 1675 that:

> God created directly only a low number of first-born species that either because of the nature of the place, or under the influence of climates, or through the succession of the various species, once scattered on the whole Earth, developed into an infinite variety of living beings.

**(My translation of Lherminier and Solignac 2005: 26)**

This is an expression of what can be called **limited transformism**, which was expressed more fully by more recent philosophers and naturalists, especially Buffon. Typically, they admitted the possibility of limited transformations, mostly to explain the diversity of species in each genus (Stemerding 1993). Some, like Buffon, thought of this as a form of degeneration caused by the environment and hybridization, and that it was reversible. Buffon shied away from the **general transformism** that would lead later systematists (starting with Lamarck) to attempt to trace affinities between higher-ranking taxa (Lherminier and Solignac 2005: 33). Nevertheless, he influenced Lamarck (Wilkins 2018: 87), who became the first true evolutionist.

Along with evolutionary theory, taxonomy was progressing. Adanson made some headway into finding a natural classification of plants. Contrary to Linnaeus, who had used mostly sexual characters, each at a given taxonomic level, Adanson thought that we should have no a priori ideas on which characters to use, and that each taxon might require a different combination of characters. Some have seen in Adanson a precursor of phenetics because he based his taxonomy on as many characters as possible and he grouped taxa "from the bottom up" (by classifying together the most similar into small groups and then lumping these small groups into increasingly larger groups). However, unlike pheneticists, he did not argue that all characters should have an equal weight in classification (Wilkins 2018: 88), and he did not compute similarity indices (Lherminier and Solignac 2005: 35). Adanson did not adhere to Buffon's reproductive definition of species because he realized that some plants and animals reproduce asexually and because he, like Linnaeus, among others, wanted to recognize mineral species as well. He may be the first systematist who realized that gaps between extant species had been created by extinction of intermediate life forms, which reconciles the reality of species with his belief that nature produces continuous variation. He also was the first to use the term "mutation" to describe taxonomic change (Wilkins 2018: 90).

### 1.4.2 Lamarck

Lamarck can be credited for popularizing the idea of organic evolution among scientists, and Mayr (1972: 61), following various French historians, called him "the founder of the theory of evolution." This is because Lamarck was the first to publish an entire book designed to present a theory of evolution, and also presented a very simple evolutionary tree. He was thus the first **general transformist** because he hypothesized not only phylogenetic relationships between species of a given genus, but also between all main animal taxa. Whereas some of his predecessors, like Buffon, admitted limited evolution that they considered a form of degradation, Lamarck viewed evolution as progress (in this, he was possibly influenced by Leibniz) and as the mechanism that yielded the most complex life forms, including Man (Lherminier and Solignac 2005: 43). Lamarck's work will be presented in greater detail here than Darwin's because it has been less well understood by many subsequent evolutionists and many misconceptions about his work remain, whereas Darwin and Alfred Russel Wallace (1823–1913), as discoverers of natural selection as the most important evolutionary mechanism, have featured prominently in subsequent evolutionary works.

Lamarck's work encompasses a great diversity of fields and schools of thought. He was initially a botanist before turning to zoology and paleontology in 1784. He was also creationist and fixist before becoming an evolutionist, around 1800 (Lherminier and Solignac 2005: 37). Lamarck was the first systematist to attribute fossils to extant species and thus to contribute to better integrate fossils into systematics; contrary to Cuvier, he did not think that the taxa represented in the fossil record were extinct (Lherminier and Solignac 2005: 41–42). Fossils thus became progressively integrated into taxonomies over several decades. By 1850, this process had progressed further, as shown by the fact that Isidore Geoffroy Saint-Hilaire (1854: 167) stated clearly that biology entailed the comparative study of all beings that live on the surface of the globe or lived there in the past, a clear allusion to fossils.

Lamarck's contributions were often downplayed, partly because he was misunderstood and partly because he failed to find a credible evolutionary mechanism. Perhaps the most notorious misunderstanding is about his proposed evolutionary mechanism. The caricatural story of the giraffe wanting to reach higher leaves and stretching its neck, thus gaining a longer neck in its lifetime and passing on to its offspring this modification, is as notorious as false. It is probably attributable to translation mistakes, notably by Charles Lyell (1797–1875) in his famous *Principles of Geology*, which Darwin read. As Mayr (1972: 58) pointed out, what Lamarck actually proposed was that changes in the environment of animals led to changes in their needs, which in turn led to changes in their activity. This, in turn, led to changes in their morphology (or other changes) through use and disuse, and the transmission of these changes to the offspring.

It is mostly this last step which has proven problematic, as we now know that acquired characters are almost never transmitted to offspring. On the contrary, Lamarck's suggestion that environmental stability promoted stability in species and that environmental change promoted evolution and could lead to speciation (Lherminier and Solignac 2005: 40) is congruent with the current evolutionary theory (but seldom attributed to Lamarck). In this respect, he can also be viewed as a precursor

of the ecological species concept (Lherminier and Solignac 2005: 48), along with É. Geoffroy Saint-Hilaire (1772–1844), who went so far as to suggest that the environment determined how animals developed (Lherminier and Solignac 2005: 52) and who is perhaps better known to have suggested that arthropods are comparable to vertebrates if you turn them upside down (because the central nerve cord is ventral to the gut in arthropods, but dorsal to it in vertebrates), a suggestion not as silly as might first seem (Minelli and Fusco 2013: 304).

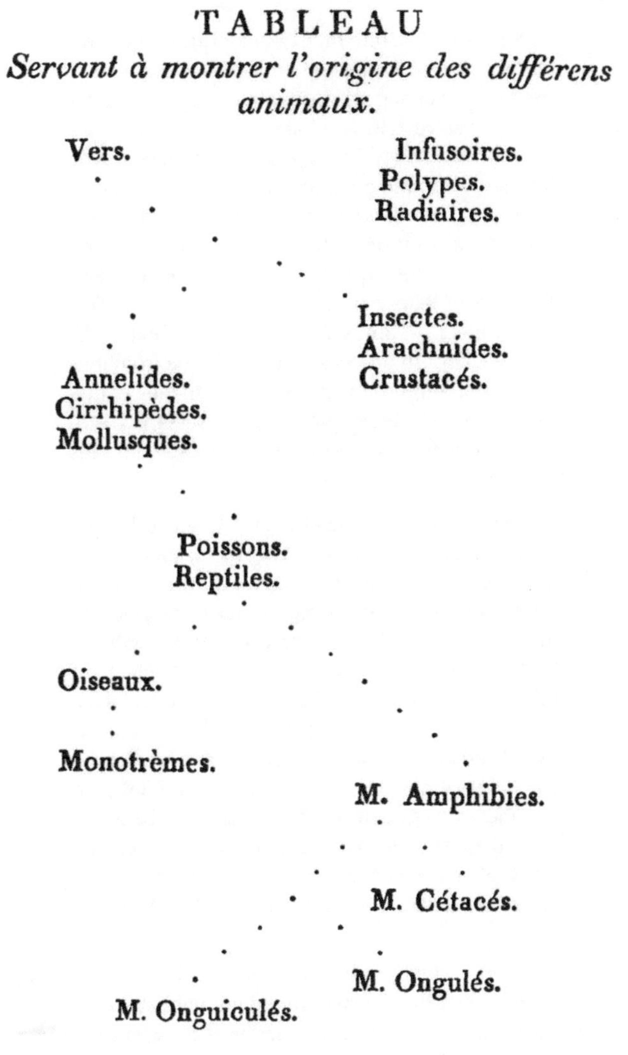

Lamarck's belief in spontaneous generation and inheritance of acquired characters was shared by most of his contemporaries. In fact, the idea of inheritance of acquired characters harks back to Antiquity (Zirkle 1946), so it really should not be attributed to Lamarck. Use and disuse played an important role in the acquisition of characters that were inherited by offspring in Lamarck's evolutionary mechanism. Darwin also accepted this mechanism, in addition to natural selection, as shown by this opening quote from the section on "Effects of Use and Disuse" in his famous book *On the Origin of Species by Means of Natural Selection Or the Preservation of Favoured Races in the Struggle for Life* (Darwin 1859: 134; abbreviated simply as *The Origin of Species* in the following): "From the facts alluded to in the first chapter, I think there can be little doubt that use in our domestic animals strengthens and enlarges certain parts, and disuse diminishes them; and that such modifications are inherited." Thus, while we must recognize that Lamarck did not discover the main evolutionary mechanism (natural selection), we can hardly blame him for accepting the effects of use and disuse in combination with inheritance of acquired characters.

Lamarck presented one of the first trees because he soon realized that the biodiversity could not be simply arranged in a gradient from simple to complex forms (Mayr 1972: 75; Wilkins 2018: 125). For the skeptics, it is worth quoting Lamarck (1809: 122–123):

## TABLEAU
### *Servant à montrer l'origine des différens animaux.*

Vers.

Infusoires.
Polypes.
Radiaires.

Insectes.
Arachnides.
Crustacés.

Annelides.
Cirrhipèdes.
Mollusques.

Poissons.
Reptiles.

Oiseaux.

Monotrèmes.

M. Amphibies.

M. Cétacés.

M. Ongulés.

M. Onguiculés.

**FIGURE 1.3** Lamarck's (1809: 463) evolutionary tree. Note the absence of evolutionary time and the ancestor-descendant relationships between extant taxa. This good scan was graciously provided by P. Tassy.

Ces variations irrégulières dans le perfectionnement et dans la dégradation des organes non essentiels tiennent à ce que ces organes sont plus soumis que les autres aux influences des circonstances extérieures; elles en entraînent de semblables dans la forme et dans l'état des parties les plus externes et donnent lieu à une diversité si considérable et si singulièrement ordonnée des espèces, qu'**au lieu de les pouvoir ranger** comme les masses, **en une série unique**, simple et linéaire, sous la forme d'une échelle régulièrement graduée, **ces mêmes espèces forment** souvent autour des masses dont elles font partie **des ramifications latérales dont les extrémités offrent des points véritablement isolés**.

These irregular variations in the improvement and degradation of non-essential organs are due to the fact that these organs are more subject than the others to the influences of external circumstances; they generate similar variations in the form and state of the most external parts and give rise to such a considerable and singularly ordered species diversity, that **instead of being able to arrange them** like the masses, **in a single, simple and linear series** in the form of a regularly graduated scale, **these same species often form**, around the masses that comprise them, **lateral ramifications whose tips offer truly isolated points**.

Although long-winded by today's standards, this text clearly states that the biodiversity cannot be arranged in a simple, linear series, and that the correct arrangement is branched. This branching is caused by the influence of the environment, which induces habits and needs (which, in an anachronistic Darwinian terminology, would be called "selective pressures"), and this leads to divergence and diversification that occurs on the branches of the Tree of Life (Lherminier and Solignac 2005: 47). Basically the same explanation, formulated slightly differently, already appeared in the slightly earlier *Discours d'ouverture* of his course entitled *Système des animaux sans vertèbres* (Lamarck 1801). In another publication, Lamarck (1802: 39) explained that this branching pattern affected the genera, and especially species, but that the more inclusive (higher-ranking) taxa formed linear series (one for animals and one for plants). Lamarck's ideas may have changed through time because the tree published in his *Philosophie zoologique* (Lamarck 1809: 463) shows that the relationships between high-ranking taxa includes both linear successions and evolutionary divergences (**Figure 1.3**). Thus, the tree suggests that "fishes" ("poissons" in the original) gave rise to reptiles, but the latter gave rise to two branches: the first gave rise to birds ("oiseaux") and the latter to monotremes, while the second gave rise to amphibious mammals ("M. Amphibies"), which then gave rise to the other mammals (but this descendance includes two more evolutionary divergences). In proposing a branched tree, Lamarck had remarkable insight and was well ahead of Chambers, an early evolutionist, who accepted multiple spontaneous creations (something widely believed, at the time) and massive parallel change (Chambers 1845; Hodge 1972: 147).

Despite Lamarck's proposal of branched trees, his evolutionary model was deterministic rather than stochastic, in this differing fundamentally from Darwin and his followers. Thus, for Lamarck, each evolutionary stage was predetermined to produce the next one, just like the terms of a mathematical series. This was based on his assumption that like causes yield like effects (Lherminier and Solignac 2005: 44). However, Lamarck was probably the first evolutionist who adopted a populational approach to the study of species. He explained the similarities between individuals of the same species by the fact that they adapted to the same environment (Lherminier and Solignac 2005: 47–48).

In addition to producing admittedly artificial classifications like many older authors, Lamarck also clearly aimed at reproducing a natural order (Lamarck 1809: chapter V, 1st page):

> Le but d'une distribution générale des animaux **n'est pas** seulement de posséder une liste **commode à consulter**, mais c'est **surtout** d'avoir dans cette liste un **ordre représentant** le plus possible **celui même de la nature**, c'est-à-dire l'ordre qu'elle a suivi dans la production des animaux et qu'elle a éminemment caractérisé par les rapports qu'elle a mis entre les uns et les autres.

The purpose of a general distribution of animals **is not only** to have a **convenient list to consult**; it is **mostly** to have in this list an **order representing** as much as possible **that of nature itself**, that is to say the order that it followed in the production of animals and that it has eminently characterized by the relationships that it established between them.

Lamarck's tree (**Figure 1.3**) was not as clearly evolutionary as those produced by more recent evolutionists, such as Darwin and Gaudry, notably because it incorporated neither geological time (or even a time dimension at all) nor the fossil record (Tassy 1981, 2011). Instead, this tree suggests that various extant taxa gave rise to other extant taxa, which is coherent with his skepticism about extinction. This, among other reasons, may explain why Darwin's tree of hypothetical taxa (which incorporated time) is mentioned much more frequently than Lamarck's tree. Nevertheless, Lamarck exposed at length in his works his views on phylogeny, without (like Linnaeus before him) shying away from placing Man close to the "quadrumanes" (primates), and even proposing a detailed scenario about the origin of Man (Lamarck 1809: 339–340):

> Effectivement, si une race quelconque de quadrumanes, surtout la plus perfectionnée d'entre elles, perdait, par la nécessité des circonstances, ou par quelque autre cause, l'habitude de grimper sur les arbres et d'en empoigner les branches avec les pieds, comme avec les mains, pour s'y accrocher, et si les individus de cette race, pendant une suite de générations, étaient forcés de ne se servir de leurs pieds que pour marcher et cessaient d'employer leurs mains comme des pieds, il n'est pas douteux, d'après les observations exposées dans le chapitre précédent, que ces quadrumanes ne fussent à la fin transformés en bimanes et que les pouces de leurs pieds ne cessassent d'être écartés des doigts, ces pieds ne leur servant plus qu'à marcher.

> Indeed, if any race of quadrumana [primates], especially the most perfected among them, should lose, by the necessity of circumstances, or by some other cause, the habit of climbing on trees and grabbing their branches with the feet (as they do with the hands), to cling to them, and if the individuals of this race, for a series of generations, were forced to use their feet only for walking and ceased to use their hands as feet, there is no doubt, according to the observations set out in the previous chapter, that these quadrumanes would eventually be transformed into bimanes [humans] and that their big toe would cease to be separated from the other toes, given that these feet would be used only to walk.

This model, which still looks essentially correct, may appear bold, given that Darwin dared not publish his ideas on evolution through natural selections until 50 years later, and only because Wallace's discovery (Darwin and Wallace 1858) forced him to do so to avoid losing priority. However, Lamarck had been preceded, to some extent, by the Genevan naturalist

Charles Bonnet (1720–1793), who had placed the orangutan (*Pongo pygmaeus*) next to man in his non-evolutionary scale of nature (Wilkins 2018: 94). To sum up all this, we may agree with Mayr (1972: 90), who suggested that Darwin had "vastly underestimated the role which Lamarck had played in preparing the intellectual climate for the subsequent Darwinian advances."

Another consideration linked to evolution must be added here. If evolution proceeds slowly and gradually, as Lamarck and (later) Darwin postulated, there should be a continuum between various life forms. Indeed, this was Lamarck's position, and it was further supported by the principle of plenitude, which many naturalists like John Ray and Lamarck (among others) accepted and which states that "the range of conceivable diversity of kinds of living things is exhaustively exemplified" (Lovejoy 1936: 52). This reflected a view, popular among deists, that Nature had a creative power that manifested itself in spontaneous generation (of fairly simple life forms) and evolution. The principle of plenitude was prevalent at the time and had been discussed earlier by philosophers such as Leibniz (Mayr 1972: 81), who coined the term "organism" (Lherminier and Solignac 2005: 26). Thus, biodiversity must form a vast continuum, an idea that was widespread at the time and inspired, to an extent, Antoine-Laurent de Jussieu's (1789) classification of plants. This conclusion is also suggested by the idea that evolution proceeds by little steps. All this implied that most units of biodiversity, as then conceptualized, from the lowermost (often called species) to the highest-ranking taxa, were artificial constructs. This is indeed what Lamarck argued. He also considered taxa ranked as families, orders and classes as artificial constructs (Lamarck 1809: 41).

Of course, we now know that extinction has created many gaps in the biodiversity, but the very existence of extinction was still doubted in Lamarck's time (though advocated strongly by Cuvier). Indeed, the principle of plenitude hardly leaves open the possibility of extinction, and Lamarck (1809: 76–77) thought that few (if any) species had become extinct. Lamarck (1809: 91) of course knew about the fossil record, but he thought that the fossilized life forms that were unknown in the extant fauna or flora simply lived in areas that had not yet been explored. This idea was much more plausible then than it is now, and John Ray had initially accepted it before subsequently arguing that fossils were not remnants of life forms, but had formed in the rocks (Wilkins 2018: 75). Lamarck also thought that many other fossils that lacked extant counterparts were simply the ancestors of extant species, but that their appearance had changed over time (Lamarck 1809: 93), a process that we now call anagenesis. He only admitted the possibility that large terrestrial animals might have become extinct (he gave several examples, all among Cenozoic mammals, including *Palaeotherium, Mastodon* and *Megatherium*) because these are more vulnerable to anthropic effects (Lamarck 1809: 92). Other species might have been lost in the great Biblical flood. Thus, if there were extinctions, this was attributed to Man or God.

Lamarck accepted Linnaean binominal nomenclature, even though this was not fully consistent with his views on plenitude and the rarity of extinction, because he realized that a worldwide consensus on how to name species would promote nomenclatural stability (Wilkins 2018: 125).

### 1.4.3 Cuvier

In the early 19th century, Cuvier (1812) and other paleontologists soon demonstrated the ubiquity of extinctions. This refuted the principle of plenitude and provided ample theoretical justification for the discontinuities observed in the biodiversity, at least in a given time slice (though not in a single lineage over time), and this further justified the practice of naming taxa. However, Cuvier's views now seem too extreme. For instance, if a given set of similar fossils was found before and after mass extinction events (which Cuvier was the first to document), he argued that mere similarity of these forms did not ensure that they belonged to the same species. In this, he would be followed by Alcide d'Orbigny.

Cuvier explained the appearance of new taxa after mass extinction events (which he called "revolutions") by immigration, rather than successive creations, as many of his critics have asserted (Lherminier and Solignac 2005: 51). This hypothesis was reasonable then because the fossil record had been studied over a tiny proportion of the Earth (and even today, this remains the case for some habitats in some geological stages). Cuvier was also a fixist, and was logically a species realist, but this partial justification (namely, extinction) for species realism faded away as the idea of biological evolution gained ground, notably with Lamarck and Darwin, and as improvement in our knowledge of the fossil record suggested that immigration could not explain all appearances of new taxa after mass extinction events.

Cuvier (1830: 123) first proposed a definition of species that many systematists would still find reasonable: "les individus qui descendent les uns des autres ou de parents communs, et ceux qui leur ressemblent autant qu'ils se ressemblent entre eux" ("individuals that descend from each other or from common parents, and those that resemble them as much as they resemble each other"). To clarify, Cuvier used three criteria to define and delimit species: similarity, descent and cross-fertilization, which are still considered important today. Cuvier's fixism should not be confused with a vaguely Aristotelian essentialism; rather, his fixism resulted from his belief that organisms were highly integrated and that, consequently, changes in one part would require changes in other parts and, as such, any change was unlikely to lead to viable results. This represents an early version of structuralism (Lherminier and Solignac 2005: 50).

### 1.4.4 The Last Stand of Scientific Fixists

The opposition between Lamarck, who viewed species as artificial because they evolve continuously, and Cuvier (and later, Lyell), who viewed species as real because they are fixed, reflects an opposition between species nominalists and realists in the first half of the 19th century. However, this simple binary opposition is a simplification. There was also a third position, advocated by the French botanists Augustin Pyramus de Candolle (1778–1841), who coined the term "taxonomy" (de Candolle 1813: 19); by his son, Alphonse de Candolle

(1806–1893); and by American botanist Asa Gray (1810–1888). They considered that species were real, but contrary to Cuvier, they accepted significant intraspecific variability that reflected environmental influences. Hybridization also created additional variability. They were species essentialists, like Cuvier, but for them, the essence was interfertility with their conspecifics rather than morphology (Wilkins 2018: 142).

By the mid-19th century, scientific fixism was grasping at straws to remain credible. The famous Scottish geologist Charles Lyell (1797–1875) made a serious attempt to defend it in his *Principles of geology* (Lyell 1832). This story is important because Lyell's influence on Darwin is well-established; Darwin took the *Principles*, including volume 2 (which describes Lamarck's work), on his long trip on the *Beagle* (Wilkins 2018: 139). He annotated it profusely and referred to it abundantly in *The Origin of Species*. Lyell thought that species were real and that they had permanent specific characters. However, he had to accommodate variations that were plain in the fossil record (and in the extant fauna). Thus, Lyell admitted that changes in their habitat resulted in changes in various attributes of species (such as color, size, shape or structure), but these changes could not exceed some bounds beyond which variations would have been fatal to the individuals. Thus, these variations did not venture too far from the species' permanent specific characters. Lyell backed up these claims partly through the absence of documented cases (at that time; this is no longer true) of hybridization to yield new species. Thus, Lyell's fixism accommodated a considerable amount of variability and it would take only a conceptual shift to view these variations in the context of transformism, a step that Darwin and Wallace would take shortly thereafter. Even Lyell reluctantly changed his mind about these points following the publication of *The Origin of Species* and discussions with Darwin (Wilkins 2018: 140).

Other contemporary fixists were less honest. Thus, one of Agassiz' pupils, Stimpson, when confronted with a mollusk that seemed to be intermediate between recognized species, ground "one of these vexatious shapes" to powder with his heel, remarking, "That's the proper way to serve a damned transitional form" (Cole 1910 quoting Nathaniel Southgate Shaler's autobiography). Clearly, evidence against fixism was growing fast and would destroy its residual scientific legitimacy in the next few decades.

### 1.4.5 Darwin, Wallace, Trees and Natural Selection

Just before Darwin's famous tree was published, Bronn (1858: 481) presented an interesting tree in a monograph that was awarded a prize by the Paris Academy of Sciences. That tree looks superficially modern. However, Tassy (2011) pointed out that its vertical axis seems to represent levels of organization rather than geological time. This theoretical tree was not meant to depict a particular taxon (no actual taxa are named). This work has a paleontological emphasis, and it contains many figures that show indented classifications to the left and stratigraphic distributions to the right, which could perhaps be viewed as a precursor of the paleontological timetree that would appear about a decade later (see Chapter 3).

Starting with Wallace (1855), who described but did not illustrate abstract evolutionary trees (only as a concept; he did not specify the affinities of actual taxa in his discussion), and more importantly Darwin (who described and illustrated this concept), the tree became the most widespread model for evolution. Darwin's famous tree incorporates geological time only in an abstract way; indeed, this tree only contains hypothetical taxa, and geological time is represented by equally abstract, hypothetical geological stages. However, his tree shows for the first time how lineages split and give rise to clades over time; it shows the diversification process as it is now conceptualized. Indeed, Bouzat (2014) argued that Darwin's tree was meant to illustrate his ideas about divergent evolution, extinction and how natural selection produced taxonomic diversification. In addition, it seems to show the evolutionary process at different timescales, with intervals XI to XIV representing a different (probably larger) timescale than intervals I-X (**Figure 1.4**; Bouzat 2014: 30). Darwin also anticipated (far ahead of cladistics and PN) that:

> Our classifications will come to be, as far as they can be so made, genealogies; and will then truly give what may be called the plan of creation. The rules for classifying will no doubt become simpler when we have a definite object in view.
>
> **(Darwin 1859: 486)**

Trees would indeed, much later, have a major impact on how taxa are discovered and delimited, but this story will be told later, in Chapter 4 on PN. However, Darwin did not reject the use of Linnaean categories for taxa; he wrote:

> Thus, on the view which I hold, the natural **system is genealogical** in its arrangement, like a pedigree; but the **degrees of modification** which the different groups have undergone, **have to be expressed by ranking** them under different so-called genera, sub-families, families, sections, orders, and classes.
>
> **(Darwin 1859: 422)**

This does not necessarily imply that he accepted paraphyletic taxa, but it is clearly an endorsement of RN (though not necessarily as it later became implemented in the rank-based codes, which are described in Chapter 2).

Another major step in our understanding of evolution was the proposal of natural selection as the main mechanism driving evolution (Darwin and Wallace 1858). The idea is simple and starts from the realization that many more offspring are produced than can survive to reproductive age in all taxa. Hence, there is a struggle for survival and reproduction. The organisms best-adapted to their environment have the best probabilities of reproducing; these are the fittest (fitness being defined by the offspring that individuals leave behind, and the proportion of these that will reproduce themselves). For this mechanism to work, we only need to postulate that at least part of the variability that affects fitness is inheritable. Darwin and Wallace did not know how such variations were transmitted; they appear not to have known (or at least, not to have considered seriously) the pioneering work of

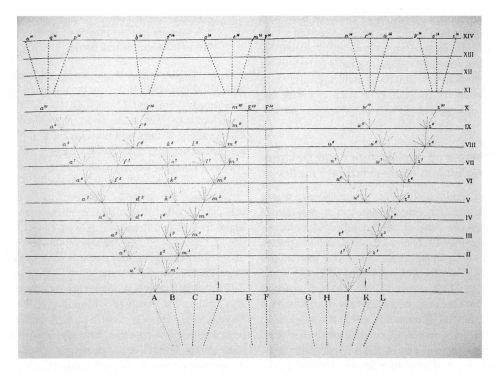

**FIGURE 1.4** Darwin's evolutionary tree published in his *Origin of species*. Image by Charles Darwin, digital picture taken by Alexei Kouprianov, Public domain, via Wikimedia Commons. Downloaded from https://commons.wikimedia.org/wiki/File:Darwin_divergence.jpg.

Gregor Mendel (1822–1884), and the rediscovery of Mendel's work in the early 20th century came after Darwin's and Wallace's work was completed. The impressive rise of genetics, with works by pioneers such as Thomas Hunt Morgan (1866–1945) and the deciphering of the structure of DNA (Watson and Crick 1953) came later still. However, the hypothesis of heritability of at least some variation (which turned out to exclude most acquired characters resulting from use or disuse during an organism's life) was sufficient. This heritability was thus supported by subsequent findings, even though it has been supplemented with others, such as genetic drift (King and Jukes 1969) and the importance of other phenomena, such as heterochrony (Gould 1977), was recognized later. But with the proposal of natural selection as an evolutionary mechanism and the tree as an evolutionary model of biodiversity, systematists had the conceptual tools to produce natural classifications of biological organisms.

It is probably no surprise that both Wallace and Darwin, the co-discoverers of natural selection, traveled extensively to study the biodiversity of distant regions. After all, Wallace had suggested to Henry Walter Bates (1825–1892) that they travel together to collect data relevant to the problem of the origin of species (Egerton 2012: 37), and it has been said that his exploration of Amazonia (where he spent about four years) facilitated his work on evolutionary theory (Egerton 2012: 42). Bates made important discoveries of his own, partly as a result of his 11-year stay in South America. Most notably, Bates discovered that some species, typically harmless and edible, mimic others that are either dangerous or inedible. This allows these harmless species to avoid predation. This phenomenon, now known as "Batesian mimicry," has long been interpreted as a spectacular confirmation

of natural selection. Indeed, "Darwin surely understood that this was the most dramatic support for his theory published by someone else" (Egerton 2012: 47). As for Darwin, the importance of his five-year trip aboard the HMS Beagle, which brought him to Brazil, Uruguay and the Galapagos Islands, among other places, is well-known. Obviously, progress in our knowledge of biodiversity triggered developments both in nomenclature (see the Preface) and in evolutionary theory.

Contrary to what has occasionally been stated, Wilkins (2018: 153) argued that Darwin was a species realist because he considered that individual species could be delimited by the way individuals were sexually attracted toward partners in nature. It was this criterion, rather than whether two captive individuals could give a viable offspring if more or less forced to mate (a criterion that had been used by Buffon), that was valid according to Darwin (Wilkins 2018: 155). However, Darwin clearly thought that the species level (the Linnaean category) was subjective (Darwin 1859: 485) and the reality and delimitation of species through evolutionary time remained problematic for Darwin (as they still are today). He recognized the key role that extinction has played in delimiting extant taxa (Darwin 1859: 432):

> Extinction has only **separated** groups: it has by no means made them; for if every form which has ever lived on this earth were suddenly to reappear, though it would be quite impossible to give definitions by which each group could be distinguished from other groups, as **all would blend together** by steps as fine as those between the finest existing varieties, nevertheless a natural classification, or at least a natural arrangement, would be possible.

This theoretical problem evoked by Darwin has become a real problem for some taxa with a rich fossil record in which anagenesis leads to significant changes that has led paleontologists to recognize two or more species in what may be a single lineage evolving through time. But these problems arose much later, in the 20th century (Nitecki 1957; Rieppel 1986), and will be discussed in Chapter 6.

### 1.4.6 Slow Progress on Phylogenetics and Delayed Incorporation into Nomenclatural Practices

Despite these important developments in evolutionary theory in the early and mid-19th century, substantial progress in phylogenetics would be delayed substantially, with an acceleration starting in the 1960s. Nomenclature started integrating phylogeny explicitly (with proposals to integrate the requirement of monophyly in nomenclatural codes) later still, in the late 1980s onwards. These stories will be told in Chapters 3 and 4, respectively. This long delay between the publication of the theory of evolution by natural selection and progress in phylogenetics may seem surprising, but it is partly explained by the fact that new theories and methods require time to be universally accepted or adopted by scientists because of the weight of tradition (Stevens 1984). For instance, some systematists remained fixists well after publication of *The Origin of Species*. This includes, notably, the Swiss and American paleontologists Louis Agassiz (1807–1873), although his views about evolution and nomenclature were not necessarily representative of paleontologists of that time (Wilkins 2018: 134). Thus, Agassiz (1859: 8) thought that taxa of various categories (including classes, orders, families and genera, in addition to species) had been created by the Creator, and that taxa of all ranks were equally real.

Another factor that explains the long delay between the rise of evolutionism and phylogenetics is that accepting the idea that there is a Tree of Life is not the same as discovering a good method to reconstruct that tree. And being able to reconstruct that tree does not necessarily imply acceptance of Nitecki's idea that this tree should be central to our taxonomy and nomenclature, even though there are compelling reasons to accept this conclusion (see Chapters 4 and 5).

Nevertheless, in the 19th century, some important methodological developments took place in phylogenetics and evolutionary biology. The most important is the proposal by anatomists and embryologists of several criteria to establish which sets of structures represented variations of the same structure. This peculiar relationship is called "homology." We now know that these structures were connected through evolution and thus formed alternative states of a given character, but this is the concept of homology held by evolutionists. This was not in Owen's formulation of homology, which is logical given that Owen did not accept the idea of evolution, as shown clearly by this brief quote (Owen 1843: 368–369):

> The testaceous Cephalopods first construct an unilocular shell, which is the common persistent form in Gasteropods, and afterwards superadd the characteristic chambers and siphon. This simple fact would of itself have **disproved the theory of**

**evolution**, if other observations of the phenomena of development had not long since rendered that **once favourite doctrine untenable**.

How ironic that a scientist so firmly opposed to the idea of biological evolution made such substantial contributions to evolutionary biology! Owen's (1843: 379) definition of a homologue is "The same organ in different animals under every variety of form and function." Owen (1843) is credited for having opposed the term "homology" to "analogy," which was an important step in the history of these terms. Owen's definition of "analogy" has aged better than his definition of "homology." He defined "analogue" as "a part or organ in one animal which has the same function as another part or organ in a different animal" (Owen 1843: 374). The meaning of "homology" and "analogy" shifted over time. In the early 19th century, various authors, notably Etienne Geoffroy Saint-Hilaire, used "homology" to designate what is now called "serial homology," and what we now call "homology" was then called "analogy" (Minelli and Fusco 2013: 291). Nevertheless, Etienne Geoffroy Saint-Hilaire's concept of homology (which he called analogy) was to an extent more modern than Owen's to the extent that it was interpreted in an evolutionary context (but like Lamarck's, his view of evolutionary mechanisms was quite different to the Neo-Darwinism that became established much later).

Commonly used criteria to establish homology (in its modern sense) are similarity in position relative to other structures, embryological origin and taxonomic continuity. This last criterion stipulates that putative homologues should be present in most closely related taxa; in this case, their most recent common ancestors can reasonably be hypothesized to have possessed one of these putative homologues. The taxonomic continuity criterion requires a good knowledge of biodiversity and of phylogeny, and progress on these fronts was made in the 19th century, notably through paleontology, as shown by key works of Cuvier (1812, 1830), Agassiz (1833–1844), Owen (1860), Gaudry (1866), Haeckel (1866) and many others, even though some of these authors were not evolutionists. Notably, Haeckel (1866) had recognized that unicellular eukaryotes (often still known as "protists") did not fit into the traditional division into the plant and animal kingdoms (Woese et al. 1990). Unfortunately, this did not deter systematists from developing rank-based codes for plants and animals in the late 19th and early 20th centuries. A code for prokaryotic organisms was only inaugurated in 1948 (see Chapter 2), and unicellular eukaryote nomenclature is still regulated by the zoological and botanical codes, even though these organisms are neither animals nor plants, in the modern sense of the word. This initial division of living organisms into animals and plants was not optimal given the state of knowledge in 1866, and is even less so now (see Chapter 3). For a more thorough treatment of this topic, see Section 3.1.5.2.

The developmental (or embryological) criterion of homology was already used early in the 19th century, as explained by Minelli and Fusco (2013: 297):

> That the study of development can provide an important and often decisive key in our attempts to establish homologies, is a notion we find already in the works

of Geoffroy Saint-Hilaire (1807), who identified the centers of ossification as the modules, or structural units, of which the skull of vertebrates consists.

However, examples of structures that appear homologous based on morphological criteria but that develop differently have been known since the late 19th century (Minelli and Fusco 2013: 297).

Using these criteria, 19th-century anatomists, embryologists and systematists established that the fins of sharks, teleosts and dipnoans are homologous with the limbs of tetrapods (e.g., Owen 1849). In the 19th century, these developments allowed more rigorous tests and more explicit formulations of evolutionary hypotheses. Much later, in the second half of the 20th century, they would form one of the foundations of cladistics, to the extent that homologues (as character states describing an aspect of a structure or organ, rather than whole organs) would become alternate states of a given character (see Chapter 3).

# REFERENCES

Adanson, M. 1763. Familles des plantes. 1ère partie. Vincent, Paris, cccxxv + 189 pp.

Agassiz, J. L. R. 1833–1844. Recherches sur les Poissons Fossiles. Imprimerie Petitpierre, Neuchâtel.

Agassiz, L. 1859. An essay on classification. Longman, Brown, Green, Longmans, & Roberts, London, viii + 381 pp.

Alves, A. S. A., L. L. dos Santos, W. S. F. Júnior, and U. P. Albuquerque. 2016. How and why should people classify natural resources?; pp. 117–121, Introduction to ethnobiology. Springer, Amsterdam.

Arabatzis, G. 2012. Michael of Ephesus and the philosophy of living things (in "De partibus animalium" 22.25–23.9); pp. 51–78 in B. Bydén and K. Ierodiakonou (eds.), The many faces of Byzantine philosophy. The Norwegian Institute at Athens, Athens.

Atran, S. 1985. The nature of folk-botanical life forms. American Anthropologist 87:298–315.

Atran, S. 1987a. The essence of folkbiology: A reply to Randall and Hunn. American Anthropologist 89:149–151.

Atran, S. 1987b. Origin of the species and genus concepts: An anthropological perspective. Journal of the History of Biology 20:195–279.

Atran, S. 1998. Folk biology and the anthropology of science: Cognitive universals and cultural particulars. Behavioral and Brain Sciences 21:547–569.

Avise, J. C. and G. C. Johns. 1999. Proposal for a standardized temporal scheme of biological classification for extant species. Proceedings of the National Academy of Sciences of the United States of America 96:7358–7363.

Balme, D. M. 1962. ΓΕΝΟΣ and ΕΙΔΟΣ in Aristotle's biology. The Classical Quarterly 12:81–98.

Berlin, B. 1973. Folk systematics in relation to biological classification and nomenclature. Annual Review of Ecology and Systematics 4:259–271.

Berlin, B. 2014. Ethnobiological classification: Principles of categorization of plants and animals in traditional societies. Princeton University Press, Princeton, NJ, 354 pp.

Berlin, B., J. S. Boster, and J. ONeill. 1981. The perceptual bases of ethnobiological classification: Evidence from Aguaruna Jivaro Ornithology. Journal of Ethnobiology 1:95–108.

Berlin, B., D. E. Breedlove, and P. H. Raven. 1966. Folk taxonomies and biological classification. Science 154:273–275.

Berlin, B., D. E. Breedlove, and P. H. Raven. 1973. General principles of classification and nomenclature in folk biology. American Anthropologist 75:214–242.

Bertrand, Y., F. Pleijel, and G. W. Rouse. 2006. Taxonomic surrogacy in biodiversity assessments, and the meaning of Linnaean ranks. Systematics and Biodiversity 4:149–159.

Bouzat, J. L. 2014. Darwin's diagram of divergence of taxa as a causal model for the origin of species. The Quarterly Review of Biology 89:21–38.

Bronn, H. G. 1858. Untersuchungen über die Entwickelungs-Gesetze der organischen Welt während der Bildungs-Zeit unserer Erd-Oberfläche. E. Schweizerbart, Stuttgart, x + 502 pp.

Buffon, G. L. L. 1749. Histoire naturelle, générale et particulière, avec la description du Cabinet du Roy. L'Imprimerie royale, Paris, 612 pp.

Bulmer, R. N. 1970. Which came first, the chicken or the egg-head?; pp. 1069–1091 in J. Pouillon and P. Maranda (eds.), Échanges et communications: Mélange offerts à Claude Lévi-Strauss à l'occasion de son 60ème anniversaire. Mouton, The Hague.

Bulmer, R. N. 1974. Folk biology in the New Guinea highlands. Social Science Information 13:9–28.

Cantino, P. D. and K. de Queiroz. 2020. International code of phylogenetic nomenclature (PhyloCode): A phylogenetic code of biological nomenclature. CRC Press, Boca Raton, Florida, xl + 149 pp.

Chambers, R. 1845. Explanations: A sequel to "vestiges of the natural history of creation." John Churchill, London, vii + 198 pp.

Choate, H. A. 1912. The origin and development of the binomial system of nomenclature. The Plant World 15:257–263.

Cole, G. A. 1910. The autobiography of Nathaniel Southgate Shaler, with a supplementary memoir by his wife. Nature 82:274–275.

Cuvier, G. 1812. Recherches sur les ossemens fossiles de quadrupèdes: où l'on rétablit les caractères de plusieurs espèces d'animaux que les révolutions du globe paroissent avoir détruites. Deterville, Paris, vi + 116 pp.

Cuvier, G. 1830. Discours sur les révolutions de la surface du globe: et sur les changements qu'elles ont produits dans le règne animal. E. d'Ocagne, Paris, ii + 408 pp.

Darwin, C. 1859. On the origin of species by means of natural selection or the preservation of favoured races in the struggle for life. John Murray, London, 502 pp.

Darwin, C. and A. Wallace. 1858. On the tendency of species to form varieties; and on the perpetuation of varieties and species by natural means of selection. Proceedings of the Linnean Society of London 3:45–62.

Davies, G. I. 2013. Introduction to the Pentateuch; pp. 12–37 in J. Barton and J. Muddiman (ed.), The Oxford Bible commentary. Oxford University Press, Oxford.

de Candolle, A.-P. 1813. Théorie élémentaire de la botanique. Deterviile, Paris, viii + 500 pp.

de Queiroz, K. 2005. Linnaean, rank-based, and phylogenetic nomenclature: Restoring primacy to the link between names and taxa. Symbolae Botanicae Upsalienses 33:127–140.

de Queiroz, K. 2012. Biological nomenclature from Linnaeus to the PhyloCode. Bibliotheca Herpetologica 9:135–145.

Diamond, J. M. 1966. Zoological classification system of a primitive people. Science 151:1102–1104.

Dubois, A. 1982. Les notions de genre, sous-genre et groupe d'espèces en zoologie à la lumière de la systématique évolutive. Monitore Zoologico Italiano-Italian Journal of Zoology 16:9–65.

Dubois, A., A. Ohler, and R. Pyron. 2021. New concepts and methods for phylogenetic taxonomy and nomenclature in zoology, exemplified by a new ranked cladonomy of recent amphibians (Lissamphibia). Megataxa 5:1–738.

Egerton, F. N. 2012. History of ecological sciences, part 41: Victorian naturalists in Amazonia-Wallace, Bates, Spruce. Bulletin of the Ecological Society of America 93:35–60.

Ellen, R. 1998. Doubts about a unified cognitive theory of taxonomic knowledge and its memic status. Behavioral and Brain Sciences 21:572–573.

Ellen, R. 2016. Is there a role for ontologies in understanding plant knowledge systems? Journal of Ethnobiology 36:10–28.

Ereshefsky, M. 2002. Linnaean ranks: Vestiges of a bygone era. Philosophy of Science 69:S305–S315.

Fögen, T. 2013. Scholarship and competitiveness: Pliny the Elder's attitude towards his predecessors in the naturalis historia; pp. 83–111 in M. Asper and A.-M. Kanthak (eds.), Science, technology, and medicine in ancient cultures, vol. 1: Writing science: Medical and mathematical authorship in Ancient Greece. Walter De Gruyter, Berlin.

Fürst von Lieven, A. and M. Humar. 2008. A cladistic analysis of Aristotle's animal groups in the "historia animalium". History and Philosophy of the Life Sciences 30:227–262.

Fürst von Lieven, A., M. Humar, and G. Scholtz. 2021. Aristotle's lobster: The image in the text. Theory in Biosciences 140:1–15.

Ganias, K., C. Mezarli, and E. Voultsiadou. 2017. Aristotle as an ichthyologist: Exploring Aegean fish diversity 2,400 years ago. Fish and Fisheries 18:1038–1055.

Gaudry, A. 1866. Considérations générales sur les animaux fossiles de Pikermi. F. Savy, Paris, 68 pp.

Geoffroy Saint-Hilaire, É. 1807. Considérations sur les pièces de la tête osseuse des animaux vertébrés, et particulièrement sur celles du crâne des oiseaux. Annales du Muséum d'histoire naturelle Paris 10:342–365.

Geoffroy Saint-Hilaire, I. 1854. Histoire naturelle générale des règnes organiques. Masson, Paris, 455 pp.

Gilmour, J. S. and S. M. Walters. 1964. Philosophy and classification; pp. 1–22 in W. B. Turrill (ed.), Vistas in botany. Pergamon Press, Oxford.

Goddard, C. 2018. A semantic menagerie: The conceptual semantics of ethnozoological categories. Russian Journal of Linguistics 22:539–559.

Goloboff, P. A. 1996. Estimating character weights during tree search. Cladistics 9:83–91.

Goloboff, P. A. and J. S. Arias. 2019. Likelihood approximations of implied weights parsimony can be selected over the Mk model by the Akaike information criterion. Cladistics 35:695–716.

González-Rodriguez, K., S. P. Applegate, and L. Espinosa-Arrubarrena. 2004. A New World macrosemiid (Pisces: Neopterygii-Halecostomi) from the Albian of Mexico. Journal of Vertebrate Paleontology 24:281–289.

Gotthelf, A. 1999. Darwin on Aristotle. Journal of the History of Biology 32:3–30.

Gould, S. J. 1977. Ontogeny and phylogeny. The Belknap Press of Harvard University Press, Cambridge, 501 pp.

Guasparri, A. 2013. Explicit nomenclature and classification in Pliny's Natural History XXXII. Studies in History and Philosophy of Science Part A 44:347–353.

Guasparri, A. 2022. The Roman classification and nomenclature of aquatic animals: An annotated checklist (with a focus on ethnobiology). Anthropozoologica 57:19–100.

Haeckel, E. 1866. Generelle Morphologie der Organismen. Reimer, Berlin, 1036 pp.

Hatano, G. 1998. Informal biology is a core domain, but its construction needs experience. Behavioral and Brain Sciences 21:575–575.

Healey, C. J. 1979. Taxonomic rigidity in biological folk classification: Some examples from the Maring of New Guinea. Ethnomedizin 5:361–383.

Hennig, W. 1950. Grundzuge einer Theorie der phylogenetischen Systematik. Deutscher Zentralverlag, Berlin, 370 pp.

Hennig, W. 1966. Phylogenetic systematics. University of Illinois Press, Urbana, Chicago, London, 263 pp.

Henry, D. 2011. Aristotle's pluralistic realism. The Monist 94:197–220.

Hillis, D. M. 1991. Discriminating between phylogenetic signal and random noise in DNA sequences; pp. 278–294 in M. M. Miyamoto and J. Cracraft (eds.), Phylogenetic analyses of DNA sequences. Oxford University Press, Oxford.

Hillis, D. M. and J. P. Huelsenbeck. 1992. Signal, noise, and reliability in molecular phylogenetic analyses. The Journal of Heredity 83:189–195.

Hodge, M. J. S. 1972. The universal gestation of nature: Chambers' "vestiges" and "explanations". Journal of the History of Biology 5:127–151.

Hoquet, T. 2007. Buffon: From natural history to the history of nature? Biological Theory 2:413–419.

Huelsenbeck, J. P. 1991. Tree-length distribution skewness: An indicator of phylogenetic information. Systematic Zoology 40:257–270.

Humar, M. 2015. The shipholder, the Remora, and the Lampreys: Studies in the identification of the Ancient Echeneis; pp. 203–220 in J. Althoff, S. Föllinger, and G. Wöhrle (eds.), Antike Naturwissenschaft und ihre Rezeption, vol. 25. WVT Wissenschaftlicher Verlag Trier, Trier.

Hunn, E. 1982. The utilitarian factor in folk biological classification. American Anthropologist 84:830–847.

Hunn, E. S. 1998. Atran's biodiversity parser: Doubts about hierarchy and autonomy. Behavioral and Brain Sciences 21:576–577.

ICZN. 1999. International code of zoological nomenclature. The International Trust for Zoological Nomenclature, London, 306 pp. www.iczn.org/the-code/the-international-code-of-zoological-nomenclature/the-code-online/

Jussieu, A.-L. de. 1789. Genera plantarum secundum ordines naturales disposita juxta methodum in horto regio parisiensi exaratam, anno 1774. veuve Herissant, Paris, 498 pp.

King, J. L. and T. H. Jukes. 1969. Non-darwinian evolution: Most evolutionary change in proteins may be due to neutral mutations and genetic drift. Science 164:788–798.

Klimis, G. 2008. Medicinal herbs and plants in Homer; pp. 283–291 in S. A. Paipetis (ed.), History of mechanism and machine science, vol. 6: Science and technology in homeric epics. Springer, Berlin.

Lamarck, J.-B. 1801. Système des animaux sans vertèbres. Deterville, Paris, viii + 432 pp.

Lamarck, J.-B. 1802. Recherches sur l'organisation des corps vivans. Maillard, Paris, viii + 216 pp.

Lamarck, J. B. 1809. Philosophie zoologique. Flammarion, Paris, 718 pp. [Reprint of the original, with a new introduction].

Laurin, M. 2005. The advantages of phylogenetic nomenclature over Linnean nomenclature; pp. 67–97 in A. Minelli, G. Ortalli, and G. Sanga (eds.), Animal names. Instituto Veneto di Scienze, Lettere ed Arti, Venice.

Laurin, M. 2008. The splendid isolation of biological nomenclature. Zoologica Scripta 37:223–233.

Laurin, M. 2010a. The subjective nature of Linnaean categories and its impact in evolutionary biology and biodiversity studies. Contributions to Zoology 79:131–146.

Laurin, M. 2010b. How vertebrates left the water. University of California Press, Berkeley, xv + 199 pp.

Laurin, M. and M. Humar. 2022. Phylogenetic signal in characters from Aristotle's History of Animals. Comptes Rendus Palevol 21:1–16.

Laurin, M. and R. Soler-Gijón. 2001. The oldest stegocephalian from the Iberian Peninsula: Evidence that temnospondyls were euryhaline. Comptes Rendus de l'Académie des Sciences de Paris, Sciences de la vie/Life Sciences 324:495–501.

Lennox, J. G. 2006. Aristotle's biology and Aristotle's philosophy; pp. 292–315 in M. L. Gill and P. Pellegrin (eds.), A companion to ancient philosophy. Blackwell, Malden, MA (USA).

Lhermin[i]er, P. and M. Solignac. 2000. L'espèce: définitions d'auteurs. Comptes Rendus de l'Académie des Sciences—Series III—Sciences de la Vie 323:153–165.

Lherminier, P. and M. Solignac. 2005. De l'espèce. Syllepse, Paris, XI + 694 pp.

Linnaeus, C. 1753. Species plantarum. Salvi, Stockholm, 1200 pp.

Linnaeus, C. 1758. Systema naturae, 10th ed. Holmiae (Laurentii Salvii), Stockholm, 824 pp.

Linné, C. V. 1751. Philosophia botanica. Salvius, Stockholm, 362 pp.

Llana, J. 2000. Natural history and the Encyclopédie. Journal of the History of Biology 33:1–25.

Lloyd, G. E. R. 1983. Science, folklore and ideology: Studies in the life sciences in ancient Greece. Cambridge University Press, Cambridge, 260 pp.

Lovejoy, A. O. 1936. The great chain of being: A study of the history of an idea. Harvard University Press, Cambridge, MA, xxv + 382 pp.

Ludwig, D. 2018. Revamping the metaphysics of ethnobiological classification. Current Anthropology 59:415–438.

Lyell, C. [1832] 1991. Principles of geology, vol. 2. [John Murry] University of Chicago Press, [London] Chicago, 330 pp.

Magnol, P. 1689. Prodromus historiae generalis plantarum in quo familiae plantarum per tabulas disponuntur. Pech, Montpellier, 79 pp.

Mayr, E. 1949. The species concept: Semantics versus semantics. Evolution 3:371–372.

Mayr, E. 1972. Lamarck revisited. Journal of the History of Biology 5:55–94.

Mayr, E. 1982. The growth of biological thought: Diversity, evolution, and inheritance. The Belknap Press of Harvard University Press, Cambridge, 974 pp.

Melville, H. 1851. Moby Dick: Or, the white whale. Harper & Brothers, New York, 1–539 pp.

Merrem, B. 1820. Tentamen systematis amphibiorum. J. C. Krieger, Marburg, xv, 191 pp.

Minelli, A. 2000. The ranks and the names of species and higher taxa, or a dangerous inertia of the language of natural history; pp. 339–351 in M. T. Ghiselin and A. E. Leviton (eds.), Cultures and institutions of natural history: Essays in the history and philosophy of science. California Academy of Sciences, San Francisco.

Minelli, A. 2022a. Species; in B. Hjørland and C. Gnoli (eds.), ISKO encyclopedia of Knowledge Organization (IEKO). www.isko.org/cyclo/species

Minelli, A. 2022b. The species before and after Linnaeus: Tension between disciplinary nomadism and conservative nomenclature; pp. 191–226 in J. Wilkins, I. Pavlinov, and F. Zachos (eds.), Species problems and beyond: Contemporary issues in philosophy and practice. CRC Press, Boca Raton, FL.

Minelli, A. and G. Fusco. 2013. Homology; pp. 289–322 in K. Kampourakis (ed.), The philosophy of biology: A companion for educators. Springer, Dordrecht.

Mishler, B. D. and J. S. Wilkins. 2018. The hunting of the SNaRC: A snarky solution to the species problem. Philosophy, Theory, and Practice in Biology 10:1–18.

Moser, B. 2013. The Roman ethnozoological tradition: Identifying exotic animals in Pliny's natural history. MA dissertation, University of Western Ontario, vii + 137 pp.

Mottequin, B., P. Edouard, and C. Prestianni. 2014. Catalogue of the types and illustrated specimens recovered from the "black marble" of Denée, a marine conservation-Lagerstätte from the Mississippian of southern Belgium. Geologica Belgica 18:1–14.

Nagaoka, L. and S. Wolverton. 2016. Archaeology as ethnobiology. Journal of Ethnobiology 36:473–475.

Newmaster, S. G., R. Subramanyam, N. C. Balasubramaniyam, and R. F. Ivanoff. 2007. The multi-mechanistic taxonomy of the Irulas in Tamil Nadu, South India. Journal of Ethnobiology 27:233–255.

Nicolson, D. H. 1991. A history of botanical nomenclature. Annals of the Missouri Botanical Garden 78:33–56.

Nitecki. 1957. What is a paleontological species? Evolution 11:378–380.

Ogilvie, B. W. 2003. The many books of nature: Renaissance naturalists and information overload. Journal of the History of Ideas 64:29–40.

Owen, R. 1843. Lectures on the comparative anatomy and physiology of the invertebrate animals, delivered at the Royal College of Surgeons, in 1843, Longman, Brown, Green, and Longmans. Longman, Brown, Green, and Longmans, London, 392 pp.

Owen, R. 1849. On the nature of limbs: A discourse delivered on Friday, February 9, at an evening meeting of the Royal Institution of Great Britain. J. van Voorst, London, 119 pp.

Owen, R. 1860. Palæontology. A and C. Black, Edinburgh, 420 pp.

Pawley, A. 2006. Wayan Fijian classification of marine animals: Some problems for lexical description and for Berlin's universals of folk taxonomies. 10th International Conference on Austronesian Linguistics, 17–20 January, Palawan, Philippines. [Unpublished draft of the talk given at the conference].

Pawley, A. 2011. Were turtles fish in Proto Oceanic? Semantic reconstruction and change in some terms for animal categories in Oceanic languages; pp. 421–452 in M. Ross, A. Pawley, and M. Osmond (eds), The lexicon of Proto

Oceanic: The culture and environment of ancestral Oceanic society. Research School of Pacific and Asian Studies, The Australian National University, Canberra.

Pellegrin, P. 1986. Aristotle's classification of animals: Biology and the conceptual unity of the Aristotelian corpus. University of California Press, Berkeley, 249 pp.

Persson, M. 2016. Building an empire in the Republic of Letters: Albrecht von Haller, Carolus Linnaeus, and the struggle for botanical sovereignty. Circumscribere: International Journal for the History of Science 17:18–40.

Pierotti, R. 2020. Historical links between ethnobiology and evolution: Conflicts and possible resolutions. Studies in History and Philosophy of Science Part C: Studies in History and Philosophy of Biological and Biomedical Sciences 81:101277.

Pleijel, F. and G. W. Rouse. 2000. Least-inclusive taxonomic unit: A new taxonomic concept for biology. Proceedings of the Royal Society of London, Series B 267:627–630.

Prance, G. T. 1994. A comparison of the efficacy of higher taxa and species numbers in the assessment of biodiversity in the Neotropics. Philosophical Transactions of the Royal Society of London, Series B 345:89–99.

Pratt, V. 1984. The essence of Ari[s]totle's zoology. Phronesis 29:267–278.

Prichard, J. C. 1813. Researches into the physical history of man. John and Arthur Arch, London, viii + 558 pp.

Ragupathy, S., S. G. Newmaster, M. Murugesan, and V. Balasubramaniam. 2009. DNA barcoding discriminates a new cryptic grass species revealed in an ethnobotany study by the hill tribes of the Western Ghats in southern India. Molecular Ecology Resources 9:164–171.

Rieppel, O. 1986. Species are individuals: A review and critique of the Argument; pp. 283–317 in M. K. Hecht, B. Wallace, and G. T. Prance (eds.), Evolutionary biology, vol. 20. Plenum Publishing Corporation, New York.

Rigato, E. and A. Minelli. 2013. The great chain of being is still here. Evolution: Education and Outreach 6:18. www.evolution-outreach.com/content/6/1/18

Romero, A. 2012. When whales became mammals: The scientific journey of cetaceans from fish to mammals in the history of science; pp. 3–30 in A. Romero and E. O. Keith (eds.), New approaches to the study of marine mammals. InTech, Rijeka, Croatia.

Romeyer-Dherbey, G. 1986. La classification des animaux chez Aristote. Statut de la Biologie et unité de l'aristotélisme. Revue de Métaphysique et de Morale 91:428–430.

Schmitz, H., N. Uddenberg, and P. Östensson. 2007. A passion for systems: Linnaeus and the dream of order in nature. Naturt och Kultur, Stockholm, 256 pp.

Sillitoe, P. 1995. An ethnobotanical account of the plant resources of the Wola region, Southern Highlands Province, Papua New Guinea. Journal of Ethnobiology 15:201–236.

Sillitoe, P. 2002. Contested knowledge, contingent classification: Animals in the highlands of Papua New Guinea. American Anthropologist 104:1162–1171.

Sillitoe, P. 2003. Managing animals in New Guinea: Preying the game in the highlands. Routledge, New York, 416 pp.

Sloan, P. R. 1972. John Locke, John Ray, and the problem of the natural system. Journal of the History of Biology 5:1–53.

Sommer, B. D. 2011. Dating pentateuchal texts and the perils of pseudo-historicism; pp. 85–108 in T. B. Dozeman, K. Schmid, and B. J. Schartz (eds.), Forschungen zum Alten Testament. Mohr Siebeck, Tübingen.

Steiner, G. 1955. The skepticism of the Elder Pliny. The Classical Weekly 48:137–143.

Stemerding, D. 1993. How to make oneself nature's spokesman? A Latourian account of classification in eighteenth-and early nineteenth-century natural history. Biology and Philosophy 8:193–223.

Stevens, P. F. 1984. Metaphors and typology in the development of botanical systematics 1690–1960, or the art of putting new wine in old bottles. Taxon 33:169–211.

Stoyles, B. J. 2013. Μέγιστα Γένη and division in Aristotle's Generation of Animals. Apeiron 46:1–25.

Strickland, H. E., J. S. Henslow, J. Phillips, W. E. Shuckard, J. B. Richardson, G. R. Waterhouse, R. Owen, W. Yarrell, L. Jenyns, C. Darwin, W. J. Broderip, and J. O. Westwood. 1843. Series of propositions for rendering the nomenclature of zoology uniform and permanent, being the Report of a Committee for the consideration of the subject appointed by the British Association for the Advancement of Science. Annals and Magazine of Natural History 11:259–275.

Sundberg, P. and F. Pleijel. 1994. Phylogenetic classification and the definition of taxon names. Zoologica Scripta 23:19–25.

Swofford, D. L. 2002. PAUP* phylogenetic analysis using parsimony (*and other methods). Version 4.0a, build 167 (updated on Feb. 1, 2020).

Tassy, P. 1981. Lamarck and systematics. Systematic Zoology 30:198–200.

Tassy, P. 2011. Trees before and after Darwin. Journal of Zoological Systematics and Evolutionary Research 49:89–101.

Thanassoulopoulos, C. 2008. Agricultural development in the Homeric era; pp. 295–301 in S. A. Paipetis (ed.), History of mechanism and machine science, vol. 6: Science and technology in Homeric Epics. Springer, Berlin.

Tournefort, J. P. de. 1694. Elemens de botanique ou methode pour connoitre les plantes. L'Imprimerie Royale, Paris, 379 pp.

Turland, N. J., J. H. Wiersema, F. R. Barrie, W. Greuter, D. Hawksworth, P. S. Herendeen, S. Knapp, W.-H. Kusber, D.-Z. Li, K. Marhold, T. May, J. McNeill, A. Monro, J. Prado, M. Price, and G. Smith. 2018. International code of nomenclature for algae, fungi, and plants (Shenzhen Code) adopted by the Nineteenth International Botanical Congress Shenzhen, China, July 2017. Koeltz Botanical Books, Glashütten, xxxviii + 254 pp.

Verne, J. 1871. Vingt mille lieues sous les mers. Ebooks libres et gratuits, 675 pp. www.ebooksgratuits.com/details.php?book=1620

Voultsiadou, E., V. Gerovasileiou, L. Vandepitte, K. Ganias, and C. Arvanitidis. 2017. Aristotle's scientific contributions to the classification, nomenclature and distribution of marine organisms. Mediterranean Marine Science 18:468–478.

Voultsiadou, E., D. Koutsoubas, and M. Achparaki. 2010. Bivalve mollusc exploitation in Mediterranean coastal communities: An historical approach. Journal of Biological Research 13:35.

Voultsiadou, E. and A. Tatolas. 2005. The fauna of Greece and adjacent areas in the age of Homer: Evidence from the first written documents of Greek literature. Journal of Biogeography 32:1875–1882.

Voultsiadou, E. and D. Vafidis. 2007. Marine invertebrate diversity in Aristotle's zoology. Contributions to Zoology 76:103–120.

Wallace, A. R. 1855. XVIII: On the law which has regulated the introduction of new species. Annals and Magazine of Natural History 16:184–196.

Watson, J. D. and F. H. Crick. 1953. A structure for deoxyribose nucleic acid. Nature 171:737–738.

Wiener, C. 2015. Dividing nature by the joints. Apeiron 48:285–326.

Wierzbicka, A. 1992. What is a life form? Conceptual issues in ethnobiology. Journal of Linguistic Anthropology 2:3–29.

Wilkins, J. S. 2018. Species: The evolution of the idea. CRC Press, Boca Raton, xxxviii + 389 pp.

Winston, J. E. 2018. Twenty-first century biological nomenclature: The enduring power of names. Integrative and Comparative Biology 58:1122–1131.

Witteveen, J. and S. Müller-Wille. 2020. Of elephants and errors: Naming and identity in Linnaean taxonomy. History and Philosophy of the Life Sciences 42:1–34.

Woese, C. R., O. Kandler, and M. L. Wheelis. 1990. Towards a natural system of organisms: Proposal for the domains Archaea, Bacteria, and Eucarya. Proceedings of the National Academy of Sciences 87:4576–4579.

Wolverton, S., A. Barker, and J. Dombrosky. 2016. Paleoethnobiology; pp. 25–32 in U. Albuquerque and R. Nóbrega Alves (eds.), Introduction to ethnobiology. Springer, Heidelberg.

Zirkle, C. 1946. The early history of the idea of the inheritance of acquired characters and of pangenesis. Transactions of the American Philosophical Society 35:91–151.

# 2

# *The Advent of the Rank-Based Codes*

## 2.1 Introduction: "At the Beginning There Was Only Chaos"[1]

We saw in Chapter 1 that Linnaeus introduced binominal nomenclature to reduce taxon names to a manageable size (two words). While this helped initially, names of taxa soon started proliferating, not only because of the discovery of new taxa, but also because many authors proposed new names for the same taxa. This practice occurred for various reasons, the most important of which was that some authors thought that a name was not appropriate because it was not descriptive enough or it did not describe the most important character of a taxon. By the early 1840s, chaos prevailed in zoological nomenclature, as aptly described and explained by Strickland et al. (1842: 4):

> It being admitted on all hands that words are only the conventional signs of ideas, it is evident that language can only attain its end effectually by being **permanently established** and **generally recognized**. This consideration ought, it would seem, to have checked those who are continually attempting to subvert the established language of zoology by **substituting terms of their own coinage**. But, forgetting the true nature of language, they persist in confounding the name of a species or group with its definition; and because the former often falls short of the fullness of expression found in the latter, they cancel it without hesitation, and introduce some new term which appears to them more characteristic, but which is utterly unknown to the science, and is therefore devoid of all authority.

> **(Bold font used for emphasis is mine, unless noted otherwise.)**

As we will see later, nomenclature was stabilized (but only to an extent) mainly through the development, promotion and enforcement of codes of biological nomenclature. However, compliance with a code is necessary, but not sufficient to ensure nomenclatural stability. Such stability also requires that the codes include clear, easy-to-follow rules to make their application objective. This was aptly summarized by Bock (1994: 11) in his monographic review of avian family-group names: "The history of zoological nomenclature clearly demonstrates that any provisions of the Code depending on the **judgment** of zoologists **destroy completely** all usefulness of those provisions of the Code and will decrease stability and universality of nomenclature." Indeed, most of the debates in the field of biological nomenclature have centered around this question: which rules will promote nomenclatural stability? And what exactly do we want to stabilize?

## 2.2 The Slow Rise of Ranks: The Example of Families

Ranks are at the core of the rank-based codes, and in zoology, rules designed to promote nomenclatural stability apply only up to the family-series level. Given that the development and gradual adoption of a family nomenclatural level preceded the advent of all rank-based codes, it needs to be covered before these codes. We saw in Chapter 1 that Linnaeus did not use families in his classification of animals and plants, but this does not mean that other systematists had not used families before. Indeed, the family rank, presumably like other "Linnaean categories," developed over an extended period and its history is both complex and difficult to document. Only a brief outline will be given here.

Magnol (1689) listed many plant families in a monograph, and indicated which lower-ranking taxa they included. Magnol (1689) is now considered by some authors to have introduced families to botanical nomenclature (Judd et al. 2008), even though his families are not established under the rank-based codes because (among other reasons) they were published well before the *Species Plantarum*. The next major step in the development of families in botany was Adanson's work (1763) entitled "Familles des Plantes" [Plant families], in which he proposed a more "natural" classification of plants. This was mentioned by de Jussieu (1789) as being one of the most successful earlier studies on the natural order of plants, and it may have inspired de Jussieu (1789) to use the family rank in botany. A concept of families as a rank between genus and order was commonly used by botanists shortly after de Jussieu (1789) started using it, but zoologists did not accept this idea for another decade.

Latreille (1797) is sometimes credited for having introduced the use of families in zoology (for instance, Bock 1994: 14). In a monograph, Latreille (1797) classified insects into classes, unnamed families (they were only described) and genera. However, the history of families in zoology seems to have started earlier. For instance, Kästner and Erxleben (1767; reprinted in Ludwig 1790) proposed a classification of amniotes in which orders were subdivided into families. Their taxa (*Monochelon, Dichelon, Trichelon* and so on) are no longer in use, and some of their "families" (such as *Testudinata*) are now ranked higher (*Testudinata*, sometimes called *Testudines* or *Chelonia*, is typically considered to be an order), but the fact that these families have names suggests that Kästner and Erxleben's concept of family was at least as modern as Latreille's. An even earlier example of the use of families in zoology is provided by Klein (1751), who used some of the same family names in synoptic taxonomic tables. For instance, Klein (1751: 3) clearly designated *Monochelon, Dichelon* and *Trichelon* as families that rank below the order *Ungulata*, to which they belong, while they clearly include taxa that have long been recognized as genera, such as *Equus, Cervus, Rhinoceros, Hippopotamus* and *Elephas*, and others that no longer designate genera but that were probably considered as such back then, such as *Porcus* and *Aries*. Other passages in the same work (pp. 40, 124–127) clearly name families in a similar context for other tetrapod taxa (notably the family *Nuda* that comprised various squamate and urodele taxa still typically ranked as genera, such as *Gekko, Lacerta* and *Salamandra*). However, these family names are not formed from the radical of a genus name followed by -idae, contrary to the practice that became established in zoological rank-based nomenclature in the 19th century (see previous section). Perhaps most surprising is that some late-18th-century authors placed the family level below the genus (Dubois 2006: 176). These authors may have been forgotten by historians of science because their work had presumably less impact than that of Latreille.

A bit later, Cuvier and Duméril (1800) used "families" in their taxonomies in tables at the end of the first volume of Cuvier's *Leçons d'Anatomie Comparée* (compiled by Duméril). Absolute ranks are not given in most of these tables, but if we attribute the ranks that were assigned by other authors to various taxa, it could be argued that there are orders and families. Paradoxically, in the only case (Table 1 for mammals) where "families" appear to be indicated (abbreviated "Fam." in the first instance and only "F." subsequently, but numbered sequentially), they designate taxa that would now be ranked at the order level or above, such as *Quadrumanes* (*Primates*), *Carnassiers* (a taxon that included bats and carnivorans, among others), *Rongeurs* (rodents) and so on. Only one (*Bimanes*, which includes only humans) might possibly match a currently recognized family (*Hominidae*). In the main text, the word "family" is clearly used in a more generic sense of "taxon." In other tables of the same work, there may plausibly be a family rank in the taxonomy. Bock (1994: 235) mentioned the table including insects (which is actually Table 8, not 5, contra Bock 1994) in this context; it includes two nested categories of taxa between the class *Insecta* and its genera. However, Table 2 on birds, and possibly others such as Table 3 on reptiles, also contain families if this criterion is adopted. In any case, these names are in vernacular French and are not based on the name of an included genus, so they are not considered available in zoological nomenclature. We could simply conclude that Cuvier and Duméril (1800) used several categories of taxa, which they did not conceive as Linnaean categories, but some of which match the family rank as it would later come to be adopted. For several decades more, French and some other European systematists continued to name families using descriptive Latinized words (rather than based on an included genus) or using vernacular, non-Latinized names (Bock 1994).

A more modern use of families is seen in Goldfuss and Schreber (1809). In this monograph, a taxonomic table (pp. XV—XIX) lists several families within orders, and each one includes genera, as in the current use. For instance, the order *Digitata* includes the family *Glires*, which includes taxa such as *Sciurus, Arctomys, Hyrax, Cavia* and *Lepus*. These are not indicated to be genera in the table itself, but this hierarchy is explained elsewhere in the text, and these names still refer to valid genera today, so the intent is clear. Other families listed in that table are *Ferae* and *Bruta*. However, the family names are not derived from an included genus, and they correspond to taxa of a higher rank in more recent taxonomies. For instance, *Glires* currently includes *Rodentia* and *Lagomorpha*, both of which are usually ranked as orders.

The British entomologist William Kirby (1813) is credited for the idea of forming family-names from the stem of an included genus, followed by the "-idae" suffix. This practice gathered sufficient support from other systematists to be incorporated into the *Strickland Code*.

To sum up, the use of families in biological nomenclature took over a century to take its current form, from Magnol (1689) to Kirby (1813). Probably, many other so-called "Linnaean categories" have similarly complex histories, but it is unnecessary to review them all here. Their development is likely to be more or less simultaneous with that of the families. Lhermin[i]er and Solignac (2000: 159) may not have been far from the truth when they declared that until the 18th century, the species was the sole category used for taxa. Entire books have been devoted to one of these categories, the species (for instance, Lherminier and Solignac 2005; Wilkins 2018) and a thorough treatment of other categories would no doubt require extensive developments.

## 2.3  The *Strickland Code*: Types, Priority, Other Innovations

### 2.3.1  Main Characteristics

Two of the central ideas developed in the previous short quote from the Strickland code (permanence of names and their general recognition) are central to biological nomenclature and effectively explain why codes were established. The very first code of biological nomenclature, the "*Strickland Code*" for short (Strickland et al. 1842), innovated by establishing a **principle of priority**, but only below the family-series level. Priority for family-level taxa started only in 1961 in zoology (Bock 1994: 8). Thus, the first published name (of a genus or species, according to the *Strickland Code*) that is proposed for a taxon is normally considered the valid one. For the *Strickland Code*, publication meant a printed work (not a label in a specimen drawer or an unpublished draft, even if its author is well respected), though in the 21st century, some electronic publications became acceptable too (typically, articles in journals with ISSN or book chapters in books with an ISBN). All subsequent names for the same taxon are either junior synonyms (different names for the same taxa) or junior homonyms (the same names applied to different taxa). This effectively prevents authors from proposing unneeded names.

The *Strickland Code* (like the more recent codes of biological nomenclature) only governed scientific names of taxa (binominal names for species and formal names of higher taxa). Application of the principle of priority requires fixing a starting point, before which names are not considered established. Given the importance of binominal nomenclature, Linnaeus' *Systema Naturae* (Linnaeus 1758) was selected as the oldest work to be considered—the starting point of priority for zoological nomenclature.

The second idea of the quote, general recognition of the names, requires compliance of the systematic community to the code. The *Strickland Code* never achieved it (Rookmaaker 2011), even though Darwin, who was a member of the commission that approved that code, tried to follow its rules and recommendations in his monograph on cirripeds (Wilkins 2018: 161). This lack of success was possibly because this code was mainly produced by Strickland and backed only by a national (British) learned society (Minelli 2022a). It was adopted in the USA and Italy, but not in most other countries (Winston 2018: 1124). All subsequent rank-based codes (which achieved global acceptance) were supported by international societies.

The *Strickland Code* (Strickland 1842) did not require illustrations of taxa that were erected, despite the obvious usefulness of illustrations in systematics; as the saying goes, "a picture is worth a thousand words." Despite this, for obvious technical reasons, antique texts (like the writings of Aristotle) were not often accompanied by illustrations. Some rare scientific illustrations were produced in the 16th and 17th century (Pyle 2000). In botany, the first woodcut illustrations appeared in *Ekphrasis*, published by Fabius Columna (1567–1650), also known as Fabio Colonna (Wilkins 2018: 63). Despite this, taxa were first described and erected (in Linnaeus' work, for instance) based only on text. Technical progress allowed illustrations to become widespread (though by no means universal) in the systematic literature in the 19th century, but the current codes still do not require illustrations (see later).

### 2.3.2 Types

The *Strickland Code* introduced a major innovation in biological nomenclature: the requirement to designate a **type**. There has been confusion in the literature about the various meanings of "type" and to which of these concepts the claim that rank-based nomenclature is typological referred. As summarized by Minelli (2019), there are four kinds of types: 1) a collection type, which is a specimen, normally stored in a museum accessible to scientists; 2) a morphological type, an abstract representation of a taxon (this is the meaning that comes closest to Platonic ideas); 3) a character-bearing (taxonomic) type or semaphoront, a term coined by Hennig (1966; see Rieppel 2003 for an analysis of this term for Hennig); and 4) a name-bearing type, or onymophoront (Dubois 2011: 74). The rank-based codes use the first and the last of these four kinds of types; the first one is used to typify species and the last one may be used to typify higher taxa (though the latter are ultimately and indirectly linked to type-specimens). Semaphoronts are useful in that they bear characters, many of which may not be noted in the first description of a given taxon but may subsequently improve our knowledge of the taxon. This is obviously very useful in taxonomy, for instance when scoring a data matrix for a phylogenetic analysis. The concept of name-bearing type centers on the fact that a name is attached to a given type (either a lower-ranking taxon or a specimen), and this allows making nomenclatural decisions. Obviously, both are related but conceptually different. The claim that rank-based nomenclature is typological has sometimes been interpreted as meaning that it implies the second meaning of the four listed previously (morphological type), but this is erroneous; clearly, rank-based nomenclature refers to types of the first and fourth kind, and taxonomy (phylogenetics) refers to types of the third kind. A specimen can be a type of the first, third and fourth kinds, whereas a type-taxon can only be a name-bearing type (onymophoront).

The *Strickland Code* recommended that a species of each genus be designated as type. In its proposition 3, we read:

> When a genus is subdivided into other genera, the original name should be retained for that portion of it which exhibits in the greatest degree its essential characters as at first defined. Authors **frequently** indicate this by selecting some one species as a fixed point of reference, which they term the "type of the genus." When they omit doing so, it may still in many cases be correctly inferred that the first species mentioned on their list, if found accurately to agree with their definition, was regarded by them as the type.

The "frequently" suggests that several authors had already adopted the habit of designating a type species, but this also implies that many more did not. That code did not formally require that type-genera be named for families, but it came close, and clarified how family names should be formed:

> It is recommended that the assemblages of genera termed families should be uniformly named by adding the termination idæ to the name of the earliest known, or most typically characterized genus in them; and that their subdivisions, termed subfamilies, should be similarly constructed, with the termination inæ.

**(Strickland 1842: 15)**

The recommended standardization of the ending of family names by the suffix "-idae" and subfamily names by "-inae" is a convention that had already been adopted by several authors by then.

For species names, the *Strickland Code* recommended the use of lower-case letters, contrary to the then-frequent habit of capitalizing the first letter when the specific epithet was derived from a proper name, but it did not mention type-specimens for species, even though Strickland argued for the use of type-specimens shortly after the *Strickland Code* was published (Strickland 1845). The formal requirement of such type-specimens was introduced much later, in the subsequent rank-based codes. Nevertheless, the designation of type-species of genera linked genus names to specimens indirectly, because labels of museum specimens typically indicated the species to which they had been attributed and the designation of a type-species provided an objective link to what each systematist who erected a genus considered to be the core component of that genus.

Linnaeus had not used types because, in any case, he thought that it was up to him to decide how taxa would be named and delimited. He did this in a very different way than current nomenclatural practice. For instance, Linnaeus continued using the name *Buchnera* to refer to an angiosperm genus even after removing all of the species that he had initially included in it (Witteveen and Müller-Wille 2020: 17)! This situation would have been prevented by the use of types, which minimally guarantee that at least the holotype of the type-species of the genus remains in the latter. But Linnaeus appears to have conceptualized taxa as "nominal spaces" in which he could put whatever he deemed appropriate (Witteveen and Müller-Wille 2020: 21). Thus, given how Linnaeus revised his nomenclature and taxonomy, and given his hegemony over systematics (in botany, in any case), there was no need for types. However, soon after Linnaeus' demise, disagreements arose among systematists about the grouping and delimitation of various taxa, and indeed, even during Linnaeus' career, several colleagues, such as Haller (Persson 2016) and Buffon (Stemerding 1993; Hoquet 2007), did not accept his taxa. The situation deteriorated further as taxonomy became decentralized, with many individuals working in several major centers (such as the Muséum National d'Histoire Naturelle in Paris and the Linnaean Society and British Museum in London) multiplied the alternative, mutually incompatible taxonomies and nomenclatures of overlapping sets of taxa, while the number of taxa was sharply increasing (Witteveen and Müller-Wille 2020: 30). As summarized by Cook (1916: 138): "The method of naming the concepts was used by Linnaeus and his followers for over a century, but had to be abandoned on account of the confusion caused by names slipping away from their original application." Time was ripe for a nomenclatural innovation that would reestablish some order.

## 2.4 The Current Codes

### 2.4.1 Basic Principles

The *Strickland Code* failed to gather sufficient support from the scientific community to have a strong stabilizing influence on biological nomenclature, but it was successful on another front: it paved the way for subsequent rank-based codes. There are five such codes: the *International Code of Zoological Nomenclature*, or *"Zoological Code"* for short (ICZN 1999); the *International Code of Nomenclature for algae, fungi and plants (Shenzhen Code)* or *"Botanical Code"* for short (Turland et al. 2018); the *International Code of Nomenclature of Prokaryotes: Prokaryotic Code (2008 revision)*, which I will simply call the *"Prokaryotic Code"* in the following (Parker et al. 2019); the *International Code of Virus Classification and Nomenclature (ICVCN)* produced by the ICTV (2018), which may be the rank-based code that has changed the most recently (Kuhn 2020); and the code for cultivated plants, formally known as the *International Code of Nomenclature for*

*Cultivated Plants*, which is the only one to deal specifically with organisms that have been modified by humans for their own use. The first edition of these codes was proposed well after the *Strickland Code*. Contrary to the *Strickland Code*, these five codes are well-accepted by the scientific community and journal editors typically try to enforce them.

The current rank-based codes extended the use of types well beyond its modest beginnings under the *Strickland Code*. Under these current codes, type specimens (or cultures, or even sequences, under some codes) are selected for species. However, for higher-ranking taxa, the type is a taxon of a lower rank. Thus, a genus has a type-species, and a family has a type-genus. In zoology, the use of types does not extend to higher levels than the family-group (which comprises subfamilies, families and superfamilies), but it does in botany (up to the class series). Since species have associated type-specimens, even higher-ranking taxa are associated with such specimens, though less directly than species.

Even for species, there are various kinds of types. In this paragraph, which only summarizes this topic without being exhaustive, the zoological terminology is used; note that other codes may use a different terminology and may not even make the same distinctions. In fact, even within a community working under a single code, additional kinds of types may be recognized (for instance, Dubois 2011; Dubois et al. 2014), but these will not be covered here. When a **single specimen** is designated as the type, it is called a **holotype**. Sometimes, holotypes have been lost; in such cases (or if no types had been designated, or if types exist but are deemed inadequate), a **neotype** is designated. Sometimes, a set of specimens is collectively designated a **type series** composed of **syntypes**. One of the syntypes can subsequently be designated the **lectotype**, which becomes the most important type specimen for that taxon; the other syntypes are then considered **paralectotypes**.

The importance of type-specimens in biological nomenclature cannot be over-emphasized. Descriptions are necessarily subjective and require interpretation, both when writing and reading them. Even drawings require some form of interpretation because they are more or less simplified compared to reality. Photographs are somewhat closer to reality, but they convey only a small part of the information included in a specimen: a given external view, at a given magnification level (microscopic details are typically not shown for large organisms), either without color (for black-and-white photos, which were the norm until the 20th century) or with colors usually limited to the visible spectrum (thus, patterns visible in other wavelengths, like ultraviolet, are lost). Without dissections (or now, virtual sections), no information about the inside of an organism is conveyed. Without histological sections or micro-tomography, microscopic details are lost. And molecular sequence data are impossible to capture by these methods. For prokaryotes, molecular data is so important that the nomenclature of these taxa had to be redone with a new start in nomenclature on January 1, 1980 (Oren 2011: 439), simply because, for many species, the original cultures were no longer available and because of the profusion of synonyms and homonyms (see the following section). Bacteriologists now use cultures of live bacteria and archaeans as types (Oren 2011: 455).

The vocabulary used differs slightly between codes, but many of the concepts are the same. In this book, unless specified otherwise, the zoological terms will be used, simply because I am more familiar with them than their botanical or prokaryote equivalents. The *Strickland Code* also used its own terminology, but this is of little relevance here.

The use of a rank (Linnaean category) and a type as basic definitions of taxon names allows determination of synonymy. In this system, two taxa are synonymous if we consider that they refer to the same organisms and have the same rank. Note that taxa that are redundant (have the same content) but have a different rank are not considered synonymous in rank-based nomenclature. Indeed, this is extremely common because many genera have only one known species, so one of the two taxa is redundant. This situation is to an extent an artefact of binominal nomenclature, given that a species must necessarily be placed in a genus. Thus, if a newly discovered species cannot be placed in an existing genus because it is too different from all existing genera or, as we would now formulate it, if it is only distantly related to clades currently ranked as genera, there is no alternative but to erect a new, redundant genus name for it.

Under rank-based nomenclature, synonymy encompasses two kinds of situations. One, called objective synonymy, is the simplest. It arises when two taxa of a same rank share the same type; this is called "**objective synonymy**" because, in this case, there is no doubt about the synonymy (it is not a matter of opinion). This situation can arise, for instance, if a systematist does not think that a name is appropriate and changes it (even though the rank-based codes have long forbidden this). Or a systematist may simply be unaware that a specimen has already been used to typify a taxon. A second, far more common situation is "**subjective synonymy.**" It arises when two taxa of a given rank that have different types are considered to be synonymous by a systematist. A difference of opinion between systematists is all that is required to generate this situation. This frequent situation is a logical outcome of the fact that that rank-based nomenclature does not delimit taxa; this is stated explicitly in Principle 2 of the introduction of the *Zoological Code*:

> Nomenclature does not determine the inclusiveness or exclusiveness of any taxon, nor the rank to be accorded to any assemblage of animals, but, rather, provides the name that is to be used for a taxon whatever taxonomic limits and rank are given to it.

> **(ICZN 1999)**

A similar statement is also found in the *Prokaryotic Code* ("General Consideration 4"; Parker et al. 2019). This feature has been claimed both to enhance nomenclatural flexibility by its proponents and to generate chaos by its detractors (Laurin 2008).

In addition to designating a type, all codes require that when erecting a taxon, its diagnostic characters be listed. This harks back to the *Strickland Code*, although that code did not distinguish between diagnosis (a short list of characters written in compact language) and a description (which is more extensive and often accompanied by illustrations), contrary to the current rank-based codes.

A recurring problem is that the nomenclatural acts are so scattered, often in journals that no longer exist (or not under their original names) or monographs that were not always widely circulated, that reviewing the history of a name or a taxon is extremely difficult. The priority of older names also means that the most important publications are also the oldest ones, which may be the most difficult to access, as anybody who has tried to consult an 18th century book in an institutional library can attest. For many taxa and for periods before registration in databases became mandatory (in some rank-based codes), the situation has become so bad that the nomenclature in use does not respect the codes. Indeed, Bock (1994: 6) stated:

> It is shown that these regulations [of the *ICZN*] are not workable without a detailed knowledge of nomenclatural history, and because such histories are not available for most groups of animals, current regulations in the Code pertaining to family-group names are not workable.

This problem is so acute that it discouraged initiatives to revise the nomenclature of some taxa. For instance, Bock (1994: 8–9) stated that an eminent zoologist (probably G. G. Simpson, a paleomammalogist):

> Decided reluctantly that he must forego an intended revision of his classification of a class of vertebrates [*Mammalia*] because although he was unwilling to transgress the *Code of Zoological Nomenclature*, he simply did not wish to devote the estimated three years of research needed to determine the history of family-group names (about 700) in this class (see also, Mayr 1969). I suspect that such a project would have taken more than three years!

Similarly, Bock (1994: 11) reported that "Most ornithologists have dealt with this problem by continuing to use the well-established family-group names, feeling that following the letter of the rules (requirements of Art. 79) was not possible without undertaking extensive historical research." Perhaps even more distressing, Bock (1994: 9) reported that:

> In their efforts to apply the new rules of zoological nomenclature (ICZN 1961, 1964, 1985a) to avian family-group names, the *ICZN* has made errors in almost every opinion issued since the first one in which *Meropidae* Rafinesque, 1815 was conserved (1954; Direction 6).

Clearly, something had to be done about this problem.

The first solution to alleviate this problem was to publish official lists of approved names, but this proved insufficient. Registration is another solution. Registration of electronic (but not paper-based) publications in an online database (zoology) or centralized listing of new taxa in a single journal (prokaryotes) is now mandatory under some codes, but not all (not for the *Botanical Code*).

To sum up, substantial differences exist between codes. The *Botanical Code* specifically requires a pdf format for electronic publications, but it does not require registration of the works. This is ironic because mycologists and phanerogamists were the first systematists to propose registration of taxon names (Ainsworth and Ciferri 1955), and many mycologists were disappointed that after approving the principle of registration in the Tokyo congress (1993), botanists rejected it at the St Louis meeting, which convened in 1999 (Hawksworth 2005). The *Zoological Code* requires registration, but only of electronic publications, not necessarily of the nomenclatural acts that they include. This is presumably because, as indicated in the "Historical background" section of the introduction of the *Zoological Code*, mandatory registration of new names was "not sufficiently acceptable to zoologists to be adopted." The *Prokaryotic Code* has a central listing of all names in a single journal and thus best centralizes nomenclatural information. Thus, each rank-based code has its own approach, which is not ideal for any user who has to deal with more than one code. Nevertheless, the preface of the current (fourth) edition of the *Zoological Code* (ICZN 1999) optimistically states:

> So far as registration of new names is concerned, this has already been introduced in bacteriology, and botanists and zoologists may come to accept it despite understandable doubts and objections. In these areas at least, the future of biological nomenclature will probably witness convergence between the various traditions which diverged during the 19th century.

Let's hope that this prediction is correct.

### 2.4.2  Zoological Code

Much of the treatment of rank-based nomenclature in this book reflects the *Zoological Code* for two reasons: first, because I am most familiar with it, given that I am a zoologist, and second, because this is the code that regulates the names of our closest known relatives, so it is presumably of greatest interest to most readers. Thus, the section on this code will be shorter than those of some of the other rank-based codes, which are discussed less frequently in subsequent chapters of this book.

Early steps in the preparation of the *Zoological Code* started in the late 19th century with preparations of new codes inspired by the *Strickland Code*. These include a code that was published by Henri Douvillé in 1881 and that was adopted internationally by geologists (ICZN 1999), as well as the American Ornithologists' Union (1886) code. Shortly thereafter, the first and second International Zoological Congresses (Paris 1889, Moscow 1892) led to publications of reports on adopted nomenclatural rules (Fischer 1894). That period also witnessed the origin of the International Commission on Zoological Nomenclature (ICZN), which updates the code. The 3rd International Congresses of Zoology (Leiden, 1895) appointed a commission of five zoologists (R. Blanchard, J.V. Carus, F.A. Jentink, P.L. Sclater and C.W. Stiles) to work on a code and to report to the 4th Congress (Cambridge, England, 1898). In due time, this became the ICZN.

A first document akin to a code, entitled *Règles internationales de la nomenclature zoologique adoptées par les congrès internationaux de zoologie*, which will be abbreviated as "*Règles*" from here on, was published a decade later (ICZN 1905). Amendments were adopted in two zoological congresses (Paris in 1948; Copenhagen in 1953), but the first genuine edition of the code, which was published in 1961, was the first complete revision of the *Règles*. The second edition was published in 1964, and the third, in 1985. The current (fourth) version (ICZN 1999) was emended slightly (articles 8, 9, 10, 21 and 78) in 2012. Some of the most recent substantial changes pertain to electronic publications and registration.

Taxa are typified from subspecies to the family-series level, as indicated by the introduction of the current *Zoological Code*. The use of type-species for genera and typification of higher-level taxa developed gradually. Thus, the *Règles* (1905) did not require that a type-species be designated for the establishment of new genus names, but this changed with the first edition of the code (1961), which applied retroactively to names established after 1930. The absence of typification and mandatory endings of names above the family-series level in the *Zoological Code* makes shifts of rank of taxa easier than under other codes (such as the *Botanical Code*), where such changes require formal nomenclatural acts (Lücking 2019: 200).

The *Zoological Code* specifies mandatory endings only for the family-series names. The suffixes are "-oidea" for superfamilies, "-idae" for families, "-inae" for subfamilies, and "-ini" for tribes and "-ina" for subtribes (Article 29.2).

Species names can be challenging to form, especially for systematists not versed in Latin grammar. Thus, Article 11.9.1. states that a species-group name, "if a Latin or latinized word must be, or be treated as" (either) "an adjective or participle in the nominative singular," "a noun in the nominative singular standing in apposition to the generic name," "a noun in the genitive case" or "an adjective used as a substantive in the genitive case and derived from the specific name of an organism with which the animal in question is associated." Furthermore, Article 11.9.2. adds that:

> An adjectival species-group name proposed in Latin text but written otherwise than in the nominative singular because of the requirements of Latin grammar is available provided that it meets the other requirements of availability, but it is to be corrected to the nominative singular if necessary.

Sometimes, strict adherence to priority would result in replacement of a well-known taxon name by another name that has long been forgotten by systematists. To deal with such cases and prevent these unfortunate consequences, appeals can be made to the ICZN to conserve a junior synonym or homonym. These "cases" are published in the *Bulletin of Zoological Nomenclature* (*BZN*) and remain open for at least eight months (longer if comments are received). Four times a year, the commission (ICZN) discusses these cases and votes on them. A two-thirds majority is required for passage. The results are published as "opinions" in the *BZN*, and corresponding changes are made in *the Official Lists and Indexes of Names and Works in Zoology* (Winston 2018: 1124).

In addition to its 90 articles, which must be followed to comply with the code, Appendix B includes a series of recommendations. Thus, Appendix B4 states: "An author establishing a new nominal taxon should clearly state the higher (more inclusive) taxa (such as family, order, class) to which the taxon is assigned." This is meant to provide information about taxonomic affinities, but in some cases, it leads to redundancy (see Chapter 3).

Appendix B6 of the *Zoological Code* states: "The scientific names of genus- or species-group taxa should be printed in a type-face (font) different from that used in the text; such names are usually printed in italics, which should not be used for names of higher taxa." As will be seen in the following, this code is isolated in treating species- and genus-series group names differently from those of other taxa in this respect (names of families and orders are not italicized). All other codes (rank-based and *PhyloCode*), even the draft *BioCode* (Greuter et al. 2011: Recommendation 4A), recommend using a different type-face for all taxon names and, usually, italics are used for that. Below the species level, the only category of names regulated by the *Zoological Code* is the subspecies.

The *Zoological Code* is available in English and French, and both versions and "are official texts and are equivalent in force, meaning and authority" (Article 86.2). Furthermore, "The Commission may authorize the publication of the Code in any language and under such conditions as it may decide" (Article 87). Availability of codes in more than one language is a goal common to most, if not all organisms that produce rank-based codes, but not all have succeeded. For instance, the *Prokaryotic Code* is currently only available in English. Parker et al. (2019: S101–S102) indicated, when discussing the Sixth International Congress of Microbiology which convened in Rome in 1953, that "Unfortunately the preparation of the text and annotations has been so time-consuming that it has not been possible to include texts of the Code in the several important languages of science," and that it was hoped that this could be done later.

The *Zoological Code* now allows erection of new taxa in electronic publications if these works are registered in *ZooBank* before they are published, if the work itself states the date of publication and evidence that it has been registered, and if the *ZooBank* registration states the name of an electronic archive where the work is intended to be preserved and the ISSN or ISBN associated with the work. To be valid, the electronic work itself must be "widely accessible electronic copies with fixed content and layout," such as a pdf (Zhang 2012). *ZooBank* is not retroactive; it thus has an exhaustive record of electronic publications erecting new taxa or including nomenclatural acts from 2012 on. Also, registration of electronic works is mandatory, but not that of printed publications. Registration of new taxon names and nomenclatural acts is not required, whether they appear in electronic or printed publications, but it is done on a voluntary basis (ICZN 2012). Thus, *Zoobank* is by no means exhaustive. Old publications are not systematically covered, even though there is an ongoing effort to cover older literature on a voluntary basis (any willing systematist can register an account in *Zoobank* and contribute to its database). Thus, this registration is only a first step toward the solution, which is to have an exhaustive database. *ZooBank* helps mostly with searches about recent nomenclatural acts, but with time, its coverage of older acts

and names should improve, without necessarily ever achieving an exhaustive coverage. In the period from March 1, 2016 to September 30, 2020, an average of 774 papers/month and 1,764 nomenclatural acts/month were registered in *Zoobank* (http://zoobank.org/Statistics, and data transmitted on October 19, 2020 by Richard Pyle, who is Architect and Manager of *Zoobank* as well as Commissioner and Councilor for the ICZN). According to Winston (2018: 1127), by May 23, 2018, 205,610 names by 55,878 authors in 88,510 publications had been registered.

### 2.4.3 Botanical Code

Many eukaryotic taxa are covered by the *International Code of Nomenclature for algae, fungi, and plants (Shenzhen Code)*, which we can call the "*Botanical Code*" for short. As its long title indicates, it covers not only photosynthetic eukaryotes (which match the traditional definition of the polyphyletic assemblage often called "plants"), but also fungi, which are more closely related to metazoans than to embryophytes (Burki et al. 2020). It also covers cyanobacteria, even though the names of these prokaryotic organisms are also regulated, to an extent, by the *Prokaryotic Code* (see Section 2.4.5). It thus covers a fairly heterogeneous, polyphyletic group of organisms.

Botany is probably the field where early nomenclature was best developed. In the *Strickland Code* (developed mostly by and for zoologists), we can read:

> The admirable rules laid down by Linnæus, Smith, Decandolle, and other botanists (to which, no less than to the works of Fabricius, Illiger, Vigors, Swainson, and other zoologists, we have been much indebted in preparing the present document), have always exercised a beneficial influence over their disciples. Hence the language of botany has attained a more perfect and stable condition than that of zoology.

> **(Strickland et al. 1842: 17)**

Nicolson (1991: 34) interpreted the *Strickland Code* as the official start of a nomenclatural split between zoology and botany, although a literature survey suggests that practices of the two communities had long diverged; the *Strickland Code* only made this more obvious and official. Indeed, Minelli (2022b: 204–205) argued that this divergence started soon after Linnaeus' death.

A first *Botanical Code* was developed by the Swiss botanist Alphonse de Candolle (1867), but according to Winston (2018: 1124), it was not accepted in all countries despite its having been adopted by an international congress that convened in Paris in 1867. Indeed, it was only adopted by the assembly (consisting of about 100 scientists from many countries) as the best guide for botanical nomenclature (de Candolle 1867: 4). It was thus not enforced to the same extent as more recent versions of the *Botanical Code* or of other rank-based codes (Nicolson 1991: 34).

Several revisions of nomenclatural rules and some codes were subsequently proposed, either by individuals, by a small group of systematists or by a large assembly in international meetings, but none of them was fully enforced worldwide. Among these initiatives, we can list (from Nicolson 1991) several of the early ones to give an idea of the frequency at which the rules changed: the "*Kew Rule*," which was produced by Henry Trimen in 1877; the "*Rochester Code*," which was produced at a meeting of a botanical club with the American Association for the Advancement of Science (AAAS) in Rochester, N.Y. in 1892; the "*Madison Rules*," which emphasized priority over use (see the following section), and which emanated from a meeting of American botanists at Madison in 1893; the "*Berlin Rules*," proposed in 1897 by Engler and his staff of the Berlin Garden and Museum; and the "*Brittonian (American) Code*," (Arthur et al. 1907), produced by a nomenclatural commission appointed at a meeting of the Botanical Club and the AAAS that convened in Rochester in 1903.

The first true code was adopted at the first meeting devoted exclusively to botanical nomenclature, which convened in Vienna in 1905. It deserves to be called a "true code" because it was the first one that was imposed internationally, though not universally. Indeed, many American botanists, unhappy with the new mandatory use of Latin to publish new names, continued using the *Brittonian Code* (Arthur et al. 1907).

A second official *Botanical Code* was adopted during the Third International Botanical meeting, which convened in Brussels in 1910. A motion to drop the requirement that the diagnosis be in Latin was defeated, but discussions at that meeting showed that there was still opposition to this requirement, especially by American botanists (Farlow and Atkinson 1910). However, important decisions were made, such as the recognition of different starting dates coinciding with important monographs for the priority of various taxa. This was to take into consideration that the earliest recognized names had to be erected in "the earliest comprehensive work treating a group, large or small, in a somewhat modern sense" (Farlow and Atkinson 1910: 221). For instance, for fungi, the chosen starting point was a series of monographs by Fries published from 1821 to 1832, but Linnaeus' *Species plantarum* (1753) was retained as the starting point of nomenclature for pteridophytes and spermatophytes. The current (Shenzhen) version of the *Botanical Code* retains different starting dates for various taxa (Article 13), although it reverted to the *Species Plantarum* (1753) for extant fungi (Article F1.1, which is superseded by a new version of chapter F presented in May et al. 2019). For long-extinct ("fossil") organisms (diatoms excepted), priority starts December 31, 1820 (Sternberg, *Flora der Vorwelt, Versuch* 1: 1–24, t. 1–13) in the *Shenzhen Code* (Article 13.1f).

Types became important in botanical nomenclature only in 1919. Hitchcock (1919) reported that a committee set up to develop regulations to fix types of genera (through type species) found a large support for this idea among 50 members of the *Botanical Society of America*. These botanists also supported the idea that the type species should be selected among the species originally included in the genus. Hitchcock (1919: 334) mentioned that, in the *Vienna Code* (from 1905), "The question of types is not touched upon." This is surprising given that the *Strickland Code* already recommended naming type species for genera, and that the formulation in that code suggests that many authors had already developed the habit of naming such species, although this habit might have been

more widespread among zoologists than among botanists in the first half of the 19th century.

As the reader probably guessed, by 1919, it was difficult for many plant systematists to keep up with nomenclatural developments. Worse, the codes quickly became too complex for most botanists to know even one of their versions well. This is shown by the fact that, in 1924, many plant systematists already felt that "The Rules are too long and complicated," and worse, "Even experts are not agreed as to the interpretation of some of the Rules." (Britten et al. 1924: 79).

These problems did not deter the development of further editions of the official code, as well as more unofficial versions that were used in some parts of the world. To name only the official codes, the third version was adopted at a congress that convened in Cambridge in 1930, and the fourth code was adopted at the Stockholm congress (1950), after a long delay explained by the major disruption caused by World War II. Further versions were adopted at the Paris congress (fifth code, 1954) and many subsequent ones: Montreal (1959), Edinburgh (1964), Seattle (1969), Leningrad (1975), Sydney (1981), Berlin (1987), Tokyo (1993), Saint Louis (1999), Vienna (2007) and Melbourne (2011). The latest one is the *Shenzhen Code*, which was adopted in 2017 (Turland et al. 2018).

On July 18, 1950, an informal meeting of about 130 taxonomists adopted a proposal to form an association with an office in Utrecht, and this association launched its official journal *Taxon* in 1951 (Nicolson 1991: 41). *Taxon* became the privileged journal in which to submit proposals to amend the code. The same association, which is now known as the International Association for Plant Taxonomy, also launched *Regnum Vegetabile*, a series of books of interest to systematic botanists, which was launched in 1952 with the publication of the *Stockholm Code* (the first publication of this series).

Recent editions of the code have been administered by *the Permanent Nomenclature Commission of the International Botanical Congress*. It was long known as the *Botanical Code* and is still sometimes informally referred to as such, but it was officially renamed in 2012 as *International Code of Nomenclature for algae, fungi, and plants*, which better reflects its coverage of several distantly related taxa.

Under this code, species are typified by a specimen, which is usually a dried organism stored in an herbarium, though metabolically inactive cultures obtained through processes such as lyophilization or deep-freezing are acceptable for algae and fungi (Oren 2019: 10). In contrast, higher-ranking taxa are typified by the same type as a specified lower-ranking taxon. Thus, each genus (or subgenus) is typified by the type of one of its species and, in practice, it is sufficient to cite that species (Article 10.1). Similarly, families and subfamilies are typified through a genus (and, ultimately, species and specimen; Article 10.9). This practice is thus similar to typification in zoology, with some differences for taxa above the family-series names. Article 7.1 states that "The application of names of taxa at the higher ranks [than family] is also determined by means of types **when the names are formed from a generic name.**" Article 10.10 explains that "The principle of **typification does not apply** to names of taxa above the rank of family, **except** for names that are **automatically typified** by being formed from generic names (see Art. 16.1(a)), the type of which is the same as that of the generic name." Article 16.1 explains that these automatically typified names are "formed from a generic name in the same way as family names" by adding the appropriate suffix. Thus, contrary to the *Zoological Code*, some orders and classes may have types, but this depends on how their names were formed. Another difference from the *Zoological Code* is that the *Botanical Code* (Article H3) allows naming hybrid species as such, with a symbol (the multiplication sign "x") or a prefix ("notho-" or simply "n-") indicating that the taxon is a hybrid (nothotaxon). Also, tautonyms (such as *Bufo bufo*), which are allowed in zoology, are forbidden in botany (Article 23.4). Finally, the *Botanical Code* recognizes several categories below the species level, contrary to the *Zoological Code;* these include subspecies, varieties, subvarieties, forms and subforms (Article 4.2).

The *Botanical Code* is more complex than the *Zoological Code* in its use of mandatory endings. Some botanical names have mandatory endings, but these are not the same as in zoology. Thus, family names end in "-aceae" (Article 18.1). Also, contrary to zoological names, some names of higher-ranking taxa (above the family group or series) also have mandatory endings. The most relevant article (16.3) about this states:

> **Automatically typified** names end as follows: the name of a division or **phylum** ends in **-phyta**, unless it is referable to the **fungi** in which case it ends in **-mycota**; the name of a subdivision or **subphylum** ends in **-phytina**, unless it is referable to the **fungi** in which case it ends in **-mycotina**; the name of a **class** in the **algae** ends in **-phyceae**, and of a **subclass** in **-phycidae**; the name of a **class** in the **fungi** ends in **-mycetes**, and of a **subclass** in **-mycetidae**; the name of a **class** in the **plants** ends in **-opsida**, and of a **subclass** in **-idae** (but not -viridae).

Thus, mandatory endings for automatically typified names in botany extends all the way up to phyla, but only a subset of the names of high-ranking botanical names are thus typified. The proper ending (suffix) depends on whether the taxon is considered to be a fungus, an alga or a plant (an embryophyte). This complexity partly reflects ancient traditions and the fact that the *Botanical Code* handles a greater variety of organisms than the *Zoological Code*.

Names of other nominal series can be tricky to form correctly. Thus, Article 23.1 indicates that the specific epithet must be "in the form of an adjective, a noun in the genitive, or a word in apposition," and Article 23.5 adds that:

> The specific epithet, when adjectival in form and not used as a noun, agrees with the gender of the generic name; when the epithet is a noun in apposition or a genitive noun, it retains its own gender and termination irrespective of the gender of the generic name. Epithets not conforming to this rule are to be corrected (see Art. 32.2) to the proper form of the termination (Latin or transcribed Greek) of the original author(s).

Besides what could be interpreted as implying that authors of species names have "proper form" and a termination (the

authors of the code obviously meant "the epithet proposed by the author," and this linguistic shortcut is not uncommon in botany but may puzzle other systematists), Article 23 shows that forming such names requires at least some basic notions of Latin grammar.

A recent development is that a special committee was set up at the San Juan meeting (which convened in Puerto Rico in 2018) to consider the possibility of using "DNA sequences as Types for Fungi" (May et al. 2019: 5). This would represent a significant departure from botanical tradition, but is in line with the increasing importance of molecular systematics.

Botanists, like zoologists, struggled to strike a good compromise between a strict observance of priority and extent of use. The former ensures fairness and objectivity in determining which name is correct, while use should be taken into consideration for the sake of stability. It is occasionally discovered that a little-used and sometimes forgotten name has priority over a frequently used, well-known name. In such a case, enforcing priority may cause unnecessary nomenclatural instability. One of the first botanical meetings where this issue was seriously considered convened in Madison, Wisconsin, in 1893. Several leading botanists participated, such as J. B. Saint-Lager, O. Kuntze and N. L. Britton. At that meeting, a strict rule of priority that admitted no exception was passed. However, this did not satisfy other botanists, and this rule was not enforced for long. Even though systematists have struggled with this issue for well over a century, the problem is not really solved. Nicolson (1991: 37) summarized the issue:

> Note that what Saint-Lager, Kuntze, Britton, and others (as radicals) emphasized [at the 1893 Madison meeting] are the rules themselves, never mind upsetting usage, which was only a short-term consequence (cost). The Candollean (conservative) concept was to try to maintain the status quo, never mind some complexity of rules, which are really of no importance. Both concepts would agree that stability is the goal of nomenclature, but one wonders if the interaction of these two schools doesn't result in a Code with **the worst of both worlds**: a complex and constantly changing Code (trying to maintain past usage) and constant conservations (to set aside rules that, despite complexity, are not maintaining the past).

This statement still largely applies today, in zoology (though that code has changed more slowly) as well as in botany.

The use of italics is more pervasive in botany than in zoology. Even though the *Botanical Code* does not require italicizing taxon names, its preface (in the current, Shenzhen edition) states:

> As in all recent editions, scientific names under the jurisdiction of the Code, **irrespective of rank**, are consistently printed in italic type. The Code sets no binding standard in this respect, as typography is a matter of editorial style and tradition, not of nomenclature. Nevertheless, editors and authors, in the interest of international uniformity, may wish to consider following the practice exemplified by the

Code, which has been well received in general and is followed in a number of botanical and mycological journals.

## 2.4.4 Code for Cultivated Plants

The first code for cultivated plants (officially called the *International Code of Nomenclature for Cultivated Plants*) arose from a need to name hybrids and taxa of subspecific and lower categories, such as varieties that are not adequately covered by the *Botanical Code* (Stearn 1953; Winston 2018: 1125). Some of these forms, such as hybrids, are not always easy to accommodate in the hierarchical arrangement of taxa assumed by most rank-based codes. In contrast, in the *Cultivated Plant Code* (as I will name it for short, from here on), as it is familiarly known, a given hybrid variety can be part of two non-nested higher taxa (called *groups*), which can thus overlap. Another way to present it is that this code provides simple, non-Latin names for plants that have been deliberately cultivated and selected by humans (Spencer and Cross 2007). These include plant varieties commercially bred for horticulture, forestry and agriculture. The scope of application of the *Cultivated Plant Code* is stated in Principle 2:

> The *International Code of Nomenclature for algae, fungi, and plants (ICN)* governs the names in Latin form for algae, fungi, and plants, **except for graft-chimaeric genera**, which are entirely governed by this *Code*.
>
> Taxa of plants whose origin or selection is primarily due to intentional human activity may be given names formed according to the provisions of this *Code*. With the exception of any Latin component within their names, the form of which is governed by the *ICN*, the nomenclature of names in the categories of cultivar, Group, and grex is governed by this *Code* alone.

This code came decades after Bailey (1918) lamented that the *Botanical Code* did not really deal with cultivated plants like corn (*Zea mays*), tuberous begonia, cultivated fuchsias and countless varieties that could not be formally named under the *Botanical Code* but that were economically important. He wrote, "The plant-breeder will bring his new groups; will taxonomy expand itself to receive them, or must they always be outcasts?" Obviously, this code proved useful because several other versions were produced; the most recent one being the ninth version (Brickell et al. 2016).

The scope of the code of nomenclature for cultivated plants is somewhat ambiguous. For instance, some species are inadvertently introduced by people to new areas, where they locally adapt, while others adapt to pesticides. Should they be covered by that code? Spencer and Cross (2007) suggested not, because only plants that are deliberately cultivated should be included.

The current *Cultivated Plant Code* recognizes three main ranks, but these are not arranged in a simple hierarchy, contrary to other RN codes. This is explained in a footnote to Article 2.1 of Chapter II: "As defined in this Code a category is a division in a system of classification. In this Code there are three categories: **cultivar, Group** and **grex**, which should **not necessarily be hierarchically limited**."

The **cultivar** is the "basic category of cultivated plants whose nomenclature is governed by this Code" (Article 2.1). The cultivar is defined in Article 2.3: "A cultivar, as a taxon, is an assemblage of plants that (a) has been **selected** for a particular character or combination of characters, and (b) remains **distinct, uniform, and stable** in these characters when propagated by appropriate means."

The Group "**may comprise cultivars**, individual plants or combinations thereof on the basis of defined character-based similarity" (Article 3.1.) Its approximate equivalency under the *Botanical Code* is evoked in Article 3.1: "A taxon previously recognized as a species or lower rank under the *ICN* [*Botanical Code*] may be designated as a Group, if such a designation is considered more appropriate and has utility." However, unlike other RN codes, taxa under this code are not strictly hierarchical. As stated in Article 3.4, "A member of one Group may also be a member of one or more other Groups if this has a practical purpose."

The grex differs rather sharply from the other categories of this code because, as indicated in Article 4.1, "The formal category for assembling plants **based solely on specified parentage** is the grex. It may **only** be used in **orchid** nomenclature." Thus, whereas other parts of this code consider taxon history irrelevant, the grex is based solely on parentage, and its use is restricted to orchids, a taxonomically lower, though commercially important, part of angiosperm biodiversity. However, other categories of taxa exist among orchids because, as indicated by Article 4.2, "One or more Groups may be formed within a grex."

Contrary to what Article 2.1 states, this code does not only recognize three categories of taxa. Thus, Article 5 recognizes the **Chimaera**, which includes the graft-chimaera (Article 5.1) and the mutation-chimaera (Article 5.2). Article 5.1 explains that:

> A graft-chimaera is a plant that results from grafting the vegetative tissues of two or more plants belonging to different taxa, and is thus not a sexual hybrid. Rules for the formation of names of graft-chimaeras **at the rank of genus** are laid out in Art. 24 of this Code. Graft-chimaeras below the rank of genus may be recognized as **cultivars**.

Thus, there are rules for genus-rank taxa, specifically in Article 24, which explains, among other things, how two genus names from the *Botanical Code* can be combined to form the name of an inter-generic graft-chimaera.

Last, but not least, this code also contains **classes** (Article 6). Their purpose is stated in Article 6.1: "A denomination class is the unit within which the use of a cultivar, Group, or grex epithet may not be duplicated except when re-use of an epithet is permitted in accordance with Art. 30." Article 6.2 further stipulates that "A denomination class under the provisions of this Code is a single genus or hybrid genus unless a special denomination class has been determined by the ISHS Special Commission for Cultivar Registration." Thus, classes under this code should not be compared with the classes in the *Botanical Code*, which are much higher-ranking taxa.

The definitions of some names regulated by the *Cultivated Plant Code* differ rather sharply from those of names regulated by other RN codes. Thus, Division V, Article 1, Note 2 of the *Cultivated Plant Code* states that "Names of graft-chimaeric genera do not have type species and therefore **no type specimen** or nomenclatural standard can be designated for them, as they are **defined solely on a statement of parentage**." However, other categories of taxa are typified, although not exactly as under RN codes. This is stated in Principle 9 of the code, which stipulates that:

> The selection, preservation, and publication of designations of **nomenclatural standards** is important in stabilizing the application of **cultivar** and **Group** names. Particular names are attached to nomenclatural standards to make clear the precise application of the names and to help avoid duplication of such names. Although **not a requirement** for the establishment of a name, the designation of such standards is **strongly encouraged**.

This code, like most other RN codes, emphasizes ranks (also called "categories" in this code) at the expense of other aspects of biological organisms that most systematists consider important, such as affinities or characters. This is stated in the first article of the Preamble (Brickell et al. 2016: 1): "The purpose of giving a name to a taxon is to supply a means of referring to it and to indicate to which **category** it is assigned, rather than to indicate its **characters** or **history**." The lack of consideration for history of taxa in this code is asserted again in Article 2.20:

> In considering whether two or more plants belong to the same or different cultivars, their **origins are irrelevant**. Plants that cannot be distinguished from others by any of the means currently adopted for cultivar determination in the group concerned are treated as one cultivar.

However, this is no doubt motivated by the utilitarian nature of this code, which regulates names of commercially important plants, and perhaps also because at the lowest taxonomic levels (often infraspecific), the history of lineages matters less than at higher levels, given that gene flux could in theory occur between adjacent cultivars (as long as this does not alter their aspect and other important properties).

### 2.4.5 Prokaryotic Code

The nomenclature of prokaryotes is hampered by the fact that "in spite of the achievements of over 200 years of research, there still is no clear concept of what a prokaryote species actually is" (Oren 2011: 438). The most widely accepted concept (the phylo-phenetic species concept) of prokaryote species is: "a **monophyletic** and genomically coherent cluster of individual organisms (strains) that show a high degree of overall similarity in many independent characteristics and is diagnosable by one or more discriminative phenotypic properties" (Oren 2011: 438). Note that monophyly is required, not explicitly by the code itself, but by the species concept used

and emphasis on phylogeny, whereas this is not the consensus in zoology and botany. The similarity criterion is common in species concepts, but the absence of an objective threshold creates delimitation problems. Thus, a pragmatic delimitation criterion was adopted, which stipulates that a species is:

> a group of strains, including the type strain, that share at least 70% total genome DNA:DNA hybridization and have less than 5°C $\Delta$Tm (=the difference in the melting temperature between the homologous and the heterologous hybrids formed under standard conditions).

**(Oren 2011: 438)**

The selected thresholds are arbitrary, but according to Oren (2011), they have proven satisfactory in most cases in prokaryotes, notably because the DNA:DNA hybridization method proved more sensitive than proportion of homology in 16SrRNA for closely related individuals and taxa and it provides a global, genome-wide measure of similarity (Stackebrandt and Goebel 1994). However, these thresholds, which correspond to about 93–94% DNA-level sequence identity (Konstantinidis and Tiedje 2005), would not work to delimit eukaryotic species; this would yield much more inclusive taxa. For instance, according to this criterion, all primates (a taxon usually ranked as an order) would belong to the same species! Furthermore, among eukaryotes, the degree of genetic divergence between taxa of a given rank is extremely variable. Thus, Avise and Liu (2011: 709) reported that "Among vertebrate animals, for example, congeneric species and confamilial genera of endotherms (birds and mammals) often show much smaller genetic distances than counterpart taxa of reptiles and amphibians." Even among prokaryotes, this criterion cannot always be followed. Thus, several strains with DNA:DNA reassociation values >70% are classified into different species, or even different genera, based on host range or pathogenicity (Konstantinidis and Tiedje 2005: 2567). In fact, some authors consider that the current criterion makes species encompass too much biodiversity and suggest raising the requirement for conspecifity of strains to >99% DNA-level sequence identity, or requiring a bit less (than 99%) sequence identity but also requiring a similar ecological niche. This would make prokaryotic species more comparable (in terms of encompassed genetic variability) to eukaryotic species, but this proposal could inflate the number of species ten-fold (Konstantinidis and Tiedje 2005: 2572). Richter and Rosselló-Móra (2009) made a strong case to replace the 70% DNA:DNA reassociation value by 95–96% DNA-level sequence identity, and showed that applying this criterion would not require complete sequences of all strains. 50% genome coverage would be sufficient, and this can even be done with as little as 20%, but less reliably.

Prokaryotes illustrate well the need for efficient nomenclatural codes. On January 1, 1980, this nomenclature made a new start (priority was reset then). It had become extremely confusing, with an abundance of synonyms, and many of the older taxa could not be properly identified because no cultures were available. Thus, a new nomenclature adopted in 1980 recognized fewer taxa. This can be illustrated by the fact that at least 28,900 bacterial names were in use in 1966, but the 1980 list included only seven classes, one subclass, 21 orders, three suborders, 66 families, 24 tribes, 290 genera, 1792 species and 131 subspecies, for a total of 2,335 names. By May 10, 2011, the number of officially recognized prokaryotic taxon names had grown to include 77 classes, 128 orders, 219 families, 2,010 genera and 10,707 species, for a total of 13,212 names, but that is still less than half as many as in 1966. Yet, this increase from the 1980 numbers reflects the fact that taxa were erected at an increasing rate; about 23 genera and 130 species per year in the 1980s, but 120 genera and 620 species per year between 2006 and 2010 (Oren 2011: 439).

The first code of nomenclature for prokaryotes was published in 1948 under the name *International Bacteriological Code of Nomenclature* (Buchanan et al. 1948). A decade later (1958), the first edition of the current code was published under the name *International Code of Nomenclature of Bacteria and Viruses*, and its scope obviously encompassed viruses, but only temporarily because a separate code for viruses was elaborated in the 1960s and 1970s (see Section 2.4.6). Thus, by the second edition of that code (1975) viruses had been taken out, as shown by the new name, *International Code of Nomenclature of Bacteria*, which was retained for the third (1990) edition. In 1999, it was decided to rename it *International Code of Nomenclature of Prokaryotes*, or *Prokaryotic Code*, to take into consideration that it also regulates *Archaea*. Its last revision is from 2008 (Parker et al. 2019), and all quotes from the code in this section are taken from this edition, unless otherwise stated.

All prokaryotic taxon names must be listed in the *International Journal of Systematic and Evolutionary Microbiology* (before 2000, the *International Journal of Systematic Bacteriology*), which is under the aegis of the International Committee on Systematics of Prokaryotes (ICSP), and an exhaustive database can be consulted at www.bacterio.net. As of September 2, 2020, it lists 20,123 species and 3,477 genera (16,763 and 3,239, respectively, if synonyms are removed), which shows that the rate of discovery of prokaryote taxa did not decrease in the last decade. This is also confirmed by the annual number of prokaryotic taxa erected available at https://lpsn.dsmz.de/archive/-number.html, even though neither list includes cyanobacterial taxa regulated by the *Botanical Code*. This suggests that many prokaryotic taxa remain to be discovered. However, none of the websites listing established prokaryotic taxa are officially endorsed by the ICSP (Oren 2019: 9); thus, the centralization of the information does not appear to be as well-ensured by this system as the zoological nomenclatural data under *Zoobank*.

The code regulates the naming at levels ranging from subspecies to phyla only. Thus, naming of infrasubspecific entities (biovars, serovars and so on) or taxa of higher categories, like kingdoms, is not regulated by the *Prokaryotic Code*. Recently, a proposal to regulate phylum names (with a mandatory ending "-ota") has been approved through a vote of the members of the International Committee on Systematics of Prokaryotes (Oren et al. 2021a), and this was followed by valid publication of the names of 42 phyla of prokaryotes (Oren and Garrity 2021). Panda et al. (2022) recognized the value of these recent changes, but also pointed out to some problems. Thus, Panda et al. (2022: 2) argued that these changes:

have led to drastically different phylum names for several widely studied and long-recognized phyla, such as *Proteobacteria* (proposed: *Pseudomonadota*), *Firmicutes* (proposed: *Bacillota*), *Actinobacteria* (proposed: *Actinomycetota*), *Tenericutes* (proposed: *Mycoplasmatota*), *Crenarchaeota* (proposed: *Thermoproteota*), and *Thaumarchaeota* (proposed: *Nitrososphaerota*).

This is no small nomenclatural problem because the names of these six phyla appeared in 89% of the PubMed entries (excluding books) and 91 genome sequences available in the NCBI Genome database (Panda et al. 2022: 2). Another problem is the disappearance of the "-archaeota" suffix that previously featured in archaeal phyla, which renders the names less informative. For instance, it is not obvious that new the name *Thermoproteota* denotes an archaean phylum, whereas this was clearly implied by the previous name *Crenarchaeota*. Panda et al. (2022: 5) thus requested that some old names be conserved to solve these problems.

Even this recently adopted change of the *Prokaryotic Code* does not mention domains, which were proposed by experts on prokaryotes to accommodate the most inclusive taxa (*Bacteria*, *Archaea* and *Eucarya*). The word "domain" is mentioned only twice in the code itself, once in an example at the end of Rule 8: "Example: Domain—*Procaryotae*; Class—*Clostridia*" and one in the bibliography on the section entitled "Seventh International Congress of Bacteriology and Applied Microbiology" (Appendix 13 of the code), in which Woese et al. (1990) is cited (the word "domain" is in the title).

As in the other rank-based codes, only species (and subspecies) have physical types (in this case, live cultures); genera, families, orders and classes are typified by lower-ranking taxa. Thus, each genus is typified by a type-species, each family by a type-genus and each class by a type-order. However, orders are typified by one of their included genera, not by a family. Names of higher taxa are composed of a genus name (the type of the higher taxon) and a mandatory ending (suffix). Mandatory endings include "-ia" for classes, "-idea" for subclasses (Parker et al. 2019: Rule 8), "-ales" for orders, "-ineae" for suborders, "-aceae" for families, "-oidea" for subfamilies, "-eae" for tribes and "-inae" for subtribes (Rule 9).

Since 2000, prokaryote nomenclature is not independent of zoological and botanical nomenclature. Thus, Principle 2 of the *Prokaryotic Code* states that:

> When naming new taxa in the rank of genus or higher, due consideration is to be given to avoiding names which are regulated by the *International Code of Zoological Nomenclature* and the *International Code of Nomenclature for algae, fungi and plants*.

This clause is aimed at preventing the erection of additional inter-code homonyms; it is not retroactive (it does not invalidate names published before November 2000). Thus, *Proteus* remains the correct name for both an amphibian and a member of the *Enterobacteriaceae* (in both cases as a genus).

Types under the *Prokaryotic Code* are not fixed specimens, contrary to the zoological and botanical standard. They are live, pure cultures (representing a single taxon) and must be deposited in at least two publicly accessible service collections in different countries (Rule 30, section 3b). This is because the phenotype of bacteria is often uncharacteristic, whereas the genome (sequences, alleles and so on) and various biochemical and metabolic features are highly diagnostic. There are a few exceptions for taxa published before 2001 (Oren 2019: 4). Cultures deposited in facilities with restricted access, such as a safe deposit (used for patent purposes, for instance), may not be used as types (Rule 30 of the code). Prokaryotes that have been well characterized but cannot presently be cultured are given a preliminary status of "*Candidatus.*" These can become full-fledged taxa later, when they can be cultured.

*Cyanobacteria* present special nomenclatural problems because they are also covered by the *Botanical Code*. Their bacterial nature has long been known (Stanier and van Niel 1962), but they started being regulated by the *Prokaryotic Code* only in 1999, and only in a limited way because the *Prokaryotic Code* is not independent of the *Botanical Code*. Thus, it is not possible to propose a bacterial name for a taxon that is already named under the *Botanical Code*. This explains why, by 2017, there were only a single family, four genera and four species, of cyanobacteria regulated by the *Prokaryotic Code* (Oren and Ventura 2017). Even recognition and delimitation of taxa are problematic because botanical types cannot be live cultures; they must be herbarium specimens, illustrations or (in some cases) metabolically inactive cultures, and this is not ideal for cyanobacteria. Such rules are poorly suited to cyanobacteria (and prokaryotes in general) because their morphological variation is much less than that of most eukaryotic organisms, especially compared with metazoans and embryophytes. Thus, cyanobacterial taxa typified by herbarium specimens create taxonomic and nomenclatural problems. The case of the genus *Nostoc*, typified under the *Botanical Code*, illustrates this problem. *Nostoc* was probably the first cyanobacterial genus to be named; it was mentioned by Linnaeus (1753), but the name harks back to Paracelsus (1493–1541) under the slightly different spelling "Nostoch." One of its species, *N. commune*, is well known and has been consumed by humans for at least 2,000 years. Molecular phylogenetic analyses of the 16S rRNA gene have shown that this genus is polyphyletic (Oren and Ventura 2017: 1264). In 1978, the ICSB (International Committee on Systematic Bacteriology, which became the ICSP) published a formal proposal to cover cyanobacteria under the *Prokaryotic Code*, but Oren (2019: 9) reported that "40 years later, the issue has not yet been solved in a satisfactory way." Three contrasting proposals were recently discussed by the ICSP in 2020: one aiming at excluding most cyanobacteria from the *Prokaryotic Code*, another one aiming at regulating all cyanobacteria through that code and a last one to modify the *Prokaryotic Code* to ensure that names of cyanobacteria validly published under the *Botanical Code* are also considered valid under the *Prokaryotic Code* (Oren 2020a), and the latter prevailed (Oren et al. 2021b).

This frustrating situation has triggered some initiatives outside the main codes. Komárek and Golubic, cited in Oren and Ventura (2017: 1261), apparently compiled a "Guide to the nomenclature and formal taxonomic treatment of oxyphototroph prokaryotes (cyanoprokaryotes)," but the URL to get this document was broken when I tried to access it (October 12, 2020). This document apparently mentioned that it was not yet an accepted code, but that its authors hoped that it could be improved and accepted in due time. Another attempt at modifying nomenclature outside of the current rules was the proposed redefinition of the genus *Planktothrix*, which was erected under the *Botanical Code*, to give it a standing under the *Prokaryotic Code* (Gaget et al. 2015a). However, the rules of both codes do not allow such a double status; each name is regulated by only one of these codes. When Oren (2015) pointed this out, Gaget et al. (2015b) replied that they "are convinced that a stepwise, and more immediate, integration of cyanobacterial taxa names under the ICNP will be more beneficial to the scientific community than waiting for the final 'big leap' in the unforeseeable future." So far, these initiatives appear to have gathered limited support. In the case of cyanobacteria, the main progress may have come from recent changes to the codes; some recently erected cyanobacterial taxa under the *Botanical Code* have both a type in a herbarium and ex-holotype cultures that are far more useful to bacteriologists. This recent practice follows Recommendation 8B of the *Botanical Code* (Oren and Ventura 2017: 1265).

The field of prokaryotic nomenclature has thus grown over the years and has reached a very high level of technical sophistication. Fortunately, this still leaves room for originality for authors who are willing to take the time to form interesting names. My favorite is the following (Oren 2020b: 4):

*Dehalogenimonas lykanthroporepellens* (Moe et al. 2009), from Gr. masc. n. *lykanthropos*, werewolf; L. pres. part. *repellens*, repelling; N.L. part. adj. *lykanthroporepellens*, repelling werewolves; the name refers to the pungent garlic aroma that is produced when these organism[s] grow in the presence of 1,2,3-trichloropropane as an electron acceptor and sulfide as a reducing agent; garlic is said to repel werewolves in some fiction literature.

As in botany, the use of italics is quite pervasive for taxon names regulated by the *Prokaryotic Code*. Advisory note A of the current version (Parker et al. 2019) states:

For scientific names of taxa, conventions shall be used which are appropriate to the language of the country and to the relevant journal and publishing house concerned. These should preferably indicate scientific names by a different type face, e.g., italic, or by some other device to distinguish them from the rest of the text.

### 2.4.6 Code for Viruses

The International *Code of Virus Classification and Nomenclature (ICVCN)* differs sharply from other rank-based codes in several key points and deserves to be presented in some detail. This code, like the others, results from a short but complex history, which is summarized nicely by Kuhn (2020). The history of viruses starts with the discovery in 1886 of the tobacco mosaic disease (which is caused by a virus) by Adolf Eduard Mayer (1843–1942). By the 1890s, it had been shown that the sap of infected plants was still infectious after passing through a filter that retained most bacteria, which pointed at a smaller entity being responsible for the disease. The first taxonomy of viruses was proposed by Bennett (1939); it was based on morphological and cytological symptoms that they caused on their hosts, as well as antigenic, chemical and physical properties of their particles. The same year, Kausche et al. (1939) published the first electron-microscope images of virus particles. Linnaean ranks were introduced in viral nomenclature by Holmes in 1948. It then included an order (*Virales*) with three sub-orders, 13 families, 32 genera and 248 species. Like the *Strickland Code*, this nomenclature did not gain general acceptance, partly because it grouped viruses based on their host tropism and ignored morphological similarities of viruses.

The first viral taxonomy that gained significant acceptance was published in a preliminary form in 1962 by Lwoff, Horne and Tournier (hence the LHT acronym by which it was known), but appeared in its definitive version four years later (Lwoff and Tournier 1966). It recognized a phylum *Vira*, two subphyla based on RNA (*Ribovira*) or DNA (*Deoxyvira*) presence in virions, classes based on the symmetry of virion capsids and lower-ranking taxa. All of its higher taxa were later dismantled, but several of its families are still accepted today (such as *Poxviridae*). Note the objective nature of subphyla and classes in this system, reminiscent of a similar objective basis for ranks in Linnaeus' botanical nomenclature. The LHT system also proposed a list of suffixes for taxa that are still in use for orders, families, subfamilies, genera and subgenera.

The virology community felt a need for an official, regulated virus nomenclature and taxonomy. Thus, at the 1966 International Congress for Microbiology in Moscow, the International Committee on Nomenclature of Viruses (ICNV) was established to "develop a globally respected and applicable virus taxonomy for all virus types of all life forms" (Kuhn 2020: 2). The ICNV became the International Committee on Taxonomy of Viruses (ICTV) in 1974, and it has been the official body to regulate virus (plus viroids and satellites) taxonomy and nomenclature ever since. Most microbiological journals request (but apparently do not enforce) that submitted drafts follow the official ICTV taxonomy and nomenclature.

As should be clear by now, the ICTV regulates not only nomenclature (as in the other rank-based codes), but also taxonomy. Also, in contrast with the *Zoological Code* and *Botanical Code*, monophyly is an important criterion for establishing taxa. These differences may reflect the greater initial difficulty in establishing affinities between major taxa of viruses (before the advent of molecular sequencing) than between animals and plants (for which the ancient Greeks had already made considerable headway) and the strong suspicion that viruses are polyphyletic. Because of these problems, the ICTV started grouping viruses from the bottom up (from species upward), in contrast to the LHT system. In its first report (published in 1971), the ICTV listed two families, 27 genera, ten subgenera and 18 "virus groups," but, by 2019, it listed one phylum, six classes, 14 orders, 143 families, 1,019 genera and 5,560 species. However, many families have still not been united into taxa of higher categories because their affinities remain unclear (Kuhn 2020: 3), except for the highest category (realms). The virus tree (Wolf et al. 2018) has five currently recognized "branches," only one of which is currently named (realm Negarnaviricota; **Figure 2.1**). The four others could be named and assigned realm rank, but note that Kuhn (2020) showed an unrooted tree, and one of these branches (number 4 on Kuhn's 2020 figure 2.1) actually comprises three adjacent branches and cannot be monophyletic under any rooting of the tree. Indeed, in the rooted trees shown by Wolf et al. (2018), branch 5 is nested within branch 4.

Not surprisingly, the species concepts accepted by the majority of virologists differ a bit from those prevailing in zoology and botany. The following definition was proposed in 1989 by Van Regenmortel and endorsed by the ICTV in 1991 (Van Regenmortel 2003: 2486): "A virus species is a polythetic class of viruses that constitute a replicating lineage and occupy a particular ecological niche." Note that this definition refers to three key ideas: "polythetic" means that no one property is essential for membership in the group, so this is reminiscent of homeostatic property clusters; the mention of "lineage" implies, in addition to common descent, that viruses change over time and, indeed, they probably evolve faster than even bacteria; and "ecological niche" is reminiscent of the ecological species concept. The currently official species definition by the ICVT (Kuhn 2020) is "the lowest taxonomic level in the hierarchy approved by the *ICTV*. A species is a monophyletic group of viruses whose properties can be distinguished from those of other species by multiple criteria." This differs from the previous definition in a few key points, including reference to the lowest taxonomic level, which suggests convergence with rank-based nomenclature as implemented in other rank-based codes, monophyly, which is still not mentioned in the *Zoological Code* and *Botanical Code* (the *Prokaryotic Code* does not mention explicitly that taxa must be monophyletic, but phylogenetic position is a criterion mentioned in Appendix 11 of that code), and an operational recognition criterion, which evokes a phenetic cluster. And, of course, there is no mention of lineages or ecological niche, which were present in the previous definition. However, Kuhn (2020: 5) noted that both definitions proved "highly controversial." Nevertheless, according to Lherminier and Solignac (2005: 441), viruses form true species, like bacteria; they are not sexual species, but regularities in their evolutionary patterns and in their pattern of gene flux suggests that the species concept can be fruitfully applied to viruses.

The concept of type is difficult to apply to virology because by 2017, the ICTV had realized that it was unlikely that we would ever be able to culture or even characterize all viral strains in a laboratory. Thus, the ICTV bases virus nomenclature on coding-complete virus genome sequences, meaning that, among other things, all open reading frames are complete; this generally means that about 90–99% of the genome has been sequenced (Ladner et al. 2014). However, this is only a minimal requirement and does not preclude using other characteristics as well, such as particle morphology, infection phenotype, and tissue and host tropism. Nevertheless, this means that the "types" in virus nomenclature are essentially nucleotide sequences and character descriptions, contrary to the physical types (preserved specimens) used in zoology and botany.

The ICTV is a large organization of over 150 members, including, among others, 42 representatives of international

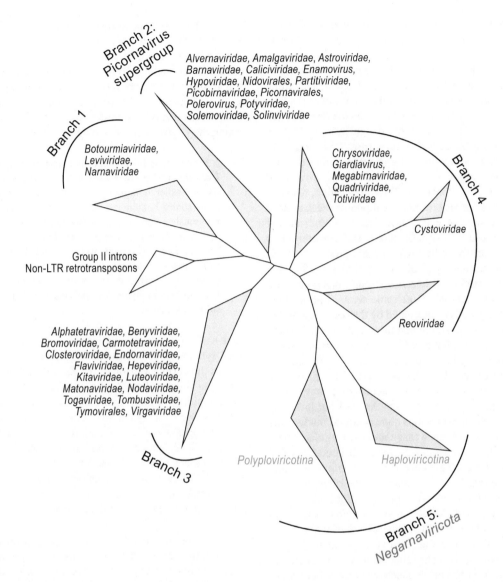

**FIGURE 2.1**   Tree of viruses produced by phylogenetic analysis of RNA. Note that "branch 4" is actually composed of three adjacent branches and cannot form a clade under any rooting of this tree. This tree had been redrawn and simplified (by Kuhn 2020) from Wolf et al. (2018), where the tree is rooted and in which branch 5 is within branch 4, as sister-group of *Reoviridae*.

*Source:* Reprinted with permission from Elsevier from Kuhn 2020: figure 1.

microbiological or virologic IUMS member societies and 101 Study Group Chairs, each of whom assembles an unlimited number of experts. Each Study Group is responsible for a major viral taxon. Thus, the number of scientists involved in regulating viral nomenclature (and taxonomy) is much greater than in the ICZN (24 members, according to www.iczn.org/about-the-iczn/commissioners/, consulted on October 2, 2020), which regulates zoological nomenclature, and also greater than in the ICSP (21 in 2020, according to www.the-icsp.org/icsp-members), which regulates prokaryotic nomenclature.

Under the ICVT, all taxon names are italicized (Kuhn 2020: 2). As for botanical and prokaryotic names, suffixes indicate the rank of taxa for a large part of the hierarchy. Thus, under the ICVT, suffixes are applied from subgenus to realm, with the particularity that the suffix ("-virus") is identical for genera and subgenera. The suffixes for other levels are (Kuhn 2020:

table 2): "-virinae" (subfamily), "-viridae" (family), "-virineae" (suborder), "-virales" (order), "-viricetidae" (subclass), "-viricetes" (class), "-viricotina" (subphylum), "-viricota" (phylum), "-virites" (subkingtom), "-virae" (kingdom), "-vira" (subrealm) and "-viria" (realm).

Contrary to other rank-based codes, names of species regulated by the ICVT are not necessarily binominal, but future revisions of the ICVT might make binominal nomenclature for these names mandatory (Kuhn 2020: 7).

### 2.4.7  Draft BioCode

By now, the reader will have realized that nomenclatural work is not only highly technical, but also complicated by problems of inter-code synonymy and homonymy and differences in rules and vocabulary. Worse, a systematist interested in distantly related taxa might have to be familiar with two or three codes, each of which is complex and takes much time to understand fully.

To remedy such problems, some systematists have tried to unify biological nomenclature into a single code, something that would facilitate work in biological nomenclature (Dubois 2005: 375). The reader may have heard about the *BioCode* initiative that was launched in the 1990s to regulate names of taxa of all groups of organisms except for viruses (Winston 2018: 1125). However, it was not the first such attempt. Shortly after the Strickland commission started to work on the "*Strickland Code*," Charles Lucien Bonaparte, Prince of Canino, who had become renowned for his ornithological publications, drafted a preliminary code that could have united zoological and botanical nomenclature (Minelli 2008). Unfortunately, this initiative was short-lived; he presented his project in a scientific meeting that convened in Padova (Italy) in 1842, but the project was abandoned because strong objections were raised at the next meeting of the same series (Lucca, Italy, 1843) against the proposals presented by the commission that had been set to develop a more definitive code on the basis of Bonaparte's draft. This initiative, like that of the Strickland commission, was a national initiative (Italian, in the case of Bonaparte's draft code). Decades later, Dall (1877) surveyed the opinions of North American systematists (many zoologists and some botanists, mostly from the USA but including a few from Canada) about various nomenclatural matters and proposed a partly unified code that could have regulated both zoological and botanical names. However, that code included separate articles for zoology and botany, so it could not really have unified the two fields. For instance, the rules that have been used to form family names differ in zoology and botany, notably the suffix used ("-idae" in zoology and "-aceae" in botany), and already by the 1870s, the weight of the tradition was such that Dall (1877) dared not envision a real unification, which would have required changing the way in which names are formed in at least one of the fields (zoology or botany), or possibly both, possibly with retroactive effects. This is only moderately surprising because nomenclatural practices in zoology and botany, which had been temporarily and partly unified by Linnaeus, started diverging again soon after his death (Minelli 2022b: 204). Dall (1877: 9) was aware of this and stated that "A serious mistake appears to have been committed at the outset by divorcing Zoological from Botanical nomenclature, as was done by the committee of the British Association [which drafted that *Strickland Code*]."

The weight of the tradition did not decrease with time, on the contrary! Thus, to again undertake an initiative to unify biological nomenclature in the 1990s was a bold enterprise. The *BioCode* initiative had slim chances of success from the outset. If it ever went into effect, the *BioCode* would regulate new names (proposed after it takes effect), while the traditional rank-based codes would continue regulating previously established names (Hawksworth 2011). The first draft *BioCode* was published in 1997 and it was abandoned four years later (ICZN 2001), "mainly owing to manifest difficulties in satisfactorily dealing with already existing names and to unwillingness of many botanists and zoologists alike to part with their traditional rules and to accept registration of new names." (Minelli

2001: 167) An attempt to revive it (Greuter et al. 2011) was short-lived (Minelli 2022a).

This project is fraught with difficulties, chiefly because each systematic community has its own culture, traditions and habits, and making these converge will be extremely difficult. In fact, there is evidence that some codes converge while others diverge. This can be exemplified by the following quote concerning the *Prokaryotic Code* (Oren 2019: 8):

> It must be noted that the current terminology differs from that of the older versions of the [Prokaryotic] Code: earlier homonyms were previously referred to as senior homonyms, and later homonyms were previously referred to as junior homonyms; earlier synonyms were previously referred to as senior synonyms, later synonyms were previously referred to as junior synonyms, homotypic synonyms were previously referred to as objective synonyms, and heterotypic synonyms were previously referred to as subjective synonyms.

The shift from "objective/subjective" to "homotypic/heterotypic" synonymy and from "senior/junior" to "earlier/later" matches the current version of the *Botanical Code*. However, the current version of the *Zoological Code* still uses the terms "senior/junior" for homonyms and synonyms, and "objective/subjective" for synonyms. Thus, the nomenclature of the *Prokaryotic Code* is now a bit more similar to that of the *Botanical Code*, but more dissimilar to the *Zoological Code*.

Development of the *BioCode* would be facilitated if the current rank-based codes converged. However, this does not appear to be the case. Indeed, the "Historical background" section of the introduction of the *Zoological Code* indicates:

> The work has revealed that it would be **premature** to introduce into this edition major changes to the established principles and structure which underlie the Code. The separate **codes have so diverged** in fundamental ways since their earliest beginnings, that the introduction of common rules today, and their application to the names established under the separate codes and which are in stable use, would result in much nomenclatural instability. Presently, a greater degree of terminological uniformity is being striven for in all codes. But the **lack of direct equivalence** in meaning of such universally applicable concepts as "availability" (zoology) and "validly published" (botany and bacteriology—where the same term has different meanings) has made the task **impossible** for the present.

Indeed, the *BioCode* is unlikely to take effect soon, if at all. The present situation was succinctly summarized by Oren (2019: 11):

> The last time the proposed *BioCode* was discussed in-depth in the meetings of the ICSP was in 2005. Then it was stated that there was little agreement among the botanists, zoologists, and microbiologist[s,] that progress was limited to attempts to maintain

a dialog, and that acceptance of a unified code of nomenclature for all living organisms **will probably not happen in our lifetime**.

This situation is unfortunate, because as illustrated by the example of cyanobacteria, the names of many organisms (called "ambiregnal") are regulated by two codes, and this creates numerous nomenclatural problems. There are many examples among unicellular eukaryotes, such as the taxon *Harosa*, which includes photosynthetic and non-photosynthetic organisms (Lücking 2019: 205) whose names are regulated by the *Zoological Code* and *Botanical Code*. If the *BioCode* were implemented, dealing with ambiregnal organisms would be much simpler.

However, the *BioCode* initiative was not a complete failure to the extent that it presumably resulted in the inclusion of Recommendation 1A in the *Zoological Code*, which states that:

> Authors intending to establish new genus-group names are urged to consult the *Index Nominum Genericorum (Plantarum)* and the *Approved List of Bacterial Names* to determine whether identical names have been established under the International Codes of Nomenclature relevant to those lists and, if so, to refrain from publishing identical zoological names.

This should slow down the appearance of cross-kingdom homonyms (at least those emanating from zoologists), even though this will not eliminate those that already exist. Another positive consequence is that "the organization of the *PhyloCode*, some of its terminology, and the wording of certain rules are derived from the *BioCode*" (Cantino and de Queiroz 2020: xv), as we will see in Chapter 4. Thus, while the *BioCode* initiative failed to convince the international community to adopt a new, united RN code, it has had positive influence on other codes.

## 2.5 Upcoming Challenges for the Rank-Based Codes

### 2.5.1 Simpler Is Better

As previously shown, correct application of nomenclature in the context of alpha-taxonomy requires a thorough knowledge of the highly technical rank-based codes. Nomenclatural work can thus be fairly time-consuming for the systematist who works only occasionally at it. A major challenge of all rank-based codes will thus be to become sufficiently streamlined to encourage systematists to follow them and formally describe taxa. If this is not done, the danger is that grey nomenclature (nomenclature that does not follow the rules of the relevant codes) of undescribed taxa (or of operational units of unknown ontology and legal status) will be increasingly prevalent (Minelli 2017). This is already a problem, and it has been promoted, to an extent, by the creation of MOTUs (molecular operational taxonomic units), even though these have been proposed as a way to quickly obtain a biodiversity inventory that may include new putative taxa (Blaxter et al. 2005). Logically,

the putative new taxa initially identified as MOTUs should subsequently be formally erected in the context of the proper rank-based code, but will they be? The alarmingly low proportion (54.3%) of Linnaean binomials found in the very popular BOLD database (Barcode of Life Data Systems; www.barcodinglife.org/) for *Clitellata* (the clade of terrestrial annelids that includes earthworms and leeches) by Minelli (2017: 660) shows that this danger is not purely hypothetical and does not concern a distant future—we are already facing this problem.

In addition to encouraging the use of grey nomenclature, the highly technical, burdensome procedures entailed by various RN codes can hamper communication. Thus, Hibbett et al. (2005: 661) lamented that:

> The constraints on taxonomic progress imposed by the [Botanical] Code (Greuter et al. 2000) are illustrated by the controversy over the classification of the genus *Coprinus*, a familiar assemblage of "inky cap" mushrooms. In 1994, and repeatedly thereafter, *Coprinus* was shown to be **polyphyletic** (Hopple and Vilgalys 1994; Redhead et al. 2001, and references therein). Redhead et al. (2001) suggested that *Coprinus* s. lat. should be divided into *Coprinus* s. str., *Coprinellus, Coprinopsis*, and *Parasola*, but this proposal has become **mired in a nomenclatural debate** concerning conservation and typification of generic names (Jørgensen et al. 2001; also see discussion at www.cbs.knaw.nl/nomenclature/index. htm [obsolete link]). Consequently, *Coprinus* s. lat. remains an "accepted" taxon (Kirk et al. 2001) fully **a decade after it was shown that it is not a clade**.

The situation is actually even worse than depicted in the quote because the nomenclatural decision that was required to answer competing proposals by Redhead et al. (2001a, 2001b) and Jørgensen et al. (2001) was only formally made and published in 2011 (Barrie 2011) because it had to be voted on by the International Botanical Congress, which meets every six years. Thus, the delay between the finding of the polyphyly of *Coprinus* by Hopple and Vilgalys (1994) and the nomenclatural decision that finally brought the nomenclature in conformity with the phylogeny spans 17 years. Admittedly, the nomenclatural debate to make *Coprinus* monophyletic took "only" 10 years, but this is still longer than it should. Given that the avowed purpose of codes of nomenclature is to improve communication by providing their users with a set of names with unambiguous meaning, this quote provides a saddening reminder of the limitations of RN codes.

The RN codes differ considerably concerning monophyly. While monophyly is required for prokaryotic taxa (though not formally by the *Prokaryotic Code*; see earlier) and the ICTV, neither the *Botanical Code* nor the *Zoological Code* have similar requirements. Given the broad consensus that taxa (at least supraspecific ones) should be clades, this aspect of these two latter codes seems anachronistic. This problem is not purely theoretical either. Thus, Hibbett et al. (2005: 661) reported that numerous fungal taxa were polyphyletic; the most problematic, the order *Polyporales*, then included "representatives of nine different clades" in the most recent edition of the *Dictionary of the Fungi*.

Will the regulating bodies of the rank-based codes rise to meet these challenges? On this front, some are better equipped than others. Thus, systematists working on viruses and (to a lesser extent) on prokaryotes have a tradition of emphasizing more molecular techniques than zoologists and botanists because phenotypic data are less abundant, and this impacts what each code accepts as nomenclatural type specimens. But even the *Code for Viruses* might need to be amended because it currently accepts coding-complete virus genome sequences as types. This is far more information than what is obtained through DNA barcoding. It would probably be unwise to allow erecting new viral taxa on the basis of DNA barcodes (which are short sequences), but perhaps an intermediate status of putative taxa could be envisioned for more exhaustive, though incomplete coding, sequences? These could perhaps be erected through a simplified procedure, potentially with temporary names or just registration numbers? The success of MOTUs and the status of *Candidatus* for prokaryotic taxa may pave the way for such developments. At the other extreme, the *Zoological Code* and *Botanical Code*, which require physical types to be deposited in collections accessible to researchers, may require more drastic changes to better accommodate the flood of molecular data that is changing our appraisal of the biodiversity of the relevant taxa, especially unicellular eukaryotes. Of course, this is a problem for the competent authorities to consider, but it may well be their greatest challenge for the 21st century.

### 2.5.2 Nomenclatural Stability at Last?

Another challenge for the rank-based codes will be to provide greater nomenclatural stability. Even some proponents of rank-based nomenclature admit that improvements are required on this front. Thus, Löbl (2015: 35) wrote: "While the [zoological] Code regulates availability of taxonomic works and acts, its effects on stability of nomina are quite limited in the practice." Yet, he is obviously a proponent of rank-based nomenclature and rather opposed to phylogenetic nomenclature, given that in the same paper (Löbl 2015: 36), he called proponents of phylogenetic nomenclature "an insignificant group of phylocodists who failed to understand the raison d'être of classification," which is both wrong and presumptuous because only the future will tell what is significant. But the occurrence of both statements in the same paper does show that at least some proponents of rank-based nomenclature opposed to phylogenetic nomenclature consider that the rank-based codes need to be amended to provide more nomenclatural stability.

This problem is acknowledged, to an extent, in the preface of the current (fourth) edition of the *Zoological Code*, where we read:

> The conventional Linnaean hierarchy will not be able to survive alone: it will have to coexist with the ideas and terminology of phylogenetic (cladistic) systematics. From a cladistic perspective, our traditional nomenclature is often perceived as too prescriptive and too permissive at the same time. **Too prescriptive**, in so far as it forces all taxa (and their names)

to fit into the **arbitrary ranks of the hierarchy; too permissive**, in so far as it may be equally **applied to paraphyletic as to monophyletic groups**. New proposals are therefore to be expected. But even in the perspective of new developments, we believe that it will never be possible or desirable to dispose of 250 years of Linnaean zoological (and botanical) taxonomy and nomenclature. One should always keep in mind that an important function of classifications is information retrieval. The Linnaean tradition will be supplemented, but not replaced, by new semantic and lexical tools.

This summarizes well the challenges facing rank-based nomenclature in the coming years to promote nomenclatural stability. The problems raised by rank-based nomenclature will be discussed in Chapter 3, and potential solutions to improve nomenclatural stability in the context of phylogenetic nomenclature are presented in Chapter 4.

## NOTE

1  Aristophanes, Birds. www.perseus.tufts.edu/hopper/text?doc=Perseus:text:1999.01.0026

## REFERENCES

Adanson, M. 1763. Familles des Plantes. 1ère partie. Vincent, Paris, cccxxv + 189 pp.

Ainsworth, G. and R. Ciferri. 1955. Mycological taxonomic literature and publication. Taxon 3–6.

American Ornithologists' Union. 1886. The code of nomenclature and check-list of North American birds adopted by the American Ornithologists' Union. American Ornithologists' Union, New York, viii + 392 pp.

Arthur, J. C., J. Barnhart, N. Britton, F. Clements, O. Cook, F. Coville, F. Earle, A. Evans, T. Hazen, and A. Hollick. 1907. American code of botanical nomenclature. Bulletin of the Torrey Botanical Club 34:167–178.

Avise, J. C. and J.-X. Liu. 2011. On the temporal inconsistencies of Linnean taxonomic ranks. Biological Journal of the Linnean Society 102:707–714.

Bailey, L. H. 1918. The indigen and cultigen. Science 47:306–308.

Barrie, F. R. 2011. Report of the general committee: 11. Taxon 60:1211–1214.

Bennett, C. 1939. The nomenclature of plant viruses. Phytopathology 29:422–430.

Blaxter, M., J. Mann, T. Chapman, F. Thomas, C. Whitton, R. FLoyd, and E. Abebe. 2005. Defining operational taxonomic units using DNA barcode data. Philosophical Transactions of the Royal Society of London, Series B 360:1935–1943.

Bock, W. J. 1994. History and nomenclature of avian family-group names. Bulletin of the American Museum of Natural History (USA) 222:1–281.

Brickell, C. D., C. Alexander, J. J. Cubey, J. C. David, M. H. A. Hoffman, A. C. Leslie, V. Malécot, and X. Jin. 2016. International code of nomenclature for cultivated plants. International Society for Horticultural Science, Katwijk aan Zee, The Netherlands, xvii + 190 pp.

Britten, J., J. Ramsbottom, T. A. Sprague, E. M. Wakefield, and A. J. Wilmott. 1924. Interim report on nomenclature. Journal of Botany 62:79–81.

Buchanan, R. E., R. St John-Brooks, and R. S. Breed. 1948. International bacteriological code of nomenclature. Journal of Bacteriology 55:287–306.

Burki, F., A. J. Roger, M. W. Brown, and A. G. Simpson. 2020. The new tree of eukaryotes. Trends in Ecology & Evolution 35:43–55.

de Candolle, A. 1867. Lois de la nomenclature botanique adoptées par le Congrès international de botanique tenu à Paris en août 1867: suivies d'une 2e édition de l'introduction historique et du commentaire qui accompagnaient la rédaction prépara-toire présentée au Congrès. H. Georg, Genève, 64 pp.

Cantino, P. D. and K. de Queiroz. 2020. International code of phylogenetic nomenclature (PhyloCode): A phylogenetic code of biological nomenclature. CRC Press, Boca Raton, Florida, xl + 149 pp.

Cook, O. 1916. Determining types of genera. Journal of the Washington Academy of Sciences 6:137–140.

Cuvier, G. and A. M. C. Duméril. 1800. Leçons d'anatomie com-parée. Baudouin, Paris, xxxi + 522 pp.

Dall, W. H. 1877. Nomenclature in zoology and botany: A report to the American Association for the Advancement of Science at the Nashville Meeting, August 31. Salem Press, Salem, 56 pp.

de Jussieu, A.-L. 1789. Genera plantarum secundum ordines natu-rales disposita juxta methodum in horto regio parisiensi exaratam, anno 1774. Veuve Herissant, 498 pp.

Dubois, A. 2005. Proposed rules for the incorporation of nomina of higher-ranked zoological taxa in the international code of zoological nomenclature. 1: Some general questions, con-cepts and terms of biological nomenclature. Zoosystema 27:365–426.

Dubois, A. 2006. Proposed rules for the incorporation of nomina of higher-ranked zoological taxa in the International code of zoological nomenclature. 2: The proposed rules and their rationale. Zoosystema 28:165–258.

Dubois, A. 2011. The rich but confusing terminology of biological nomenclature: A first step towards a comprehensive glos-sary. Bionomina 3:71–76.

Dubois, A., A. Nemésio, and R. Bour. 2014. Primary, second-ary and tertiary syntypes and virtual lectotype designation in zoological nomenclature, with comments on the recent designation of a lectotype for *Elephas maximus* Linnaeus, 1758. Bionomina 7:45–64.

Farlow, W. G. and G. F. Atkinson. 1910. The botanical congress at Brussels. Botanical Gazette 50:220–225.

Fischer, M. 1894. Rules of nomenclature adopted by the interna-tional zoological congress, held in Moscow, Russia, 1892. Part II. The American Naturalist 28:929–933.

Gaget, V., M. Welker, R. Rippka, and N. Tandeau de Marsac. 2015a. A polyphasic approach leading to the revision of the genus *Planktothrix* (Cyanobacteria) and its type species, *P. agardhii*, and proposal for integrating the emended valid botanical taxa, as well as three new species, *Planktothrix paucivesiculata* sp. nov.[ICNP], *Planktothrix tepida* sp. nov.[ICNP], and *Planktothrix serta* sp. nov.[ICNP], as genus and species names with nomenclatural standing under the ICNP. Systematic and Applied Microbiology 38:141–158.

Gaget, V., M. Welker, R. Rippka, and N. Tandeau de Marsac. 2015b. Response to: "Comments on: A polyphasic approach lead-ing to the revision of the genus *Planktothrix* (Cyanobacteria) and its type species, *P. agardhii*, and proposal for integrating the emended valid botanical taxa, as well as three new spe-cies, *Planktothrix paucivesiculata* sp. nov.[ICNP], *Planktothrix tepida* sp. nov.[ICNP], and *Planktothrix serta* sp. nov.[ICNP], as genus and species names with nomenclatural standing under the ICNP," by V. Gaget, M. Welker, R. Rippka, and N. Tandeau de Marsac, Syst Appl Microbiol (2015). http://dx.doi.org/10.1016/j.syapm.2015.02.004, by A. Oren [Syst Appl Microbiol (2015). doi:10.1016/j.syapm.2015.03.002]. Systematic and Applied Microbiology 38:368–370.

Goldfuss, G. A. and J. C. D. Schreber. 1809. Vergleichende Naturbeschreibung der Säugethiere. Waltherschen Kunst-und Buchhandlung, xix + 314 + XXXVI (plates) pp.

Greuter, W., G. Garrity, D. L. Hawskworth, R. Jahn, P. M. Kirk, S. Knapp, J. McNeill, E. Michel, D. J. Patterson, R. L. Pyle, and B. J. Tindall. 2011. Draft BioCode (2011): Principles and rules regulating the naming of organisms. New draft, revised in November 2010. Bionomina 3:26–44.

Hawksworth, D. L. 2005. Universal fungus register offers pattern for zoology. Nature 438:24.

Hawksworth, D. L. 2011. Introducing the Draft BioCode (2011). Bionomina 3:24–25.

Hennig, W. 1966. Phylogenetic systematics. University of Illinois Press, Urbana, Chicago, London, 263 pp.

Hibbett, D. S., H. R. Nilsson, M. Snyder, M. Fonseca, J. Costanzo, and M. Shonfeld. 2005. Automated phylogenetic taxonomy: An example in the homobasidiomycetes (mushroom-form-ing Fungi). Systematic Biology 54:660–667.

Hitchcock, A. S. 1919. Report of the committee on generic types of the Botanical Society of America. Science 49:333–336.

Hopple Jr, J. S. and R. Vilgalys. 1994. Phylogenetic relationships among coprinoid taxa and allies based on data from restric-tion site mapping of nuclear rDNA. Mycologia 86:96–107.

Hoquet, T. 2007. Buffon: From natural history to the history of nature? Biological Theory 2:413–419.

ICTV. 2018. The international code of virus classification and nomenclature. https://talk.ictvonline.org/information/w/ictv-information/383/ictv-code

ICZN. 1905. Règles internationales de la nomenclature zoologique adoptées par les congrès internationaux de zoologie: International rules of zoological nomenclature. Internationale Regeln der zoologischen Nomenklatur. FR de Rudeval, Paris, 57 pp.

ICZN. 1999. International code of zoological nomen-clature. The International Trust for Zoological Nomenclature, London, 306 pp. www.iczn.org/the-code/the-international-code-of-zoological-nomenclature/the-code-online/

ICZN. 2001. International committee on bionomenclature. Bulletin of Zoological Nomenclature 58:6–7.

ICZN. 2012. Amendment of articles 8, 9, 10, 21 and 78 of the international code of zoological nomenclature to expand and refine methods of publication. ZooKeys 219:1–10.

Jørgensen, P. M., S. Ryman, W. Gams, and J. Stalpers. [1486] 2001. Proposal to conserve the name *Coprinus* Pers. (*Basidiomycota*) with a conserved type. Taxon 50:909–910.

Judd, W., C. Campbell, E. Kellogg, P. Stevens, and M. Donoghue. 2008. Plant systematics: A phylogenetic approach. Sunderland, Sinauer Associates, Sunderland, MA, xv + 611 pp.

Kästner, A. G. and J. C. P. Erxleben. 1767. Dissertatio inauguralis physica sistens diivdicationem systematum animalium mammalium. Rosenbusch, 14 pp.

Kausche, G. A., E. Pfankuch, and H. Ruska. 1939. Die Sichtbarmachung von pflanzlichem Virus im Übermikroskop. Naturwissenschaften 27:292–299.

Kirby, W. 1813. VI. Strepsiptera, a new order of insects proposed; and the characters of the order, with those of its genera, laid down. Transactions of the Linnean Society of London 11:86–122.

Klein, Jacob Theodor. 1751. Quadrupedum dispositio brevisque historia naturalis. Ionam Schmidt, Lipsiae.

Konstantinidis, K. T. and J. M. Tiedje. 2005. Genomic insights that advance the species definition for prokaryotes. Proceedings of the National Academy of Sciences 102:2567–2572.

Kuhn, J. H. 2020. Virus taxonomy. Reference Module in Life Sciences B978-0-12-809633-8.21231–4.

Ladner, J. T., B. Beitzel, P. S. Chain, M. G. Davenport, E. Donaldson, M. Frieman, J. Kugelman, J. H. Kuhn, J. O'Rear, and P. C. Sabeti. 2014. Standards for sequencing viral genomes in the era of high-throughput sequencing. mBio 5:1–5. https://doi.org/10.1128/mBio.01360-14

Latreille, P. A. 1797. Précis des caractères génériques des insectes disposés dans un ordre naturel. Prévôt, Paris, xiii + 201 pp.

Laurin, M. 2008. The splendid isolation of biological nomenclature. Zoologica Scripta 37:223–233.

Lhermin[i]er, P. and M. Solignac. 2000. L'espèce: définitions d'auteurs. Comptes Rendus de l'Académie des Sciences—Series III—Sciences de la Vie 323:153–165.

Lherminier, P. and M. Solignac. 2005. De l'espèce. Syllepse, Paris, XI + 694 pp.

Linnaeus, C. 1753. Species Plantarum. Salvi, Stockholm, 1200 pp.

Linnaeus, C. 1758. Systema naturae. Holmiae (Laurentii Salvii), Stockholm, 824 pp.

Löbl, I. 2015. Stability under the international code of zoological nomenclature: A bag of problems affecting nomenclature and taxonomy. Bionomina 9:35–40.

Lücking, R. 2019. Stop the abuse of time! Strict temporal banding is not the future of rank-based classifications in fungi (including lichens) and other organisms. Critical Reviews in Plant Sciences 38:199–253.

Ludwig, C. F. 1790. Delectus opusculorum ad scientiam naturalem spectantium. Lipsiae, VIII + 560 pp.

Lwoff, A. and P. Tournier. 1966. The classification of viruses. Annual Reviews in Microbiology 20:45–74.

Magnol, P. 1689. Prodromus historiae generalis plantarum in quo familiae plantarum per tabulas disponuntur. Pech, Montpellier, 79 pp.

May, T. W., S. A. Redhead, K. Bensch, D. L. Hawksworth, J. Lendemer, L. Lombard, and N. J. Turland. 2019. Chapter F of the international code of nomenclature for algae, fungi, and plants as approved by the 11th International Mycological Congress, San Juan, Puerto Rico, July 2018. IMA Fungus 10:21.

Minelli, A. 2001. Zoological nomenclature: Reflections on the recent past and ideas for our future agenda. Bulletin of Zoological Nomenclature 58:164–169.

Minelli, A. 2008. Zoological vs. botanical nomenclature: A forgotten "BioCode" experiment from the times of the Strickland Code. Zootaxa 1950:21–38.

Minelli, A. 2017. Grey nomenclature needs rules. Ecologica Montenegrina 7:654–666.

Minelli, A. 2019. Zoological nomina with typus or typicus as the specific epithet. Bionomina 16:1–21.

Minelli, A. 2022a. Species; in B. Hjørland and C. Gnoli (eds.), ISKO Encyclopedia of Knowledge Organization (IEKO). www.isko.org/cyclo/species

Minelli, A. 2022b. The species before and after Linnaeus: Tension between disciplinary nomadism and conservative nomenclature; pp. 191–226 in J. Wilkins, I. Pavlinov, and F. Zachos (eds.), Species problems and beyond: Contemporary issues in philosophy and practice. CRC Press, Boca Raton, FL.

Nicolson, D. H. 1991. A history of botanical nomenclature. Annals of the Missouri Botanical Garden 78:33–56.

Oren, A. 2011. How to name new genera and species of prokaryotes?; pp. 437–463 in F. Rainey and A. Oren (eds.), Taxonomy of prokaryotes. Elsevier, Amsterdam.

Oren, A. 2015. Comments on: "A polyphasic approach leading to the revision of the genus *Planktothrix* (Cyanobacteria) and its type species, *P. agardhii*, and proposal for integrating the emended valid botanical taxa, as well as three new species, *Planktothrix paucivesiculata* sp. nov.[ICNP], *Planktothrix tepida* sp. nov.[ICNP], and *Planktothrix serta* sp. nov.[ICNP], as genus and species names with nomenclature standing under the ICNP," by V. Gaget, M. Welker, R. Rippka, and N. Tandeau de Marsac, Syst. Appl. Microbiol. (2015), http://dx.doi.org/10.1016/j.syapm.2015.02.004. Systematic and Applied Microbiology 38:159.

Oren, A. 2019. Prokaryotic nomenclature; pp. 1–12 in W. B. Whitman (ed.), Bergey's manual of systematics of Archaea and Bacteria. John Wiley & Sons, Inc., in association with Bergey's Manual Trust, New York. DOI: 10.1002/9781118960608.bm00004.pub2.

Oren, A. 2020a. Three alternative proposals to emend the rules of the international code of nomenclature of prokaryotes to resolve the status of the cyanobacteria in the prokaryotic nomenclature. International Journal of Systematic and Evolutionary Microbiology 70:4406–4408.

Oren, A. 2020b. Prokaryotic names: The bold and the beautiful. FEMS Microbiology Letters 367:fnaa096.

Oren, A., D. R. Arahal, R. Rosselló-Móra, I. C. Sutcliffe, and E. R. Moore. 2021a. Emendation of rules 5b, 8, 15 and 22 of the international code of nomenclature of prokaryotes to include the rank of phylum. International Journal of Systematic and Evolutionary Microbiology 71:004851.

Oren, A., D. R. Arahal, R. Rosselló-Móra, I. C. Sutcliffe, and E. R. Moore. 2021b. Emendation of general consideration 5 and rules 18a, 24a and 30 of the international code of nomenclature of prokaryotes to resolve the status of the cyanobacteria in the prokaryotic nomenclature. International Journal of Systematic and Evolutionary Microbiology 71:004939.

Oren, A. and G. M. Garrity. 2021. Valid publication of the names of forty-two phyla of prokaryotes. International Journal of Systematic and Evolutionary Microbiology 71:005056.

Oren, A. and S. Ventura. 2017. The current status of cyanobacterial nomenclature under the "prokaryotic" and the "botanical" code. Antonie van Leeuwenhoek 110:1257–1269.

Panda, A., S. T. Islam, and G. Sharma. 2022. Harmonizing prokaryotic nomenclature: Fixing the fuss over phylum name flipping. Mbio 13:e00970–22.

Parker, C. T., B. J. Tindall, and G. M. Garrity. 2019. International code of nomenclature of prokaryotes: Prokaryotic code (2008 revision). International Journal of Systematic and Evolutionary Microbiology 69:S1–S111. https://doi.org/10.1099/ijsem.0.000778

Persson, M. 2016. Building an empire in the republic of letters: Albrecht von Haller, Carolus Linnaeus, and the struggle for botanical sovereignty. Circumscribere: International Journal for the History of Science 17:18–40.

Pyle, C. M. 2000. Art as science: Scientific illustration, 1490–1670 in drawing, woodcut and copper plate. Endeavour 24:69–75.

Redhead, S. A., R. Vilgalys, J. M. Moncalvo, J. Johnson, and J. S. Hopple Jr. 2001a. *Coprinus* Pers. and the disposition of Coprinus species sensu lato. Taxon 50:203–241.

Redhead, S. A., R. Vilgalys, J.-M. Moncalvo, J. Johnson, and J. S. Hopple Jr. 2001b. (1473–1474) Proposals to conserve the name *Psathyrella* (Fr.) Quél. with a conserved type and to reject the name *Pselliophora* P. Karst. (*Basidiomycetes: Psathyrellaceae*). Taxon 50:275–277.

Richter, M. and R. Rosselló-Móra. 2009. Shifting the genomic gold standard for the prokaryotic species definition. Proceedings of the National Academy of Sciences 106:19126–19131.

Rieppel, O. 2003. Semaphoronts, cladograms and the roots of total evidence. Biological Journal of the Linnean Society 80:167–186.

Rookmaaker, L. 2011. The early endeavours by Hugh Edwin Strickland to establish a code for zoological nomenclature in 1842–1843. Bulletin of Zoological Nomenclature 68:29–40.

Spencer, R. D. and R. G. Cross. 2007. The international code of botanical nomenclature (ICBN), the international code of nomenclature for cultivated plants (ICNCP), and the cultigen. Taxon 56:938–940.

Stackebrandt, E. and B. M. Goebel. 1994. A place for DNA-DNA reassociation and 16S rRNA sequence analysis in the present species definition in bacteriology. International Journal of Systematic Bacteriology 44:846–849.

Stanier, R. Y. and C. van Niel. 1962. The concept of a bacterium. Archiv für Mikrobiologie 42:17–35.

Stearn, W. T. 1953. International code of nomenclature for cultivated plants. Royal Horticultural Society London, London, 29 pp.

Stemerding, D. 1993. How to make oneself nature's spokesman? A Latourian account of classification in eighteenth-and early nineteenth-century natural history. Biology and Philosophy 8:193–223.

Strickland, H. E. 1845. Report on the recent progress and present state of ornithology; pp. 170–221 in Report of the fourteenth meeting of the British Association for the Advancement of Science held at York in September 1844. Pichard and John E. Taylor, London.

Strickland, H. E., J. S. Henslow, J. Phillips, W. E. Shuckard, J. B. Richardson, G. R. Waterhouse, R. Owen, W. Yarrell, L. Jenyns, C. Darwin, W. J. Broderip, and J. O. Westwood. 1842. Report of a committee appointed "to consider of the rules by which the nomenclature of zoology may be established on a uniform and permanent basis. Annals and Magazine of Natural History 11:1–17.

Turland, N. J., J. H. Wiersema, F. R. Barrie, W. Greuter, D. Hawksworth, P. S. Herendeen, S. Knapp, W.-H. Kusber, D.-Z. Li, K. Marhold, T. May, J. McNeill, A. Monro, J. Prado, M. Price, and G. Smith. 2018. International code of nomenclature for algae, fungi, and plants (Shenzhen Code) adopted by the Nineteenth International Botanical Congress Shenzhen, China, July 2017. Koeltz Botanical Books, Glashütten, xxxviii + 254 pp.

Van Regenmortel, M. H. 2003. Viruses are real, virus species are man-made, taxonomic constructions. Archives of Virology 148:2481–2488.

Wilkins, J. S. 2018. Species: The evolution of the idea. CRC Press, Boca Raton, xxxviii + 389 pp.

Winston, J. E. 2018. Twenty-first century biological nomenclature: The enduring power of names. Integrative and Comparative Biology 58:1122–1131.

Witteveen, J. and S. Müller-Wille. 2020. Of elephants and errors: Naming and identity in Linnaean taxonomy. History and Philosophy of the Life Sciences 42:1–34.

Woese, C. R., O. Kandler, and M. L. Wheelis. 1990. Towards a natural system of organisms: Proposal for the domains Archaea, Bacteria, and Eucarya. Proceedings of the National Academy of Sciences 87:4576–4579.

Wolf, Y. I., D. Kazlauskas, J. Iranzo, A. Lucía-Sanz, J. H. Kuhn, M. Krupovic, V. V. Dolja, and E. V. Koonin. 2018. Origins and evolution of the global RNA virome. MBio 9:e02329–18.

Zhang, Z.-Q. 2012. A new era in zoological nomenclature and taxonomy: ICZN accepts e-publication and launches ZooBank. Zootaxa 3450:8.

# 3

## *Rank-Based Nomenclature and Evolution*

### 3.1 The Rise of Phylogenetics

We saw in Chapter 1 that the development of evolutionary biology in the 19th century led systematists to draw the first evolutionary trees. However, phylogenetics at that time, and until the middle of the 20th century, was qualitative and fairly subjective because it basically consisted of comparing presumably related taxa, assessing their similarities and differences, and guessing how they might be related to each other. Evolutionary trees produced that way were essentially opinions, hence irrefutable. Well into the middle of the 20th century, attempts at inferring the phylogeny relied mostly on the fossil record (Cain 1959: 241), as implied by this brief quote from Sneath and Sokal (1962: 856): "the available fossil record is so fragmentary that the phylogeny of the vast majority of taxa is unknown." When fossils were available, they were often considered potential ancestors of extant taxa (e.g., Romer 1966). This slow rise of phylogenetics allowed the use of rank-based nomenclature to become firmly entrenched for many decades before alternatives that emphasize more phylogeny could be considered, as explained by Papavero et al. (2001: 5):

> Lack of well founded knowledge of the sistergroup relationships during the late 19th and early 20th centuries had allowed further use of the Linnaean hierarchy which should expressly not reflect 'speculations' about phylogenetic relationships. This resulted in **more securely establishing the use of Linnaean categories** until recently.
>
> **(Bold font used for emphasis is mine, unless noted otherwise.)**

The advent of more explicit phylogenetic inference methods that allow replication and refutation, as well as a greater resolving power, was thus a prerequisite for the development of phylogenetic nomenclature (abbreviated PN from here on). A brief history of the development of these methods, in approximate chronological order, follows.

#### 3.1.1 Phenetics

Numerical taxonomy, also called phenetics, sought to classify organisms objectively, but contrary to cladistics and other more recent methods, it does not attempt to infer evolutionary patterns, even though Sneath and Sokal (1962: 860) envisioned this as a potential distant goal: "Has numerical taxonomy nothing to say regarding phylogenetics? We believe that it will have, though this

is still an unexplored field." In fact, phenetics developed partly as a reaction against the "phylogenetic speculations" that prevailed in taxonomy when phenetics was developed with the explicit purpose of providing greater objectivity to assess "affinities" (envisioned as mere similarities) between organisms (Sneath and Sokal 1962). Phenetics thus attempts at forming ontologically agnostic groups of similar organisms. Thus, contrary to cladistics, characters need not be (and even, cannot be) polarized. Some consider that the development of phenetics harks back, in its simplest form, to Michel Adanson (1727–1806), to the extent that Adanson tried to base his classifications on many characters (Lherminier and Solignac 2005: 35; see Section 1.4.1).

Phenetics has been abandoned by most systematists to classify large groups of organisms, partly because some of its assumptions are problematic and partly because it is not ambitious enough, given the strong growth of evolutionary biology (assessing similarity is not enough; systematists and evolutionary biologists want to unravel evolutionary history). Thus, one of the main tenets of phenetics is that if you measure (for continuous characters) or score (for discrete characters) a random sample of all possible characters, you will get an objective measure of similarity that can be used to group specimens into low-ranking taxa and the latter into progressively more inclusive taxa. This premise was criticized long ago by Griffiths (1974: 103):

> The introduction of a statistical rule of construction [phenetics] achieves nothing, because the **incorrect premise that taxa are classes** (= sets) constructed out of the attributes of individuals still remains. In modern biological theory species and "higher" taxa are postulated to be physical systems at a higher level of organization with irreducible attributes of their own, **not classes of individuals**.

Another problem is that phenetics assumes that there is a finite number of characters, and that the latter can be broken down into "atomic statements," also called "unit characters" (Sneath and Sokal 1962: 858), which are basic characters that cannot be broken down into something simpler. But, as pointed out by Griffiths (1974: 110):

> Nobody ever succeeded in finding convincing examples of atomic statements and in demonstrating how more complex statements (such as those involving universals) can be reduced to these. Furthermore, the epistemological content of the doctrine conflicts with modern theories of perception, notably Gestalt theory. . . . Nowadays the theory of logical atomism is only of historical interest to philosophers.

DOI: 10.1201/9781003092827-4

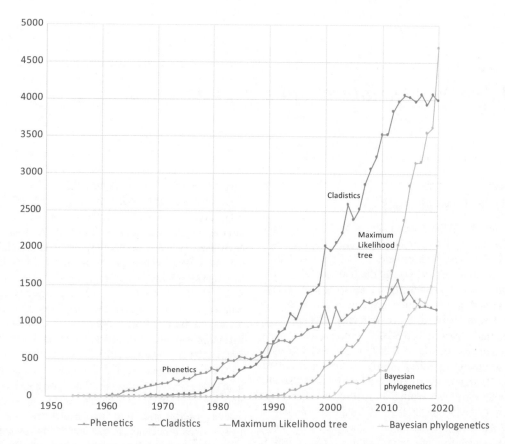

**FIGURE 3.1** Development of systematic methods (phenetics, cladistics, maximum likelihood, and Bayesian phylogenetics) to classify organisms as shown by the number of papers mentioning four methods each year. The data for this figure are based on a Google Scholar search carried out on June 3, 2021, using, respectively, the words or expressions "phenetics," "cladistics," "maximum likelihood tree" and "Bayesian phylogenetics." Note that for the last two, using a shorter expression would not have been a viable option because both maximum likelihood and Bayesian methods are used in many other (non-phylogenetic) contexts.

These reasons may explain why phenetics was progressively abandoned to classify higher-ranking taxa in the 1980s and 1990s. Consequently, this method will not be detailed further here. Phenetics remains in use to recognize and delimit small groups, such as subspecies and species, which explains its continued growth (in number of published papers per year) until 2012 (**Figure 3.1**).

### 3.1.2 Parsimony

#### 3.1.2.1 Character Polarization

The German entomologist Willi Hennig (1913–1976) provided a far more objective phylogenetic inference method. The parsimony criterion, which spread in phylogenetics with the advent of cladistics, stipulates that the tree that requires the fewest character transformations must be preferred because any other tree requires more ad hoc hypotheses of character transformations. Hennig (1950, 1966) pointed out that systematists had used primitive as well as "derived" (new) characters in assessing relationships, but that their evolutionary significance was not comparable. For instance, among vertebrates, many taxa (sharks, skates, gars, salmons, the coelacanth and lungfishes to name a few) possess paired fins, whereas others

(salamanders, frogs, turtles, crocodiles and so on) possess paired limbs with digits. However, various paleontological, anatomical and developmental data (Hall 2007; Laurin 2010a) suggest one of these structures (the fin) gave rise to the other (the limb). Thus, the presence of fins is primitive, whereas the presence of limbs is derived. Hennig (1950, 1966) pointed out that primitive characters like the fin (in this context) do not indicate close relationships, only derived characters (here, the limb) do. Consequently, the presence of fins does not indicate that sharks, salmons and lungfishes are closely related. On the contrary, the presence of limbs does indicate close affinities between the taxa that possess it. Recent phylogenetic research supports this point and shows that lungfishes and the coelacanth, despite their fins, are more closely related to the taxa that possess limbs (the tetrapods) than to other finned taxa (such as sharks and salmons). Together, they form the taxon *Sarcopterygii* (Janvier 1996). This is why tetrapods form a clade (a monophyletic group), whereas fishes do not, and thus, the taxon *Pisces*, recognized in ancient studies (and unfortunately a few recent ones as well), is no longer recognized by systematists who accept Hennig's (1966) conclusions.

How can we determine which state of a character is primitive? In early parsimony analyses, this had to be determined before performing the rest of the analysis, as Hennig (1966) did. For instance, the character "appendage type" in the example given can have two states: "fin" and "limb," and in the data matrices that are compiled for such analyses (these consist of cells representing character states in taxa), these states are represented by a simple code, most frequently an integer (0, 1, 2 and so on). This was only a slight reformulation of the principle of homology that was developed in the 19th century (see Chapter 1); only homologous structures or traits can be alternate states of a given character.

Character polarization (determining which of two or more states appeared first) is an old problem that has been thoroughly discussed through a vast body of literature; only a very brief, simple account needs to be given here because this step is no longer required by recent phylogenetic analysis methods, at least not as a preliminary step in the analysis (see the following). However, O'Hara (1997) listed "the distinction between ancestral and derived character states" as one of the most important concepts associated with the advent of cladistics around the 1960s. The most frequently used criterion and the most generally valid is outgroup comparison (Stevens 1980: 351); ontogeny and stratigraphy (geological age) are additional, less frequently used criteria. Early studies in this field used up to seven methods to polarize characters (Stevens 1980), but several of these soon fell into disrepute (for good reasons) and were soon discarded by systematists. These discarded criteria include: 1) ingroup analysis, the principle that the most common state in the ingroup must be primitive; 2) notions about character evolution, using inferred function or biogeographic distribution; 3) character correlation, based on the inference that if a taxon possesses many primitive states, for a character of unknown polarity, that taxon probably retains the primitive condition; and 4) inferred evolutionary trends. This leaves three polarization principles that remain valid: outgroup comparison, the ontogenetic criterion and stratigraphy, which are detailed in the following.

The principle of outgroup comparison stipulates that the state present in most outgroups is probably primitive for the ingroup. This requires defining two key terms: the ingroup is the group of interest, whose phylogeny is under study; the outgroups are taxa that do not belong to the ingroup, but that are hypothesized to be closely related to it. Note that this criterion rests on parsimony because if state 0 occurs in the outgroup, and if the ingroup contains both taxa that display state 0 and others with state 1, in the absence of any phylogenetic information within the ingroup, this distribution can be most parsimoniously interpreted as a single gain of state 1 in a subset of the ingroup from the primitive condition (state 0). Any alternative hypothesis requires at least two steps (evolutionary transformations) rather than one. For instance, if state 1 is hypothesized to be primitive, the outgroup and a subset of the ingroup both have to (convergently) acquire state 0.

Note that in some cases, the outgroup criterion can fail, for instance, if the ingroup has ancestrally acquired state 1 (at its basal node), with part of it reverting to state 0. This can be detected if enough characters are correctly polarized and if the phylogenetic analysis recovers the correct tree, but this can be established only after the analysis has been completed.

Another situation that can lead to failure of the outgroup criterion is if the outgroup acquires a derived condition convergently with part of the ingroup. To prevent this problem,

multiple outgroups can be used, typically successive sister-groups of the ingroup. For instance, in a study of extant tetrapods, instead of only using *Dipnoi* (lungfishes) as the outgroup, we could add *Actinistia* (coelacanths) and, eventually, *Actinopterygii*. Closer extinct outgroups could also be used, but I mentioned better known, extant taxa for the sake of simplicity, and if the analysis emphasized characters that do not fossilize, such as molecular sequences or behavior, long-extinct outgroups would not be suitable. This lowers the probability that all outgroups convergently acquire a derived condition that also occurs in part of the outgroup. However, using multiple outgroups also raises other problems. First, if they display different states, which one will be taken as primitive? Perhaps the state present in the majority of outgroups, but if an even number of outgroups is present, a tie could be encountered. And if the character is not binary, more than two states could occur in the outgroups. To accurately tackle a variety of situations, an algorithm was presented by Maddison et al. (1984).

Outgroups need to be fairly closely related to the ingroup to adequately polarize many characters. To take the tetrapod example given, suppose that one of the characters used to study tetrapod phylogeny is the number of digits in the hand, with the following states: 0, five digits (present in many amniotes, including turtles, crocodilians, many squamates and mammals); 1, four digits (present in anurans and urodeles); 2, one to three digits (present in some taxa with limb reduction, notably among some urodeles); and 3, no digits (present in limbless taxa, such as gymnophionans and snakes). Given that the extant sister-groups of *Tetrapoda* possesses fins and hence lack digits (which may be neomorphs; Laurin 2010a), they do not allow polarizing this character, or they would lead to the wrong conclusion that the absence of digits is primitive for tetrapods. In this case, only extinct outgroups would be useful. Using *Seymouriamorpha* (for instance) as the outgroup would show that five digits were present ancestrally in *Tetrapoda* (defined as a crown group; that is, the smallest clade that includes all extant tetrapods).

The ontogenetic criterion uses the various states present in ontogeny to provide additional data to polarize characters. Its earliest uses rested on a parallel with the now-discredited law of recapitulation (Stevens 1980), which stipulated that if two or more states occur successively in ontogeny, the earliest-occurring one is probably primitive. For instance, the ontogeny of gymnophionans (apodan amphibians) documents the fusion of maxilla and palatine into a compound maxillopalatine (Wake and Hanken 1982). The ontogenetic criterion suggests that the retention of two separate bones is primitive compared to their fusion in extant gymnophionans (but of course, outgroup comparison yields the same conclusion without the need to have extensive growth series).

However, two kinds of exceptions hamper widespread application of the ontogenetic order in which states appear as a polarization criterion. The first is paedomorphism, the retention of a larval or immature phenotype in the adult, as shown (Vance 2017) by the axolotl, *Ambystoma mexicanum*, an endangered urodele that retains external gills and an aquatic lifestyle in the adult. Use of the ontogenetic order criterion for the presence/absence of gills in the adult of urodeles would be misleading because the presence of gilled larvae in most uro-deles (exceptions are direct-developing urodeles) would lead to the wrong conclusion that the presence of external gills is primitive for adults, and absence, derived. In this case, out-group comparison and the fossil record show that the adult ancestral urodeles probably never possessed external gills (Laurin 2010a). On the contrary, paedomorphosis allowed some urodele taxa to become fully aquatic at the adult stage, often through retention of external gills, and this appears to have happened repeatedly among urodeles, probably as early as in the Jurassic (Evans et al. 1988; Buffrénil et al. 2015) and as late as in the Pleistocene (Shaffer and McKnight 1996).

The second kind of exceptions that raise doubts about the usefulness of the ontogenetic order criterion to polarize charac-ters is modification of early developmental stages through time, such as larval specialization. For instance, the anuran tadpole possesses various specialized structures (such as the soft oper-culum that covers the external gills and a cement gland that allows it to become attached to various large objects) that were probably never present in the adults of anuran ancestors. Larval development is probably primitive for lissamphibians, but the ancestral larva probably looked more like a urodele larva (with free external gills) than like tadpoles (Laurin 2010a; Pough et al. 2018). For these various reasons, Stevens (1980: 342) con-cluded that this criterion was not very helpful.

More recent uses of the ontogenetic criterion ignore ontoge-netic order; they only use the distribution of states in taxa to determine the most general condition, which is inferred to be primitive (de Queiroz 1985; Bryant 2001). Consider an exam-ple of five taxa (two outgroups and three ingroup taxa) and two states, at two developmental stages, X (early stage) and Y (later stage), and a character that displays two successive states (A and B) in a subset of the taxa (Table 3.1).

In this case, both ingroup and outgroup display both states in ontogenetic stage X. Thus, applying the outgroup criterion to this stage would not allow polarizing the character. Using the ontogenetic generality criterion, we would conclude that state A is primitive because all taxa display state A at some stage of their ontogeny, whereas state B occurs only in three taxa.

This example also highlights the important difference between the two interpretations of the ontogenetic criterion. Given that state B occurs before state A in ontogeny in the taxa that display it, the ontogenetic order criterion would have led to the conclusion that state B is primitive.

The stratigraphic criterion states that the geologically oldest of the states is probably primitive. This reflects to an extent the belief (evoked above) of some botanists (Stevens 1980: 342),

microbiologists and zoologists (Sneath and Sokal 1962: 856) that fossils were so important that phylogeny could not be done without them. It is intuitively obvious that this criterion must be correct most of the times, but it is useful only if taxa of different geological ages are compared, or if the character history is docu-mented in the fossil record, even if the relevant extinct taxa are not directly included in the analysis. In the tetrapod example, recall that the dipnoan and actinistian outgroup does not allow polar-ization of the character about digit number. However, the strati-graphic criterion would show that among Permo-Carboniferous taxa, the hand included four or five digits, depending on the taxa, and that Devonian taxa included more than five digits (a condi-tion called polydactyly). The stratigraphic criterion would then suggest a morphocline (gradational variation) and a trend (direc-tional change) of decreasing number of digits; hence the tetrapod crown-group would be correctly inferred to have had five digits ancestrally, and state 0 is primitive (Laurin 1998, 2010a). The reliability of this criterion depends on the completeness of the fossil record and on our ability to incorporate the fossils into a phylogeny. Thus, the fossil record of organisms with a complex skeleton that fossilizes well (as for vertebrates, arthropods and echinoderms, among others) is most useful, but it has also proven useful for gasteropod and bivalve mollusks (Vermeij 1999).

Early studies (before the age of computer-assisted phy-logenetics) thus relied on a priori polarization to manually compute most parsimonious trees. With the advent of phy-logenetic analysis software in the 1970s and 1980s (Luckow and Pimentel 1985; Platnick 1987; Felsenstein 1993; Swofford 2002; Goloboff et al. 2008), characters no longer need to be polarized a priori (although some early software allowed or required this preliminary step). This is fortunate because, as previously shown, polarity cannot be known with certainty. Instead, the outgroup (or multiple outgroups) is used to root the tree. Character polarity can then be assessed a posteriori by reconstructing the history of a character on a tree, a procedure known as **character optimization**.

### 3.1.2.2 Character Optimization

Exact algorithms for character optimization have been imple-mented in all recent phylogenetic analysis programs (e.g., Swofford 2002; Goloboff et al. 2008). In parsimony, these algorithms simply ensure that given a topology (tree shape, disregarding branch lengths, which express either the evo-lutionary time or the amount of change on each branch), the number of character transformations required to produce the observed character state distribution (of terminal taxa) is mini-mized. These algorithms allow assigning ancestral states to nodes of the tree, although, in some cases, multiple solutions exist; this phenomenon is called "ambiguous optimization" (Swofford and Maddison 1987).

### 3.1.2.3 To Order or Not to Order, That Is the Question

If more than two states exist, they may need to be ordered; this consists of determining a sequence or a more complex pattern according to which transitions are thought to have occurred. The simplest case is a linear sequence, like 0–1–2, which

**TABLE 3.1**

Ontogenetic polarization criterion using only state distribution (not ontogenetic order). Stage X precedes Y in ontogeny. Taxa 1 and 2 are outgroups; taxa 3 to 5 form the ingroup. Modified from Bryant (2001).

| | Taxa | | | | |
|---|---|---|---|---|---|
| Ontogenetic stage | 1 | 2 | 3 | 4 | 5 |
| X | A | B | A | B | B |
| Y | A | A | A | A | A |

means that transitions between states 0 and 2 necessarily go through the intermediate state 1. Alternatively, if there are no good grounds to infer such a sequence or pattern, states can be left unordered, and all transitions between two different states are possible without intermediate steps. This is a key issue in phylogenetic analysis because each transformation "costs" a step, and the total number of steps is minimized in cladistics (parsimony). Thus, under the simple ordering sequence 0–1–2, a transition from 0 to 2 costs two steps, whereas under unordered states, it costs a single step. This explains why, for a given matrix, the optimal (most parsimonious) trees can differ, depending on the ordering hypotheses made.

There has been a long debate about whether or not, and how, character states should be ordered (e.g., Mickevich 1982; Hauser and Presch 1991). In most parsimony-based phylogenetic studies published in the 1970s and 1980s, multi-state characters were generally ordered (Slowinski 1993), but later, the most common approach became not to order any character. It is neither possible nor desirable to give an exhaustive history of this question, but a few key points should be useful. First, many of the discrete character states used in phylogenetic analyses represent discretization of inherently continuous variation. This is the case of all size and shape characters, for instance (but not of molecular sequence data, which are genuinely discrete). States like "small," "mid-sized" and "large" need to be defined quantitatively to be meaningful. Let's suppose that in a given taxonomic sample, size varies between 1 kg and 1,000 kg. We might score as "small" taxa of less than 50 kg, "mid-sized" those of 50 kg–200 kg and "large" those weighting more than 200 kg. Note that these boundaries are arbitrary (perhaps 100 kg and 300 kg, or 10 kg and 150 kg might have been more appropriate), and that discretization results in information loss. Such characters were initially discretized because many popular phylogenetic analysis computer programs were unable to use quantitative information, and because such precise information was not always available. However, software is no longer a constraint because TNT (tree analysis using new technology) can use such quantitative phenotypic characters (Goloboff et al. 2006).

It should be intuitively obvious that discrete states of continuous traits should be ordered because the logic that allows such discretization is precisely the same as the one that dictates the need for ordering (Wiens 2001). Namely, in both cases, systematists need to assume that taxa that display similar but not identical character values are more closely related to each other than taxa that display very different character values (near the opposite ends of the observed range); this simply assumes gradual evolution rather than evolution through large leaps. Simulations of quantitative characters on known phylogenies, followed by their discretization and parsimony analysis show, unsurprisingly, that by ordering them, better results are obtained; more correct clades and fewer incorrect are recovered (Rineau et al. 2015). This remains true even in the presence of a few, minor ordering errors (Rineau et al. 2018).

### 3.1.2.4 Clade Support Measures

Another important advancement in phylogenetics was the development of clade support indices. Two main types will be presented here, in the context of parsimony: the Bremer (decay) index and bootstrap. The decay index was developed by Bremer (1988), hence its first name. It simply consists of measuring the number of extra steps required to find trees in which the clade of interest is absent. Bremer (1988) did this by looking at strict consensus of trees of increasing length, but, subsequently, scripts were developed to perform this task more efficiently (Eriksson 1998; Müller 2004). An advantage of the Bremer index is that its significance is intuitively obvious because it is expressed in terms of number of steps. Its main flaw is that it does not inform on character conflict. Thus, a clade supported by five uncontradicted synapomorphies has a decay index of 5, but another clade supported by ten synapomorphies contradicted by five synapomorphies that support an incompatible clade also has a decay index of 5. Yet, it could be argued that the first of these two clades is better supported because no character conflict suggests an alternative topology.

Contrary to the decay index, bootstrap addresses character conflict in assessing clade support. This support value, developed by Felsenstein (1985), consists in resampling with replacement n times the characters of a matrix of n characters to produce new matrices. Thus, each new matrix contains as many characters as the original matrix, but some of the characters have been omitted, many occur once and a decreasing number occurs twice, three times or more. Each such matrix obtained by resampling is analyzed to obtain the most parsimonious trees. This procedure (obtaining a resampled matrix and analyzing it) is called a replicate. A large number of replicates needs to be performed; early analyses relied on 20 or more because of computing time, but now it is customary to perform 100, 1,000 or even more bootstrap replicates. The bootstrap frequency of a clade is the proportion (typically expressed as a percentage) of most parsimonious trees in which this clade occurs. Often, clades with a bootstrap frequency greater than 80% or 95% are considered robust, but this is only a rule of thumb; bootstrap frequencies are not probabilities that the clade really exists. To return to the previous example, a clade supported by five uncontradicted synapomorphies should have a greater bootstrap value than another clade supported by 10 synapomorphies but contradicted by five synapomorphies that favor an alternative topology, simply because some of the bootstrap trees could recover the alternative topology.

### 3.1.2.5 The Slow Rise of Cladistics

The adoption of cladistics by the systematic community was slow. The first version of Hennig's (1950) book on this topic was largely ignored, perhaps because it was written in a rather difficult German, and, even after the English translation was published (Hennig 1966), the number of studies using this new method rose slowly; only by the mid-1980s had cladistics established a firm foothold in systematics (**Figure 3.1**). For a vivid and humoristic account of this phase of the history of systematics, see Felsenstein (2001).

### 3.1.3 Maximum Likelihood

Maximum likelihood phylogenetic inference developed simultaneously with cladistics and phenetics, probably because all three methods benefited from progress in computer science.

Maximum likelihood can be defined as the goodness of fit of a model to a dataset. In phylogenetics, likelihood is estimated using evolutionary models. There is a large body of literature on this topic, especially about evolutionary models for molecular data; thus, only the basics need to be covered here. We will take the example of phenotypic data because this will facilitate comparisons with cladistics and phenetics. For phenotypic characters, the simplest model (with as single parameter) estimates the transition rate between two or more states. Such a model (Lewis 2001) is symmetrical if the transition rate from 0 to 1 is constrained to be equal to reversions from 1 to 0 (to take the example of a binary character, but this holds for any number of states ordered linearly); this is the simplest, one-parameter Markov model (MK1 from here on). A slightly more complex model would include separate rates for transitions from 0 to 1 (which would also be the transition rate from states 1 to 2, and so on, if more than two states exist), and from 1 to 0 (that rate would also apply to transitions from state 2 to 1, if three or more states were present); hence, it requires estimating two parameters and is often called the MK2 model.

These very simple considerations highlight a very basic difference between likelihood and parsimony in phylogenetic inference. While parsimony simply minimizes the number of changes required to explain the taxonomic distribution of character-states of taxa on a given tree, likelihood maximizes the fit of a given model on a tree (topology and branch lengths) to explain the data (observations, in terms of character-state distribution on taxa). Thus, in maximum likelihood phylogenetic inference, branch lengths are taken into consideration, contrary to parsimony. As a simple, intuitive example (which may not be the most frequent situation), suppose that a data matrix includes extant and extinct taxa. For a given tree, the character-state present in an old, extinct terminal taxon will have greater weight than that of its extant sister-group to determine the state of their last common ancestor (**Figure 3.2B**). Another practical difference from parsimony is that under maximum likelihood, characters that appear on a single terminal branch in a dataset (autapomorphies of a single terminal taxon) are phylogenetically informative because such autapomorphies are more likely to appear on a long branch than on a short one. Such characters thus play a role in assessing branch lengths and, through this, topology can also be affected, for instance if a tree is constrained to be ultrametric (as is appropriate in a tree of extant taxa, if branch lengths reflect evolutionary time), or in a tree that incorporates extinct taxa if their geological age is used to place the terminal taxa at the correct relative height.

Maximum likelihood phylogenetic inference is much more computer-intensive than parsimony-based phylogenetic inference (cladistics) because not only do the parameters of the model need to be estimated (or even the kind of model, when applicable, as for molecular data for which many models exist), but the topology and branch lengths need to be estimated as well. Note that these branch lengths represent amount of change in this context, not evolutionary time, and that given the heterogeneity in such rates between taxa and through time, estimating these lengths is no simple task!

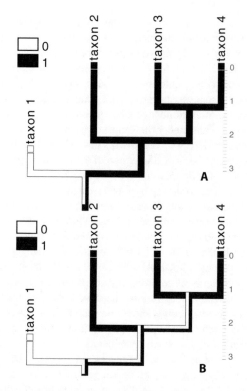

**FIGURE 3.2** Comparison between parsimony (cladistic) analysis and maximum likelihood. Simple example of a four-taxon case with one extinct (taxon 1) and three extant taxa (2–4). In this case, the fossil is a fairly old representative of the clade given that it is separated from the last common ancestor (LCU from here on) of all four taxa by a branch measuring only 0.5 units (which could be millions of years, or multiples thereof, for instance), and that LCU (represented by the basal node) is separated from extant taxa by 3 units. (A) Parsimony analysis yields ambiguous optimization; the last common ancestor could have possessed states 0 or 1 (both alternatives are equally parsimonious). (B) Maximum likelihood (here, using a symmetrical Markov model, abbreviated MK1 from here on) also gives uncertain state assignments for the LCU, but both alternatives are not equally probable. State 0 is assigned a probability of 0.7615, whereas state 1 has a modeled probability of 0.2385. This is because taxon 1 is old, hence linked to the LCU by a short branch (0.5 units), along which probability of change is lower than in the other branch (one unit long) leading to taxa 2–4. The model also considers the lower probability of parallel changes from 0 to 1 on the branches leading to taxon 2 and that leading to the LCU of taxa 3 and 4.

### 3.1.4 Bayesian Phylogenetics

Bayesian phylogenetics shares many similarities with maximum likelihood-based phylogenetics to the extent that both are model-based methods. However, whereas maximum likelihood is used to estimate the best evolutionary model, the best values of the evolutionary model parameters, and the best topology and set of branch lengths to yield a single tree, Bayesian phylogenetics estimates support for various values of model parameters and uses all possible parameter values (weighted by their support, expressed as posterior probability) to produce a population of trees differing in both topology and branch lengths. This method is thus ideally suitable to produce credibility intervals of various attributes of trees, such as node height (distance from the root, in evolutionary time or amount of change), which can be translated into divergence ages through molecular dating. A drawback is that this method is even more computationally intensive than maximum likelihood. This is due to the fact that topology and branch lengths are evaluated over a great number of parameter values, and that the evolutionary model parameters, topologies and branch lengths need to be estimated in millions of samples, typically performed through Markov chain Monte Carlo (MCMC) methods (Buckley et al. 2002). Nevertheless, the great progress in computer science has now made Bayesian phylogenetics practicable on a wide range of computers and datasets, and this method has risen sharply in popularity since the 2000s (**Figure 3.1**).

Model-based phylogenetics (through maximum likelihood and Bayesian methods) is now well-established in molecular systematics, but its use for phenotypic characters is more contentious. Adoption of these methods to analyze phenotypic data has been hampered by the paucity of evolutionary models implemented in phylogenetic analysis software. So far, only the Markov model for discrete characters (binary, linearly ordered or unordered) is widely available (Lewis 2001), whereas many evolutionary models are available for molecular characters (Jayaswal et al. 2007; Su et al. 2014; Ayres et al. 2019), and methods to assess support for alternative molecular evolution models have been developed (e.g., Posada 2008).

Yet, phenotypic characters are more complex than molecular characters, given that many phenotypic features are influenced by several genes and that the details of this influence remain to be determined in many cases. It is thus not entirely surprising that Goloboff et al. (2019) found that most phenotypic datasets that have been used by systematists (among those that were sampled in their study) do not support the hypothesis of a "common mechanism." Such a "common mechanism" is assumed by most evolutionary models, and this implies that a single set of branch lengths that determines the "probability of change for all characters increases or decreases at the same tree branches by the same exponential factor" (Goloboff et al. 2019: 494). Indeed, morphologists have long known that many phenotypic characters do not behave in this way. For instance, Farris (1983: 15) suggested that the number of teeth varies slowly in mammals, but fast in actinopterygians and, more recently, Ascarrunz et al. (2016) demonstrated through model-based analyses, that even within amphibians, several highly significant shifts occurred in the evolutionary rates of number of presacral vertebrae and, to a lesser extent, in body proportions (ratio between the length of the presacral region of the vertebral column and the maximum width of the skull). An even more spectacular example is provided by the appendix of the caecum, which never appeared in Laurasiatheria (a large clade of placental mammals), whereas 36 transitions (mostly gains) occur in its sister-group, the Eurachontoglires; the probability of such a pattern arising randomly is inferior to $10^{-5}$ (Smith et al. 2017: 48). This non-conformity of phenotypic characters to a "common mechanism" is probably the greatest challenge facing model-based phylogenetics for phenotypic characters, and this problem will need to be solved if we wish to use such methods in paleontology, given that except for the most recent fossils (mid- to late Pleistocene), only phenotypic characters are available. This may also explain why Goloboff

et al. (2018) reported that parsimony outperformed model-based phylogenetics on most such datasets.

Another feature of Bayesian phylogenetics that may not be optimal for phenotypic characters is its emphasis on characters that are scored in most taxa of a matrix. King (2019) reported that this led Bayesian analyses to converge on trees that reflect a strong influence of two linked characters (an endoskeletal and a dermal cranial joint) that were scored in most taxa in his data matrix on early osteichthyans, whereas other characters that he thought were important, such as the presence of tooth enamel and eyestalk attachment, had little influence on the Bayesian trees because they were scored in fewer taxa. Thus, while it seems likely that the popularity of Bayesian phylogenetics will continue to grow, even among paleontologists and other systematists who use phenotypic characters, future growth of this method should entail the development of evolutionary models more appropriate for phenotypic characters.

### 3.1.5 Timetree Construction

#### 3.1.5.1 Timetrees, Branch Lengths and Variable Rates

So far, we have only tackled the problem of inferring tree topology, even though model-based phylogenetics also requires estimating branch lengths, which are initially assessed in terms of amount of evolutionary change (number of evolutionary transformations, in phenotypic characters or nucleotides, typically). However, each character evolves at its own speed. For instance, for molecular data, it is well known that each gene evolves at its own pace, and even within a given gene, introns may evolve faster than exons, and even codon position matters; the third nucleotide of each triplet tends to evolve faster than the first two. Needless to say, phenotypic characters also evolve at highly variable rates, and these may be poorly correlated with rates of molecular evolution. Thus, the most generally useful branch length is evolutionary time, which is useful to reconstruct history, if absolute time rather than a relative chronology within a small clade is of interest.

#### 3.1.5.2 Early Paleontological Timetrees

Paleontologists have been time-calibrating their trees in a crude way since the middle of the 19th century, a time at which the French paleontologist Albert Gaudry (1866) and the German evolutionary morphologist Ernst Haeckel (1866: plate 7) published some trees in which various extinct and extant vertebrate taxa were placed in a tree and in the stratigraphy. Gaudry's (1866) trees (**Figure 3.3**) showed placental mammals, mostly from the Miocene locality of Pikermi (Greece). Haeckel's (1866) tree included all vertebrates placed in a stratigraphy that ranged from the Silurian to the present. According to Tassy (2011), both scientists were influenced by Darwin's (1859) *Origin of Species*, which featured a tree with a theoretical timescale, but Haeckel's work is more often remembered than Gaudry's in this context. For instance, Avise (2008) stated that the metaphor of the phylogenetic tree had been popularized by Haeckel; Gaudry is not mentioned in that paper.

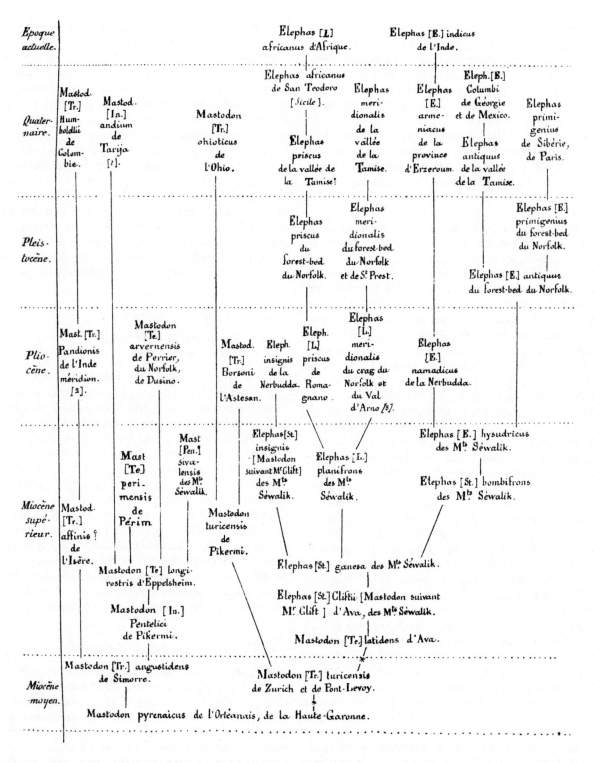

**FIGURE 3.3**  Proboscidian tree from Gaudry (1866: 38), one of the first true evolutionary trees ever published. In this monograph, Gaudry published several such timetrees of various mammalian clades from the Miocene locality of Pikermi (Greece). This scan was kindly provided by P. Tassy.

Fossils thus played a critical role in producing the first time-trees, but early paleontologists also used the fossil record to search for ancestors. Indeed, the temporal calibration of early paleontological timetrees depended directly on the identification of actual ancestors to identify divergence times with any degree of precision. In their absence, all that could be inferred

is that the last common ancestor lived before both of the taxa to which it gave rise. All this changed with the advent of cladistics, which shifted the focus to determining sister-group relationships. Worse, Hennig (1966) argued quite convincingly that it was impossible to be sure to have identified the correct ancestor. This is because the identification of ancestors relies both on positive evidence (the presumed ancestor must be geologically older than its descendants, and share apomorphies with them) and negative evidence (the absence of autapomorphies that would suggest that it is not the ancestor, but instead, a related but distinct evolutionary lineage). Reliance on negative evidence is always problematic, but this is worse in paleontology than in other fields. Indeed, all that could be claimed is that we have not found autapomorphies. In extant taxa, in this situation, if skeletal morphology has not yielded autapomorphies, additional searches can be conducted in other types of characters, such as soft anatomy, behavior, physiological attributes and molecular characters. But in paleontology, all we typically have is the skeleton and, often, only a small part of it. In this latter case, for instance, in a mammalian species known only from the teeth, claiming an ancestral status of a species because the dentition failed to yield autapomorphies seems difficult to justify! Then, how could divergence events be dated using the fossil record? This question is not entirely solved, even though much progress has been made in the last decade, but the immediate consequence has been that paleontologists focused on inferring the minimal (rather than most probable) age of taxa.

The recognition that the fossil record yields minimal, rather than actual, divergence dates has had a deep impact on molecular dating. The wide availability of minimal divergence ages and the paucity of maximal divergence dates based on the fossil record resulted in most molecular studies incorporating many minimal age constraints, but very few maximal age constraints or best point estimates of these divergences. For instance, Roelants et al. (2007) used 22 minimal age constraints but a single maximal age constraint to date the lissamphibian tree. This bias may explain part of the discrepancy between paleontological and molecular age estimates of clades (Marjanović and Laurin 2007), in addition to the simple fact that paleontologists have focused on minimal ages. The problem with such a highly skewed set of dating constraints can be illustrated by a simple analogy with curve-fitting: what would be the result of minimizing the square of the distance between the points and the curve only in one direction? The curve would of course be shifted to the level of the highest (or lowest) points.

### 3.1.5.3 Paleontological Dating

#### 3.1.5.3.1 First Quantitative Methods

The fossil record is notoriously incomplete, and the fossil recovery rate varies in time and space (Foote 2001; Lu et al. 2006; Benton et al. 2011; Smith and Barrett 2012; Warnock et al. 2020). These factors make it challenging to obtain precise estimates of divergence times from fossil evidence alone, given that a literal reading of the fossil record will always underestimate the true age of taxa (Marshall 2008; Laurin 2012; Heath et al. 2014; Didier and Laurin 2020). Fortunately, the combined effort of estimating maximal ages of clades based on their fossil record and the development of more sophisticated software to date trees have led to changes in practices in dating the Tree of Life using molecular or paleontological data. Strauss and Sadler (1989) pioneered the efforts to produce unbiased estimates (along with confidence intervals) of the age

of taxa based on the fossil record. Their method was applicable in a limited number of situations because it assumed a uniform fossil record through time and it was designed to apply to evolutionary lineages rather than clades, but it was developed in a series of subsequent studies to be applicable in an increasingly broad range of situations (Marshall 1994, 1997, 2008; Marjanović and Laurin 2008). See Laurin (2012) and Marshall (2019) for recent reviews of these methods.

#### 3.1.5.3.2 Tip Dating

More recently, great progress has been made in dating trees using the fossil record through the development of two methods: tip dating and the fossilized birth-death process (FBD from here on). In tip dating, fossils are not used to specify nodal ages a priori. They are integrated into a data matrix that may also (optionally) include extant taxa. When the latter are represented by molecular data, in addition to the morphological data required to assess the relationships between extant and extinct taxa, this is also called "total evidence dating," but tip dating can also be performed using only phenotypic data. The morphological data are used to estimate branch lengths and topology. These, along with the age of extinct taxa represented by fossils, allows dating nodes. This method assumes that the branch tip represents the age of the fossils that have been used to score the matrix. In extinct taxa with a long stratigraphic range, this may be slightly problematic because the best-preserved fossil that has been used to score the matrix may not be one of the geologically youngest of its lineage. However, the main (and most problematic) assumption of the method is that it assumes some kind of morphological clock. In other words, phenotypic characters (the only ones that can be scored for the vast majority of fossils given that all ancient DNA documented so far is less than 2 Ma old) are assumed to evolve according to a simple evolutionary model, typically the Markov model (Lewis 2001) in all branches and from the root of the tree to the tips. This does not appear to be realistic, as shown by the empirical study of Goloboff et al. (2019) mentioned previously. However, this method has the advantage that only the extinct taxa best represented in the fossil record need to be integrated; problematic taxa, or those known by fragmentary remains, can simply be left out. The studies in which this method was initially developed used it to date lissamphibian (Pyron 2011; **Figure 3.4**) and hymenopteran (Ronquist et al. 2012b) phylogeny.

#### 3.1.5.3.3 The Fossilized Birth-Death Process

The fossilized birth-death process (FBD) was originally developed by Stadler (2010) as a modification of the birth-death process that has been used to study taxonomic diversification from extant taxa alone. In the absence of fossil data, it was soon realized that the extinction rate could be severely underestimated, and that fluctuations through time in speciation (here equated with cladogenesis, meaning lineage splitting) and extinction rates could not be assessed properly (Höhna et al. 2011: 2577). The FBD can overcome these limitations by integrating the fossil record, and it is complementary with tip dating because it can also be used to date cladogeneses (Heath et al. 2014; Didier and Laurin 2020) and even extinction times (Didier and Laurin 2021). Like tip dating, it assumes a clock,

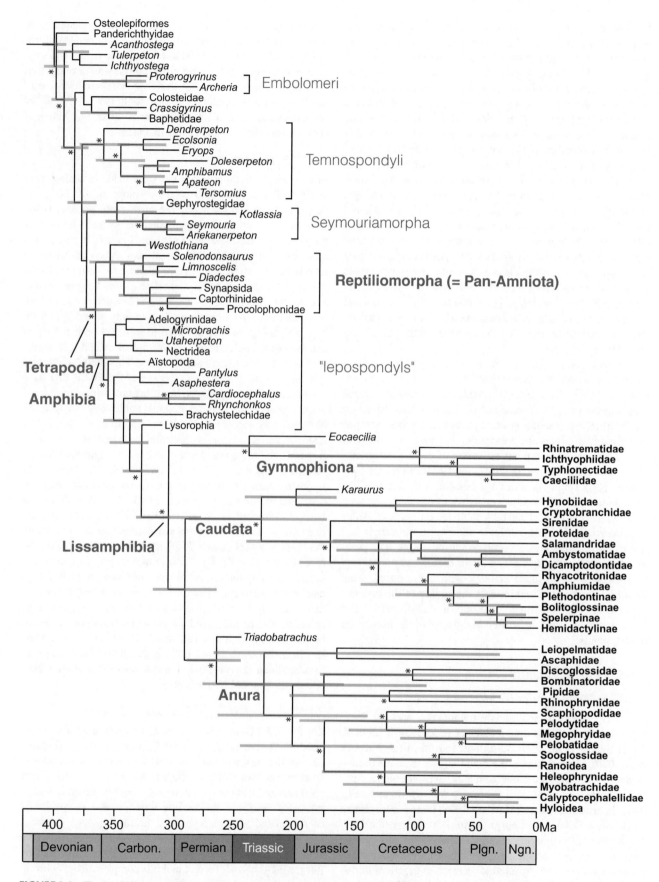

**FIGURE 3.4** Tip-dated phylogeny of lissamphibians and early stegocephalians initially published by Pyron (2011). Updated from Laurin (2012). The age of each extinct taxon represented in the analysis as a branch tip is considered to be known without error, but the age of each node is associated with a 95% credibility interval (gray bars).

but in the diversification (cladogenesis and extinction) and fossilization process, rather than in the evolution of phenotypic characters. To my knowledge, whether or not this is a realistic assumption in most cases has not yet been tested. Indeed, testing the clock-like nature of the cladogenetic process is difficult because divergence times are notoriously difficult to infer, methods to estimate extinction times have received less attention and the age of fossiliferous strata is seldom known with sufficient precision for a meaningful test. A corollary of the assumption of the FBD clock is that a representative sample of the studied clade must be included, from its origin to a given point in time (not necessarily the present), and all fossiliferous horizons of a given taxon (or a representative sample thereof) is required, rather than only the age of the fossils scored in the matrix (as in tip dating). This may seem limiting, but this representative sample may consist in fossils that are sufficiently well-preserved to be placed reliably in a phylogeny. Also, dating with FBD does not necessarily require incorporating the morphological data from the relevant fossils; instead, a tree (or set of trees) from a previous study can be used. Nevertheless, the FBD imposes more constraints (it may require greater efforts to compile datasets) than tip dating, in which the choice of extinct taxa to include can be purely arbitrary and a single geological age (of the fossils scored in the matrix) for the extinct terminal taxa needs to be determined. The FBD has been used in various implementations (reviewed in Gavryushkina and Zhang 2020), which differ in the way the fossil record is used, and either alone (such as Silvestro et al. 2014) or in the context of node dating to provide node age priors and to date various clades, such as bears (Heath et al. 2014) and tetraodontiform teleosts (Arcila et al. 2015).

Zhang et al. (2016) further developed the FBD by accounting for "diversified sampling," which is a rather well-established practice in many comparative studies. Random sampling, which is assumed by default by many statistical methods (including the FBD) consists in sampling species (tips of the tree) randomly, but many systematists try to include representatives of all higher-ranking taxa, and this "diversified sampling" is not random. Zhang et al. (2016) applied their method to a hymenopteran dataset (which includes data on extant and extinct wasps, bees and ants) initially described by Ronquist et al. (2012b). They showed that the initial radiation of extant Hymenoptera dated from close to the Permo/Triassic boundary (about 252 Ma) rather than the late Carboniferous (about 309 Ma), a rather sizeable difference (Zhang et al. 2016: table 5).

Stadler et al. (2018) added flexibility to the FBD by accounting for stratigraphic ranges of "species" (not conceptualized as in most systematic studies) and three speciation modes: budding (or asymmetric), bifurcating (or symmetric) and anagenetic (which involves phenotypic or genetic change, but not cladogenesis). This more complex speciation model reflects long-established ideas among systematists and paleontologists, but it is also applicable to neontological datasets of fast-evolving organisms. Stadler et al. (2018) provide the example of epidemiological data in which stratigraphic range is the duration of an infection of a patient, budding is transmission of the pathogen to another patient, death (or extinction) is patient recovery and sampling is sequencing of the genome of a pathogen in a patient.

A series of papers have provided various exact calculations for the FBD, especially aimed at better exploiting the fossil record; these methods are applicable to extant as well as entirely extinct clades (**Figure 3.5**). Thus, Didier et al. (2017) provided equations to compute the likelihood of a tree with fossils without requiring divergence time estimates, which are always poorly constrained. This improves estimates of the parameters of the FBD model. Didier and Laurin (2020) showed how exact distribution of divergence times could be computed from trees with fossil occurrence data. This method was used to date the evolutionary radiation of early amniotes based on a dataset that includes 109 terminal taxa. Didier and Laurin (2020) concluded that Amniota (defined as the smallest clade that includes extant mammals and reptiles) originated around 330 Ma. Given that the oldest fossil in that clade (*Hylonomus*) dates from 317–319 Ma, this suggests a gap in the fossil record covering the first 12 Ma of amniote evolution.

The fact that numerous molecular dating studies have assumed an age of 305 to 330 Ma for this divergence (338–288 Ma in San Mauro et al. 2005; 315 Ma for Hugall et al. 2007, at least 312.3 Ma for Irisarri et al. 2012 and 312–330 Ma for Shen et al. 2016) may seem almost reassuring because the difference between the constraints used and what now seems optimal is slight. However, for other events, the discrepancy between molecular and paleontological estimates can differ much more. For instance, three studies had dated the divergence between Bactrian camel and dromedary at 4 Ma (Wu et al. 2014; Heintzman et al. 2015) to 8 Ma (Cui et al. 2007), but the FBD suggests that event took place about 1 Ma, and the probability that it dates from more than 2 Ma is negligible (Geraads et al. 2020). This difference apparently results mostly from an improper interpretation of the fossil record that led to the use of erroneous paleontological calibration constraints. In this instance, use of the FBD proved critical to assess the timing of diversification of the camelids.

Didier and Laurin (2021) showed how extinction times could be computed. Using a dataset of 50 eupelycosaur taxa with 179 fossil occurrences, they showed that the extinction of the various lineages (nominal species) of ophiacodontids, edaphosaurids and sphenacodontids occurred gradually over a period of several millions of years, mostly in the late Cisuralian, which suggests a gradual decline in diversity rather than a mass extinction event near the Cisuralian/Guadalupian (early/middle Permian) boundary (about 272.3 Ma).

Many methods based on the FBD are still in early stages of development because they currently assume a single set of rates (of cladogenesis, extinction and fossilization events) per tree, but they could be developed further to allow variations in rates over time, just like molecular dating has become increasingly sophisticated over time by allowing variations in evolutionary rates over the tree.

### 3.1.5.3.4 Incorporating Uncertainties about Affinities of Extinct Taxa

Both tip dating and the FBD can (and typically do) incorporate phylogenetic uncertainty about the systematic position of fossils to the extent that such analyses are typically carried out on populations of trees (Pyron 2011; Ronquist et al. 2012b;

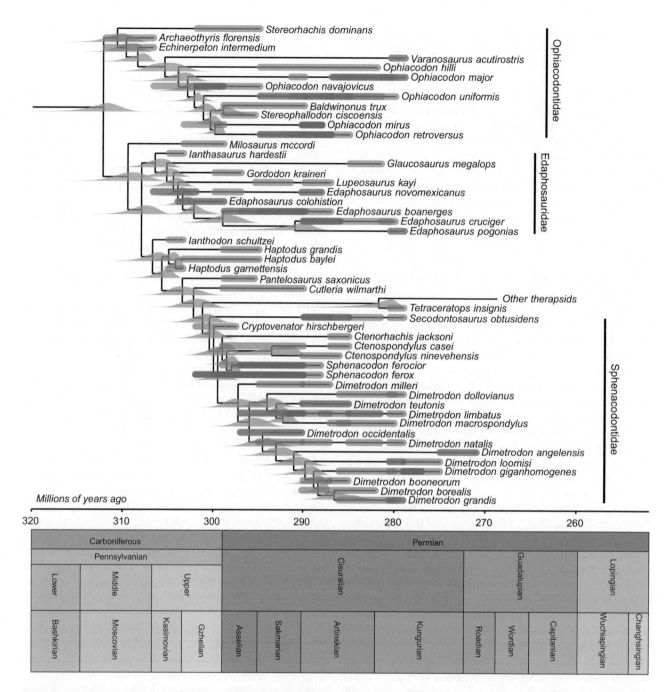

**FIGURE 3.5** Timetree of permo-carboniferous synapsids produced through the fossilized birth-death process (FBD). Slightly modified from Didier and Laurin (2021). Each brown (in the electronic version) bar represents the uncertainty associated with the age of each fossiliferous horizon. Darker brown bars represent the overlap of two or more fossiliferous horizons. The blue (electronic version) curves represent the probability densities of divergence times of each node. The taxon "other therapsids" includes mammals.

Didier and Laurin 2020). Typically, molecular node dating (see Section 3.1.5.4) has ignored this kind of uncertainty, or has considered it indirectly (and arguably inadequately) by setting "soft" minimal age constraints (which are linked to a statistical distribution that lets a certain proportion of sampled ages exceed the bounds of confidence intervals).

However, molecular node dating can take into consideration uncertainty about the systematic affinities of fossils. Sterli et al. (2013) showed how bootstrap analysis could be used in the context of a parsimony analysis (which is still the most

frequently used phylogenetic inference method in paleontology) to quantify the uncertainty about minimal age constraints of various clades to reflect uncertainty on the systematic position of fossils. This was performed on a dataset of extant and extinct turtles (**Figure 3.6**) to assess the minimal age of three **crown-clades**: *Testudines* (the turtle crown-clade), *Pleurodira* and *Cryptodira*.

Sterli et al. (2013) found that such uncertainty greatly exceeded the bounds of the 95% confidence intervals of the age of the same nodes estimated by previous molecular dating

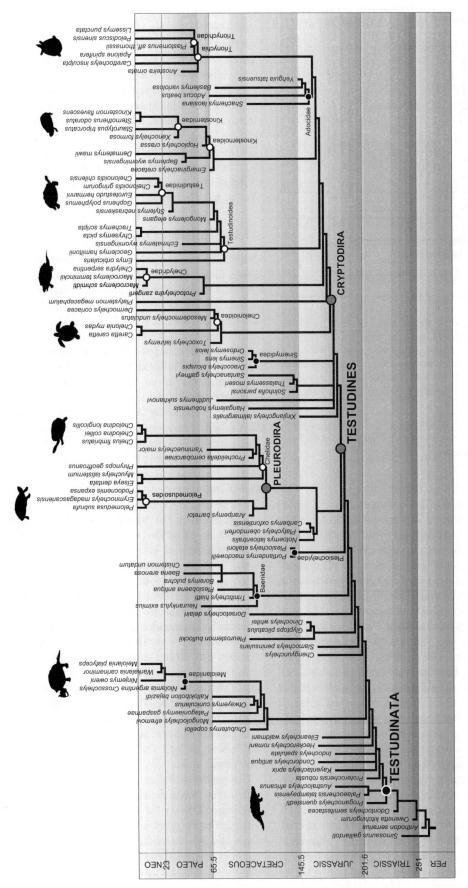

**FIGURE 3.6** Tree of Testudinata (turtles and tortoises) temporally calibrated with the fossil record. The age of three taxa (*Testudines*, *Pleurodira* and *Cryptodira*; nodes in red in the ebook version; grey in the print version) was investigated by looking into the effect of uncertainty in the systematic position of various extinct taxa through bootstrap. Ages are in million years. Abbreviations: NEO, Neogene; PALEO, Paleogene; PER, Permian.

*Source:* Modified from Sterli et al. (2013: figure 1).

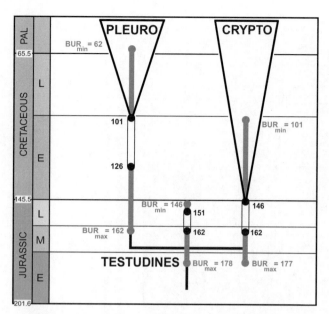

**FIGURE 3.7** Simplified tree (see Figure 3.6 for the whole tree) showing the uncertainty associated with the age of three clades (*Testudines, Pleurodira* and *Cryptodira*) reflecting the uncertainty of the age of fossils (in black) and the much greater uncertainty (BUR for Bootstrap Uncertainty Range, in grey) linked with uncertainty about the systematic position of extinct taxa. Abbreviations: E, Early; M, Middle; L, Late; PAL, Paleogene.

*Source:* Modified from Sterli et al. (2013: figure 2). Colors of geochronological units follow the Commission for the Geological Map of the World (CGMW), located in Paris (France).

studies of turtles (**Figure 3.7**). This was especially true for the minimal age of clades because in some bootstrap trees, the oldest fossils that have been attributed to various clades were located outside these clades and, in some cases, the next oldest fossil was younger by several (sometimes tens of) millions of years. This effect was particularly severe for *Pleurodira*; its minimal age is generally considered to be more than 100 Ma (for instance, Joyce et al. 2013; Marjanović 2021) because parsimony analyses suggest that early Cretaceous taxa like *Prochelidella cerrobarcinae* and *Araripemys barretoi* are included in this crown-group. Nevertheless, Sterli et al. (2013) found that in 4% of the bootstrap replicates, these taxa and even the Paleocene taxon *Yaminuechelys major* were excluded from this crown-group, and the oldest fossil attributed to that clade in these rare bootstrap replicates is barely 1 Ma (Pleistocene). Thus, dating constraints for this clade, as typically used in molecular node dating, should somehow incorporate a low probability (about 0.04) that this taxon may have appeared much more recently than previously assumed, possibly only 1 Ma. A similar phenomenon concerns the maximal estimate of the minimal appearance time because in 21% of the bootstrap replicates, several Late Jurassic taxa that are typically interpreted as stem *Pleurodira* (that is, *Notoemys laticentralis, Caribemys oxfordiensis* and *Platychelys oberndorferi*) are placed within the crown group, and this increases its estimated minimal age to 162 Ma (Callovian, Middle Jurassic). Thus, the estimate of the minimal age of *Pleurodira*

ranges from 1 to 162 Ma, and this interval does not tackle the maximal plausible age of this taxon, which is (as for nearly all taxa) much more difficult to assess. Sterli et al. (2013) showed that the confidence intervals published by previous molecular dating studies were typically too narrow because they were often much narrower than those that incorporate uncertainty in the systematic position of various extinct taxa (**Figure 3.8**). Conversely, some molecular estimates have surprisingly wide 95% credibility intervals of nodal ages. Thus, Dornburg et al. (2011) estimated that for *Testudines*, this interval extended into the late Carboniferous (**Figure 3.8**), even though the fossil record of this clade starts in the Late Jurassic and that several Triassic to Late Jurassic stem-trutles are known.

### 3.1.5.4 Molecular Dating

Despite the exciting recent developments in paleontological dating, most timetrees produced in the last four decades have used molecular (rather than paleontological) node dating. This method consists of scaling molecular branch lengths through calibration constraints typically derived from the fossil record or from geological events (Sauquet 2013; Magallón 2020).

Initially, molecular node dating required assuming a strict molecular clock (Zuckerkandl and Pauling 1965), which implies that the rate of sequence evolution is approximately constant over time and among lineages. This assumption was suspected from the start not to be realistic, but soon, more flexible methods aimed at relaxing this assumption were developed. These included, in approximate chronological order, quartet dating (Cooper and Penny 1997), which estimated one evolutionary rate per quartet (set of four terminal taxa) but could use a set of taxa to produce many quartets to estimate divergence times; penalized likelihood (Sanderson 2002), which allows rate variations across lineages but penalizes these changes, with a cross-validation procedure to select the right smoothing parameter; and uncorrelated methods (e.g., Drummond et al. 2006), in which rates are not correlated between adjacent branches. Instead, they are drawn from some distributions (for instance, lognormal or exponential). Furthermore, Ho et al. (2005) demonstrated that molecular evolution is much faster at short timescales (as assessed through population-level studies) than at longer timescales (studied through comparisons between various evolutionary lineages, often separated by millions of years of evolution), and this implies that temporal calibration of a molecular timetree must be done at the appropriate timescale. This list is not exhaustive and developments in molecular dating continue, among other things by the incorporation of Bayesian methods (e.g., Ronquist et al. 2012a, 2012b; Sauquet 2013; Guindon 2020) and studies on molecular evolutionary rates.

The molecular dating methods mentioned here are collectively called "node dating" because they require specifying a priori constraints on the age of some nodes, typically through geological events, such as the separation of continental plates, which may provide minimal divergence dates between taxa that have poor overseas dispersal abilities, or the fossil record, which may provide minimal divergence dates between clades that have an adequate fossil record. However, in both cases, the actual divergence date, or even a plausible maximal divergence date,

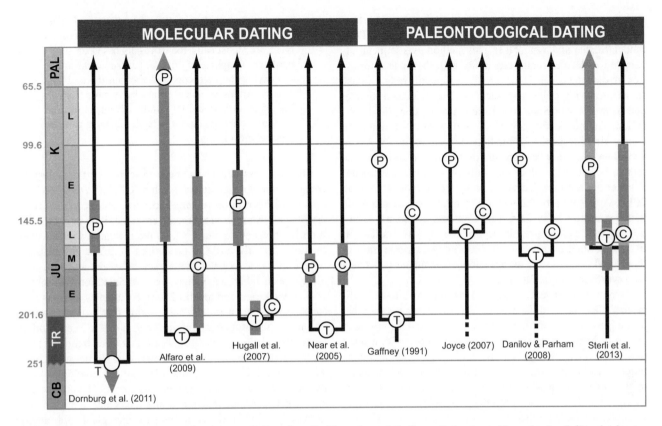

**FIGURE 3.8** Comparison of the age of three taxa (T: *Testudines*, P: *Pleurodira* and C: *Cryptodira*) estimated by molecular (left) and paleonto-logical (right) studies. Dark grey bars indicate, for molecular studies, the 95% confidence intervals of node ages. For this study (right), the dark grey bars indicate the Bootstrap Uncertainty Range; the light grey bars represent only the uncertainty linked with the age of fossils. Abbreviations: CB, Carboniferous; E, Early; K, Cretaceous; JU, Jurassic; L, Late; M, Middle; PAL, Paleogene; and TR, Triassic.

*Source:* Modified from Sterli et al. (2013: figure 3). Colors of geochronological units follow the Commission for the Geological Map of the World (CGMW), located in Paris (France).

are difficult to estimate. This is by far the greatest challenge to effective molecular dating because various distributions (flat, log-normal, truncated Cauchy and so on) are available in molecular dating software, and the impact on divergence date estimates of the choice of such distributions, as well as their exact parameter settings, has been demonstrated empirically (Warnock et al. 2012) along with the importance in setting appropriate bounds, espe-cially maximal ages (Marjanović and Laurin 2007). However, we are barely getting a first glimpse at the actual probability densi-ties of divergence dates suggested by the fossil record (Didier and Laurin 2020, 2021).

This is the weakness in molecular dating analyses because these dating constraints are always subject to various sources of uncertainty. Thus, not only is the date of separation of con-tinents difficult to date precisely, but this is a gradual process, and various taxa may cease to be able to disperse across such separating continents at different times. For instance, when South America separated from Africa in the Cretaceous (Arai 2014), taxa composed of flightless, salt-water intolerant organ-isms, like most lissamphibians, may have lost the ability to disperse between both continents fairly early, though possi-bly well after initial separation of these continents because dispersal on natural rafts (e.g., Egerton 2012: 50) may have remained possible. Taxa capable of aerial dispersal, such

as birds, bats and insects, may have continued to disperse between Africa and South America until much later. In addi-tion, the assumption that the geographic distribution of extant taxa reflects vicariance is not always justified. For instance, the distribution of extant paleognaths (the flightless ratites and the tinamu) suggests a Gondwanan distribution, in which case the timing of breakup of Gondwana could provide sepa-ration times between the main taxa. However, Phillips et al. (2010) questioned this assumption and concluded that capacity for flight may have been lost independently in various ratite lineages, which may have enabled overseas dispersal, at least when the separation between continents was still recent and distances were low. Indeed, late Miocene ostrich fossils are known from Eurasia (Boev et al. 2009), which can be rec-onciled with a Gondwanan distribution only if we postulate dispersal from Africa to Eurasia in the Cenozoic. Another spectacular counter-example is provided by the distribution of extant gymnophionans, which also suggests a Gondwanan distribution, yet the oldest stem-gymnophionan is from North America (Jenkins et al. 2007).

Tip dating, which we have briefly reviewed (Section 3.1.5.3.2), does not require making assumptions about the age of some ("calibrating") nodes. However, like node dating, it requires a correct assessment of the taxonomic affinities of extinct taxa. One molecular dating method that does not relies on fossils uses a birth–death-sequential-sampling (BDSS) model (Stadler and Yang 2013). This method uses sequentially sampled molecular sequences, along with their age information, to date the nodes. However, this is typically only availabe for fast-evolving organisms, such as viruses. Indeed, Stadler and Yang (2013) demonstrated the method using SIV/HIV-2 (simian immunodeficiency virus and human immunodeficiency virus, respectively; the latter is responsible for AIDS) gene data. A further development of the method allows ancestors to be sampled, which can improve parameter estimates if ancestors are included in the sample (Gavryushkina et al. 2014). Application of the method to the bear dataset (which includes ancient bear DNA used to assess systematic affinities, as well as stratigraphic occurrence of bear fossils) initially described by Heath et al. (2014) suggests that most of their fossil samples were the ancestors of extant or extinct taxa (Gavryushkina et al. 2014: 11).

Despite these caveats, molecular dating has allowed calibrating many parts of the Tree of Life, including for taxa that lack a fossil record, and it has arguably improved the precision of age estimates of some taxa, especially those with a scant fossil record (e.g., Hedges and Kumar 2009). This, of course, requires calibrating the tree using data from more distantly related taxa with an adequate fossil record, and this necessarily results in less precise or reliable estimates. In the extreme example of bacteria, other types of time constraints, like the appearance in the fossil record of some biomarkers; the rise in the level of oxygen, which can be inferred to have been triggered by cyanobacteria; or the maximal age of terrestrial life based on the age of the oldest continents (Battistuzzi et al. 2009) can be used. Nevertheless, the resulting approximate ages are very useful because bacterial morphology is overall fairly conservative, and the fossil record could thus provide little useful data to date the spectacular evolutionary radiations of bacteria.

Even for taxa with a rich fossil record, molecular dating has proven extremely useful. For instance, mammals have a rich fossil record that spans the Jurassic to the Recent (even though it is only rich starting in the Paleocene), but dating the early evolutionary radiation of placental mammals has proven challenging because the affinities of many Mesozoic and Paleocene mammals are contentious and the Mesozoic fossil record is relatively poor (Springer et al. 2019). Fortunately, progress has been made recently on this front (Halliday et al. 2017). Resolving the phylogeny of Cretaceous and Paleogene mammals is important because it is critical to determine if this evolutionary radiation started well before the end of the Cretaceous, as many molecular dating (Springer et al. 2017, 2019) and some paleontological studies (such as Archibald 1999) suggest, or at the base of the Paleocene, as most paleontologists maintain (such as Benton 1999; Davies et al. 2017; Halliday et al. 2017), sometimes forcefully. For instance, Alroy (1999: 116–117) stated:

> there are few ways to avoid this study's major conclusions: In terms of both taxonomic diversity and body mass distributions, the single most important radiation of mammals occurred not during the

Cretaceous, but during the early Paleocene. Far from being a "literal reading of the fossil record," these are statistically robust results based on standardized sampling regimes. They show the folly of denying the Paleocene radiation on the basis of loosely calibrated molecular clocks. A more fruitful line of inquiry would be to take advantage of this obviously important event by exploring its impact on molecular evolution.

However, this debate has been obscured by the failure in some paleontological studies (such as O'Leary et al. 2013) of distinguishing minimal divergence dates, which can readily be established from the fossil record from maximal or even most probable divergence dates, which require more sophisticated methods (like the FBD or tip dating) to estimate. Similarly, uncertainties associated with molecular dating have not always been taken adequately into consideration (Bromham et al. 1999; Graur and Martin 2004). Fortunately, there is also substantial progress in molecular dating that brings both sets of dates (paleontological and molecular) closer to each other (Phillips and Fruciano 2018). There is a clear trend for both sets of methods to increasingly converge on similar nodal ages when both kinds of data (molecular and paleontological) are used correctly with adequate methods. Recent blatant discrepancies between paleontological and molecular age estimates frequently result from erroneous paleontological calibrations (e.g., Geraads et al. 2020).

One of the latest additions to molecular dating (in the broad sense) is the use of lateral gene transfer (LGT) to obtain relative ages between branches (Davín et al. 2018). This method is based on the simple principle that the donor clade must have existed before the receiving clade, and this allows for establishing relative node ages even between distant parts of a tree. This provides additional information that can be combined with classical molecular timetrees calibrated with fossil calibrations to better constrain the age of various nodes. This is especially useful in parts of the tree with a poor fossil record (either because there are few fossils, or because they cannot be placed confidently in reasonably precise parts of the tree). Davín et al. (2018) demonstrated the advantages of this method with three trees that cover three such clades: *Cyanobacteria*, *Archaea* and *Fungi*. They proposed a relatively recent age (860 Ma) for the *Prochlorococcus–Synechococcus Cyanobacteria* clade. This method is especially appropriate to date this divergence event because *Prochlorococcus* experienced genome reduction and fast evolution, both of which create problems with classical molecular dating methods. Among *Archaea*, Davín et al. (2018) proposed an early diversification of methanogens, which started around 3 Ga, and is consistent with evidence of biogenic methane as early as 3.5 Ga. Davín et al. (2018: 906) also proposed that "The relative order of appearance of archaeal energy metabolisms corresponds to increasing energy yield, with methanogenesis evolving before sulfate reduction, and the oxidative metabolisms of Thaumarchaeota and Haloarchaea evolving most recently." Among *Fungi*, Davín et al. (2018) improved age estimates for many of the deepest nodes, such as *Zoopagomycota*, which appears to have differentiated around 712 Ma.

### 3.1.5.5 Timetrees Beyond Biological Nomenclature

Timetrees are useful in various branches of biology, and they are part of the alternatives that are required to replace rank-based nomenclature by more appropriate methods. A good example is conservation biology, which often requires biodiversity assessments of various regions to determine conservation priorities, such as which region to preserve when it is not possible or too costly to preserve two or more pristine areas. Traditionally, these assessments have used taxon counts at a given level (often supraspecific taxa). This raises various problems (because of lack of equivalence of taxa of a given rank; see Section 3.4.2), but what is the alternative? Faith (1992) developed the "phylogenetic diversity" index, which uses branch lengths (ideally scaled by evolutionary time, even though amount of change may also be used), which has been increasingly used in conservation biology, as shown by 3,859 Google Scholar citations of that paper (as of September 25, 2021). Thus, timetrees are an essential tool to make PN a viable alternative to rank-based nomenclature among other scientists who are not directly involved in biological nomenclature (or even in systematics), but who use nomenclatural and taxonomic information in their research.

Other uses of timetrees have more tenuous links with biological nomenclature but deserve to be mentioned here briefly for their relevance to other scientific fields. Indeed, timetrees are required to address many macroevolutionary topics, such as the dynamics and geological age of evolutionary radiations (e.g., Brocklehurst 2017; Ascarrunz et al. 2019; Didier and Laurin 2020) or past biological crises, also called mass extinction events (e.g., Ruta et al. 2011; Sidor et al. 2013; Brocklehurst 2018; Didier and Laurin 2021) and their putative causes (Arens and West 2008; Wignall et al. 2009; Tabor et al. 2020).

### 3.1.6 Molecular Evolution

So far, evolution has been discussed as if molecular characters reflected a simple history. However, the phenomenon of homoplasy, which has long been known among systematists, also affects molecular data through various evolutionary mechanisms that have been documented in the last few decades. One of them is gene duplication in a given lineage and subsequent loss in a subset of its daughter-lineages. Another is hybridization, including introgressive hybridization, which produces lineages that have genes derived from two contemporary ancestors. A third is incomplete lineage sorting, which occurs when an ancestral polymorphism persists across cladogenetic events; the subsequent fixation of a given allele may occur convergently, which yields a phylogenetic signal that is different from that of the lineages. To take a simple example (**Figure 3.9**), suppose that an ancestral lineage has two alleles (x and y) of a given gene and that allele x appeared before y. This lineage then splits into A and its sister-groupe that then splits into B and C, and the latter splits into D and E. Allele x is lost (y is fixed) in B. Lineage C retains both alleles and passes them into the basal part of lineages D and E, but D loses allele y whereas E loses allele x. By looking at these alleles, in the absence of external information, we would be misled into thinking that lineages B and E are sister-taxa because they

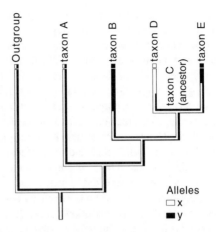

**FIGURE 3.9** Hypothetical example showing how incomplete lineage sorting can generate a misleading phylogenetic signal.

share the derived allele y, but this does not reflect the actual history of these lineages. These phenomena have been well-documented in East African cichlids (Salzburger 2018). This is why it has become customary, in molecular phylogenetics, to distinguish species trees from gene trees.

### 3.1.7 Relevant Progress in Evolutionary Biology

Homology is important for phylogenetics to the extent that homologous features must be compared to obtain a reliable phylogeny. This is intuitively most obvious with discrete phenotypic characters. For instance, the now-consensual hypothesis that tetrapods form a clade whereas "fishes" do not is supported (among others) by the conclusion that limbs and fins are homologous structures and that fins preceded limbs in vertebrate evolution. However, homology is also important to analyze molecular sequence data, as we will see in the following. The concept of homology now appears much more complex than the popular formulation by Owen in the 19th century. Several criteria have been proposed to assess it, and several kinds of homology have been recognized. However, this trend may have gone too far; Minelli and Fusco (2013: 290) found unfortunate that simple structural similarity, especially as seen in nucleotide or aminoacid sequences, are called "homology." The reasons for this are obvious, especially in the case of nucleotides, given that only four nucleotides (A, T, C and G) can be present in most positions. Similarly, in the context of morphometrics, calling "homologous" the reference points used to quantify shape through geometric morphometrics is unwarranted (Bookstein 1994).

Minelli and Fusco (2013) recognized three main concepts of homology that are still valid. One is historical homology, which can be conceived as a reinterpretation of Owen's "sameness" concept of homology in an evolutionary perspective. Various authors have proposed definitions of historical homology, including Mayr (1969: 85, cited in Minelli and Fusco 2013: 292), who defined it thus: "homologous features (or states of features) in two or more organisms are those that can be traced back to the same feature (or states) in the common ancestor of those organisms." From this historical perspective, similarity that appeared convergently is called "homoplasy."

A second set of current concepts of homology is called "proximal-cause concepts of homology." This considers homologous traits or structures that share the same developmental cause or generative mechanism. These causes can be genetic or epigenetic. The ancient (but still useful) notion of serial homology in body parts is a kind of proximal-cause homology because these serially repeated structures share a similar development. This includes several classical examples, such as the visceral arches, which include, in all primitively aquatic vertebrates, the gill arches and, in gnathostomes, the hyoid and mandibular arch. Other examples include the vertebrae and ribs, which are repeated in many body segments of many ganthostomes, but also body segments and appendages (including buccal parts) of arthropods, among others. The previous examples all feature linearly aligned structures, but the pattern may differ. For instance, echinoderms have radial symmetry and radially repeated body segments. Note that even within such homologous structures, development needs not be conserved completely, as the frequent cases of heterochrony attest (Minelli 2016: 49).

Note that both sets of homology concepts (historical and proximal-cause) do not define congruent sets of homologous traits. Thus, traits that are obviously homologous in the historical sense may be controlled by different genes or different ontogenetic processes. Such considerations led Minelli (1998) to propose a "combinatorial (or factorial) concept of homology," a third, more complex concept of homology. Minelli and Fusco (2013: 294) justified it thus:

> It has become clear that units (modules) ascribed to different levels of biological organization (such as genes, mechanisms of development, morphological structures) evolve to an extent largely independent from each other, sometimes providing conflicting pictures of homology relationships.

They provided two examples. One is the vertebrate alimentary canal, which must obviously be homologous in the historical sense of the term, but which, depending on the taxon, derives from at least three different precursors. Another example is provided by the songs of acridid grasshoppers, which appear to be homologous through the clade based on their taxonomic distribution. However, the mechanism used to produce sounds varies unexpectedly. Most species sing by rubbing the femur of their hindlegs against the forewings, but one species (*Calliptamus italicus*) sings by rubbing the mandibles against each other. Under this factorial concept, homology should be treated as relative or partial.

Homology also needs to be semantically circumscribed. Thus, bird and bat wings are homologous as forelimbs, but they are convergent (homoplastic) as wings, given that the last common ancestor of both taxa used its forelimb to walk, rather than to fly, and that limb lacked any adaptation to flight.

Homology is important in the context of phylogenetic analyses; in this context, two concepts of homology must be distinguished. Primary homology, based on structure, position and eventually development, is a hypothesis that must be made to score characters in taxa because only homologous traits should be compared; it must be used when scoring a phylogenetic data matrix. Secondary homology is a validation of historical homology obtained at the end of the analysis, as shown by a reconstruction of character history on a tree (or set of trees).

Molecular biology developed its own concepts and nomenclature of homology. Thus, a gene present in various taxa is "orthologous" if all these genes derive from an ancestral gene that was present in the last common ancestor of these taxa. However, genes can be duplicated, with two or more copies being present in a given taxon. These genes are "paralogous". *Hox* genes and collagen genes are present in multiple copies and thus include many paralogues. Finally, a gene can be acquired through interspecific (horizontal) transfer; this gives rise to "xenologous" genes. These distinctions are important because molecular phylogenies must be based on comparisons of orthologous genes. Comparing paralogous or xenologous genes would yield a mix of species trees and gene trees, which would not reflect the evolutionary history of the organisms.

So far, we have discussed homology of structures and molecules, but gene expression patterns and developmental stages can also be homologous (or not). These are not trivial questions. For instance, are larvae of various taxa homologous? Is an anuran tadpole homologous as a larva with a caterpillar (larval butterfly) and with an annelid trochophore? Even at a lower scale, homology is sometimes contentious. Thus, Hanken (1999) suggested that the homology between the anuran tadpole and the urodele larva is uncertain (but likely). And within anurans, the tadpole may have reappeared from a direct-developing ancestor as many as four times (Hanken 1999: 66), in which case not all anuran tadpoles are homologous to each other in the historical sense of homology. A similar phenomenon apparently occurred in plethodontid salamanders (Chippindale et al. 2004).

Evolutionary biology progressed on many other fronts after the 19th century, but most other developments, while interesting, are less relevant to phylogenetics and nomenclature and will not be developed here. An example is that organisms do not simply occupy an ecological niche; sometimes, they construct it. Beavers are a prime example because the dams that they build modify ecosystems by creating lakes, and many other taxa benefit from this (Pierotti 2020: 4). Other examples include termites, who build large nests; earthworms that increase porosity and aeration in the soil; and, in a distant past, cyanobacteria who, by releasing molecular oxygen ($O_2$), created major changes in the biosphere. All these changes, by modifying the environment, create new evolutionary pressures and shift phenotypic optima toward which various taxa evolve.

## 3.2 Evolution, Delimitation and Ranks

### 3.2.1 Ranks before and after Evolution

Rank-based nomenclature was developed in the context of a non-evolutionary worldview (de Queiroz and Gauthier 1992; Mishler 1999). Indeed, when Linnaeus wrote his main works that laid the foundation of rank-based nomenclature (RN hereafter) and were later selected to start nomenclatural priority by the rank-based codes, most biologists were fixists. However, this changed through the 19th century, and this created problems for rank-based nomenclature. We can read, in the code adopted by the American Ornithologists' Union (1886: 5) that:

> No one appears to have suspected, in 1842 [when the Strickland code was elaborated], that the Linnaean system was **not the permanent heritage of science**, or that in a few years a **theory of evolution was to sap its very foundations**, by radically changing men's conceptions of those things to which names were to be furnished.

Proponents of RN still have to fully integrate this fact. As de Queiroz and Gauthier (1992: 472) argued:

> The taxonomic system developed by Linnaeus, and formalized in the various codes of biological nomenclature, has governed taxonomic practices admirably for over 200 years. Indeed, it is a tribute to a taxonomic system based on non-evolutionary principles that it has persisted for well over 100 years into the era dominated by an evolutionary world view—an era in which taxonomy is purported to be evolutionary. But biological taxonomy must eventually outgrow the Linnaean system, for that system derives from an inappropriate theoretical context. Modern comparative biology requires a taxonomic system based on evolutionary principles.

Abandoning Linnaean ranks (absolute categories) was advocated by many other authors, not only among proponents of PN (such as Papavero et al. 2001: 5). The use of ranks creates two distinct sets of problems: the incoherence that arises when phylogeny changes and results in taxa of a given rank becoming nested in a taxon of an equal or lesser rank, and the lack of delimitation that creates nomenclatural instability even when the phylogeny does not change. Let's look first at the problems created by changes in our ideas about the phylogeny.

### 3.2.2 Rank Changes Required by New Trees

The rise of phylogenetics and of molecular systematics has had major taxonomic and hence, nomenclatural implications. This effect has been of uneven intensity depending on the taxa. A well-known and spectacular example has been the discovery (by paleontologists) that the class *Aves* is deeply nested within the dinosaurian suborder *Theropoda*, which contradicts the basic principles of rank-based nomenclature, if taxa are to be monophyletic (which is not a requirement of the *Zoological Code*). But such examples need not involve long-extinct taxa. The ancient class *Pisces* had to be dismantled because it is paraphyletic and includes *Tetrapoda* (to mention only extant taxa), which was typically unranked, but which includes classes *Amphibia, Reptilia, Aves* (itself in *Reptila*) and *Mammalia* (Janvier 1996; Laurin 2010a). Similarly, the mammalian order *Cetacea* (whales and dolphins) is now thought to be included in the order *Artiodactyla*, which resulted in the erection of a new taxon, *Cetartiodactyla*, for this large clade (Montgelard et al. 1997; Hassanin et al. 2012). There are also many botanical examples. For instance, all angiosperms used

to be divided into monocots and dicots, but it is now well-established that monocots are deeply nested within dicots (Cantino et al. 2007). Less well-known but more problematic examples are found in other taxa regulated by other codes. For instance, in a discussion of this problem among Fungi, Lücking (2019: 200) wrote:

> Given the considerable changes in our understanding of evolutionary relationships in Fungi including lichens and the fact that, in contrast to Metazoa, higher classification and nomenclature of the fungi up to the phylum level is governed by a strict and formal Code, the **integration of Linnean principles into molecular phylogenetic classifications has caused serious conflicts** in this kingdom. Traditionally delimited taxa were often found nested within each other, at different hierarchical levels, a problem already discussed by Hennig (1950) under the concept of apomorphic subgroups (Figure 3.1).

However, ranks also create problems in many other situations. This is explained in the preface of the *PhyloCode* (Cantino and de Queiroz 2020: xi):

> In a group in which the standard ranks are already in use, naming a newly discovered clade requires either the use of an **unconventional intermediate rank** (e.g., supersubfamily) [which have no official standing in the rank-based codes] or the shifting of less or more inclusive clades to lower or higher ranks, thus causing a **cascade of name changes**. This situation **discourages systematists from naming clades** until an entire classification is developed. In the meanwhile, well-supported clades are **left unnamed**, and taxonomy falls progressively farther behind knowledge of phylogeny. This is a particularly serious drawback at the present time, when advances in molecular and computational biology have led to a burst of new information about phylogeny, much of which is not being incorporated into taxonomy.

Thus, ranks create serious communication problems in biological nomenclature even in routine taxonomic works. This is shown, for instance, by the appendix entitled "Classification of the vertebrates" that appears toward the end of a vertebrate paleontology textbook (Benton 2005). It includes a class Agnatha within which the Infraphylum Gnathostomata occurs (first inversion), within which the class Osteichthyes includes the subclass Sarcopterygii that contains the infraclass Tetrapodomorpha and the superorder Osteolepidida that includes the superclass Tetrapoda (second inversion), an order Pelycosauria (which includes only the basalmost synapsids) that should include (if the phylogeny were translated correctly into the taxonomy) the order Therapsida and within the latter, a suborder Cynodontia within which the class Mammalia occurs (third inversion), a subclass Diapsida that includes a superorder Dinosauria and a suborder Theropoda, all of which include the class Aves (fourth inversion). This indented, ranked classification of vertebrates nicely shows the problems entailed by retaining rank assignments that were coined long ago, when we had very vague notions about the affinities of the taxa, in a 21st-century taxonomy that incorporates recent phylogenetic progress.

### 3.2.3  Absolute Ranks and Taxon Delimitation

Under RN, taxa are not delimited. The definition consists of a type, which is objective, and a rank, which is subjective. The only constraint provided by such definitions is that the type must be included in the taxon. Each systematist delimits the taxon as he wishes, with the consequence that closely related taxa can be synonymized or a taxon can be split, simply because of personal preferences. Strangely, not all systematists, and not even all experts on biological nomenclature, realize this. Thus, Dubois (2005a: 370) wrote:

> In zoology, the current taxonomic system allows to place all the known species unambiguously in a hierarchic arrangement of taxa (genus, family, order, class, etc.), and **the nomenclatural system allows us to designate unambiguously any of these taxa by a single nomen**, that should be agreed upon and used by all zoologists worldwide.

It is easy to demonstrate that the second part of this quote (in bold type) is far from correct; RN allows to unambiguously attribute the name of a given taxon if it is assigned a given rank, but given that ranks do not exist in nature, this assignment is subjective, and hence, name attribution is subjective too. This was acknowledged more recently by Dubois et al. (2021: 14):

> This is not to say that these ranks by themselves are biologically meaningful, which they generally are not, but that their application in amphibians (as in most groups) is based neither on a robustly defined historical tradition or on recent conventions, nor on a meaningful division of the taxonomic hierarchy to reflect the structure of the tree. Thus, the current amphibian taxonomy in many ways represents **the worst of all possible worlds**.

These problems occur at all levels, from infraspecific taxa to high-ranking supraspecific taxa. They can be illustrated with a simple hypothetical example (first proposed by Laurin 2008) with four species (named j, k, m and n), two genera (O and P) and a single family (Oidae) recognized in the context of rank-based nomenclature (**Figure 3.10A**). In this demonstration, we will assume that the phylogeny is correct, that it is accepted by all practicing systematists and that no new factual data about these taxa are discovered. Even in this situation, which should be ideal to promote nomenclatural stability in this small clade of four terminal taxa, rank-based nomenclature allows considerable instability.

Several nomenclatural changes reflecting subjective opinions are allowed by the rank-based codes. One possibility is synonymizing one of the two genera (**Figure 3.10B**). Another is to recognize additional genera among the four nominal species (**Figure 3.10C**). A third set of changes consists in recognizing additional families, each of which may have one or two genera (**Figure 3.10D**). Just with these few mechanisms, 12 additional nomenclatures can be recognized while maintaining monophyly of all taxa (which is not required by the *Zoological Code* and *Botanical Code*) and without changing species delimitation. This also means that the meaning of taxon names is unclear without additional information. For instance, Oidae could include, depending on nomenclatural preferences, species j–n (**Figure 3.10A**), only species j and k (**Figure 3.10D**, first nomenclature), or even only species O (**Figure 3.10D**, second nomenclature). The contents of genus O can vary to the same extent. Imagine the possibilities afforded by this nomenclatural system with over 1.5 million named species! Under PN, which will be presented in Chapter 4, no alternative is allowed; additional (redundant) taxa could be erected, but the names of all taxa initially defined would still apply to the same sets of organisms.

Before going further, the notion of redundancy should be explained. In RN, two taxa are redundant if they refer to the same set of organisms and if they have been assigned different ranks (taxa of the same ranks that refer to the same organisms are considered synonyms. In RN, redundant taxa are extremely common simply because of nomenclatural constraints. Thus, binominal nomenclature requires that every species be placed in a genus. If a new nominal species is a sister-group of a clade composed of two or more genera, a new redundant genus must be erected if we wish taxa to be monophyletic. The genus contains exactly the same organisms as the species; thus, the genus is redundant with the species. This problem can happen at higher nomenclatural levels too. If the new nominal species is a sister-group to a clade that includes at least two families, it is typically included into a new redundant family, and this can extend in theory to any level, like the phylum (**Figure 3.11**). Note that such redundancy problems (beyond the trivial case of monotypic genera) are as old as RN itself, as pointed out by Papavero et al. (2001: 26). For instance, in the first edition of Linnaeus' (1735) *Systema Naturae*, Amphibia (a class) included a single order (*Serpentia*). Class *Amphibia* was no longer redundant in the 10th edition of the *Systema Naturae*, in which Linnaeus (1758) removed most redundancy, although some rare redundant taxa remained, as for elephants, for which Linnaeus recognized only one species, *Elephas maximus*, which now designates only the Asian elephant.

### 3.2.4  Evolutionary Models That Would Justify Absolute Ranks

The problems with nomenclatural instability evoked here occur because rank allocation of taxa is subjective. No objective property of taxa allows unambiguous ranking,

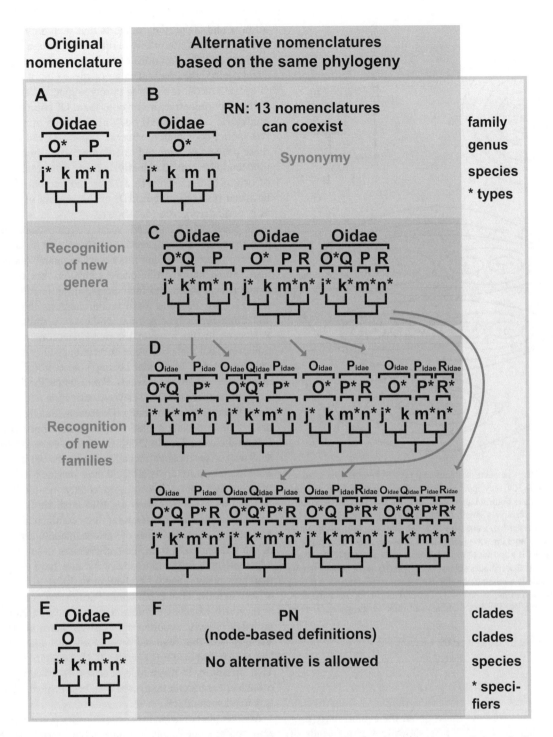

**FIGURE 3.10** Hypothetical example showing some of the nomenclatural instability allowed by rank-based nomenclature as implemented in the *Zoological Code* and *Botanical Code*, and assuming that systematists require monophyly of taxa (which is not required by these codes), that the phylogeny is stable and that species delimitation is not contested. Types are marked by an asterisk (*). The initially proposed nomenclature (A) can be modified by synonymizing one of the two initially recognized genera (B), and by erecting additional genera (C) and new families (D). Under phylogenetic nomenclature (see Chapter 4), no alternative is allowed (E, F); additional taxa could be erected, but the names of all taxa initially defined would still apply to the same sets of organisms.

*Source:* Modified from Laurin (2008: figure 1).

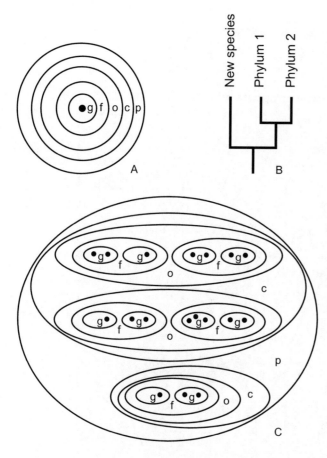

FIGURE. 3.11 Redundancy. Redundant taxa (A) contain the same set of organisms. In RN, they appear to be nested within each other, and they differ only in their nomenclatural rank. The black solid circle represents the species; successively larger, hollow circles represent a genus (g), a family (f), an order (o), a class (c) and a phylum (p). In this hypothetical example, a phylum was erected for a single species. This could be required by RN if a species (new or not) were discovered to be the sister-group of a clade that includes at least two phyla (B). Most nomenclatures under RN include a mixture of redundant and non-redudtant taxa (C). In this hypothetical example, one of the two classes (bottom) is redundant, as is one of the three orders (bottom) and three of the genera (one in each order).

*Source:* Modified from Laurin (2005: figures 2, 3).

and this reflects what we know about the evolutionary process. This can be explained by showing various kinds of evolutionary models that would allow such unambiguous, objective rank allocation (**Figure 3.12**). In an ideal case for rank-based nomenclature, the tree would be completely symmetrical (with each daughter-branch of a given node having the same number of tips at a given time), cladogeneses would be simultaneous across clades (preferably for the whole Tree of Life) and the rate of divergence (phenotypic or genetic) per time unit would be homogeneous across taxa (**Figure 3.12A**). In this case, three criteria (number of included lineages, age of origin of taxa or degree of phenotypic divergence) could be used to rank taxa (**Figure 3.12A**). However, what we know about the evolutionary process for

all taxa and all periods suggests that such a model is unrealistic and does not reflect reality, even though some accelerations and decelerations of the process can be induced by special circumstances. If none of these conditions are met (**Figure 3.12B**), ranking is purely subjective and does not reflect any objective property of taxa. Of course, these two models do not reflect all possibilities (even though the second model does match what we know). If even only one of these processes occurred in an unrealistically regular way, taxa could be ranked objectively using (for example) the age of origin of taxa (**Figure 3.12C**), the number of included terminal taxa (**Figure 3.12D**) or the amount of phenotypic or genetic divergence (not shown).

As previously mentioned, special circumstances probably affect the rate of cladogenesis and phenotypic divergence, but we have no evidence that this allows to unambiguously rank taxa. Thus, many evolutionary radiations are documented in the fossil record, but their exact timing is still controversial. For instance, the fossil record suggests that mammals and birds underwent a great evolutionary radiation in the Paleocene in the wake of the end-Cretaceous mass extinction event, but their timing is uncertain, as already discussed (Section 3.1.5.4 "Molecular Dating"). Also, not all clades are affected equally by such events. For instance, the fossil record suggests neither especially elevated extinction rates among lissamphibians around the K/Pg (Cretaceous/Paleogene) boundary nor an important evolutionary radiation shortly thereafter (Marjanović and Laurin 2014). Of course, there are good reasons why crises do not affect all taxa equally. For instance, the Messinian salinity crisis affected most marine Mediterrenean taxa because that crisis affected chiefly the Mediterranean Sea and its basin (Roveri et al. 2014) and, similarly, marine anoxic events should affect far less continental taxa than marine ones. Finally, crises of various magnitudes are known in the Phanerozoic, from the end-Permian event, which was apparently one of the most severe but may have lasted longer than previously thought (Viglietti et al. 2021), to much milder crises, like the Grande Coupure (Eocene/Oligocene boundary), which affected continental vertebrates in Europe and probably reflects a cooling event, given that it affected most strongly taxa that required a warm climate, many of which had an African origin (Legendre 1987; Lemierre et al. 2021). This variability in magnitude of crises would complicate use of such crises to rank taxa, even if all taxa responded the same way to all such crises.

Another phenomenon in the evolutionary process could possibly confer some reality to Linnaean categories, namely, if a special process could give rise to taxa of a given rank directly, rather than to new lineages that are typically ranked as species. Such a proposal will surely seem strange to most readers, but it has been proposed by Dubois (1981), who suggested that new genera arose through a more drastic genetic revolution than he envisioned for species, and that this "geniation" process was accompanied by an ecological niche shift that facilitated subsequent evolutionary radiations (within each new genus). This proposal does not seem to have gained much support, simply because it does not appear to reflect a biological reality. However, a similar proposal was made recently by Lücking (2019: 224), who argued that genera were composed

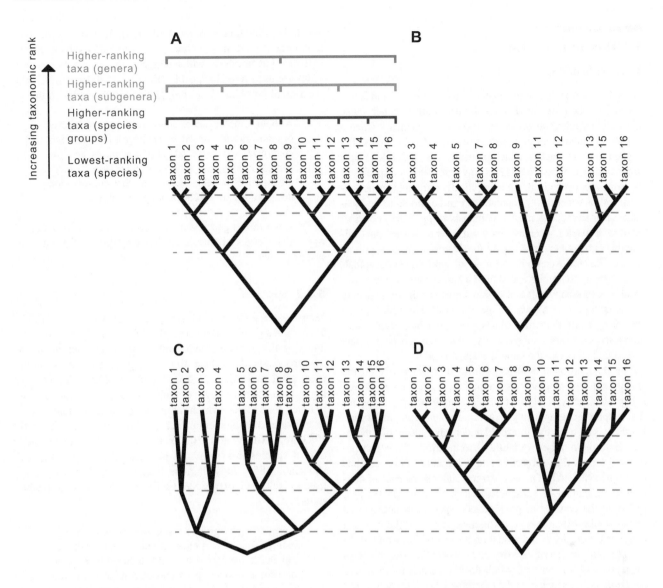

**FIGURE 3.12** Evolutionary models and the subjective nature of Linnaean categories (absolute ranks). If the tree were completely symmetrical, if cladogeneses were simultaneous (but occurred at several discrete periods of a clade's evolutionary history) and if the rate of phenotypic (or genetic) divergence per time unit (represented by the slope of the branches, rather than the spacing of taxa) was regular, three criteria (number of included lineages, age of origin of taxa or degree of phenotypic divergence) could be used to rank taxa (A). However, what we know of the evolutionary process suggests that such a model is highly unrealistic; instead, the tree is not completely symmetrical, cladogeneses can occur at any time and the rate of phenotypic (and genetic) divergence varies in time and across lineages (B). Trees with only one property that could allow objective ranking: simultaneous divergences across clades (C), and symmetrical tree (D). Grey dashed lines indicate times of cladogeneses in trees (A, C) in which these occur in simultaneous bursts.

*Source:* Modified from Laurin (2010b: figure 2).

of sets of similar species, but that genera were separated from each other by marked discontinuities:

> Thus, there are objective traits bound by evolutionary constraints that allow us to determine the next rank above the species level: the **switch of an evolutionary pattern** or theme, with the repetition of internal, analogous patterns of diversification.

However, these apparent gaps in the extant biodiversity, which have been used to delimit taxa (Michener 1963: 153), presumably simply reflect the extinction of intermediate forms,

as suggested by an examination of the fossil record, which documents much more continuous variation in the biodiversity (Carroll 1997; Laurin 2010a).

As previously mentioned, some rank-based codes (those used by zoologists and botanists, for instance) do not require taxa to be monophyletic, even for higher taxa where this requirement has been endorsed by a majority of systematists. This may be because there is still a fairly large proportion of systematists who wish to continue using paraphyletic taxa (such as Benoit et al. 2016: 779). Another reason may be that requiring monophyly would upset the Linnaean hierarchy of many long-established taxa (see Section 3.2.2 "Rank Changes Required by New Trees").

## 3.3  How to Rank Taxa?

### 3.3.1  Subspecies

The subspecies rank is immediately below the species and is recognized by the rank-based codes. Its definition is rather vague and polymorphic; subspecies are typically recognized when a species is found to exhibit fairly strong geographic variability in phenotypic characters, although ideally, it is desirable to have some evidence of restricted gene flow from other subspecies of the same species. Indeed, if geographically different (phenotypically and/or genetically) populations can be shown to be incapable of interbreeding, most systematists will probably consider them to be species, whereas if they can interbreed when put (artificially) into contact, it is logical to consider them to represent subspecies. However, no matter how subspecies are conceptualized, contrary to species and taxa of a higher rank, a single subspecies cannot be recognized within a taxon of the immediately higher rank (a species); when a species is found to exhibit strong variability, a minimum of two subspecies must be erected simultaneously. Of course, subsequently, additional subspecies may be recognized within the same species, if need be.

The recognition of subspecies also depends partly on the species concept adopted by systematists. As mentioned earlier, subspecies are typically recognized among polytypic species that display geographically structured phenotypic variability, but there is no objective threshold under this concept of subspecies as to the required variability. Application of this subspecies concept is thus subjective and is a matter of convenience (Cracraft 1983: 167). On the contrary, under some species concept, there cannot be subspecies. Thus, Cracraft (1983: 170) proposed the concept of phylogenetic species, which consist of the smallest diagnosable taxonomic units, or "the smallest diagnosable cluster of individual organisms within which there is a parental pattern of ancestry and descent." Under this phylogenetic species concept, which does not entail reproductive isolation, no subspecies can be recognized because they would be based on variability within the lowest-ranking taxa and hence, on taxonomically irrelevant data (Cracraft 1983: 171). Cracraft viewed these phylogenetic species (rather than biological species defined as reproductive communities) as the most important units of evolution. However, the species thus defined could turn out to be trivial if subsequently they fused with each other following some environmental changes that leads them to become sympatric and if they remained interfertile.

The prevalence of recognized subspecies varies between taxa, with the most intensively studied taxa having more subspecies than poorly studied taxa (Haig et al. 2006: 1586). Unsurprisingly, Haig et al. (2006: 1588) reported a wide disparity in molecular divergence between recognized subspecies. It has even been argued that teleost subspecies should not be recognized because under a:

> strict subspecies definition (e.g., a population in a particular region that is genetically distinguishable from other such populations and is capable of interbreeding with them), every isolated creek and pond could have a unique subspecies or species of fish.

> **(Haig et al. 2006: 1588)**

Similarly problematic is the finding, from a survey of 104 monographs, that most authors recognize either subspecies or varieties, but not both, which suggests that in practice, these ranks are used interchangeably (Haig et al. 2006: 1589). This is no small problem; Haig et al. (2006: 1590) concluded that

> Although the scientific community has some level of comfort with the subjective nature of subspecies classification (Hey et al. 2003), agencies and the general public want subspecies criteria to be more quantitative or better defined so that conservation designations are applied more predictably.

Thus, the ranking problems for subspecific taxa affect biology well beyond systematics, as shown by this brief discussion of subspecies.

### 3.3.2  Species

Many systematists have long hoped for a taxonomy in which all taxa assigned to the same absolute rank (Linnaean category) would be equivalent (Dubois 1982: 11), but despite a few proposals along those lines, this goal still seems out of reach (Dubois et al. 2021: 26). An objective basis for assigning taxa to a given category would seem, at first sight, easiest to reach for species, which many authors and agencies seem to consider the elementary unit of biodiversity (Cracraft 1983: 160–161; Minelli 2022b: 192, 200). The central position of species in systematics is exemplified by the title of Darwin's (1859) most famous book, and it harks back at least to Isidore Geoffroy Saint-Hilaire (1859: 365–366), who stated:

> L'espèce est le groupe **fondamental** donné par la nature. Tout en part ou y aboutit; comme la variété qui en est une dérivation accidentelle; et la race une dérivation devenue permanente; comme la famille ou compagnie, la société, l'agrégat et la communauté, qui en sont des subdivisions naturelles; comme le **genre** qui est la collection des espèces qui se ressemblent le plus; comme les **groupes supérieurs** eux-mêmes qui sont des collections de genres, par conséquent, médiatement, d'espèces. Si cela est, s'il n'y a dans la nature que des espèces diversement considérées, tellement qu'il ne reste, en dehors d'elles, que " des ombres"; on ne s'étonnera pas de voir la définition de l'espèce placée par les maîtres de la science au nombre des **plus grands problèmes** dont l'esprit humain ait à se préoccuper.

> The species is the **fundamental** group given by nature. Everything starts or ends from it; like the variety, which is an accidental derivation from it; and the race is a derivation that has become permanent; like the family or company, the society, the aggregate and the community, which are natural subdivisions of it; like the **genus**, which is the collection of the species that most resemble each other; like the **higher groups** themselves, which are collections of genera, and therefore, mediately, of species. If this is the case, there is in nature only species that are considered in different ways, so that apart

from them, only "shadows" remain; it will not be surprising to see the definition of species placed by the masters of science among the **greatest problems** with which the human mind has to concern itself.

This view arguably still prevails today to the extent that species are considered somehow more important than taxa of other ranks. This is shown, for instance, by the mediatic lists of threatened species published by the IUCN (e.g., Betts et al. 2020). Its importance is shown, among other things, by the fact that the international community started engaging conservation efforts for amphibians shortly after the red-listing process for this taxon was started, even though the decline of amphibians had been noticed 15 years earlier (Betts et al. 2020: 638). Note that there is no dedicated list of threatened genera, families or orders (although these can be listed if all their species are already listed). Subspecies have been listed by other institutions (under the US Endangered Species Act of 1973, for instance, abbreviated ESA in the following), but this raises a number of problems, the most acute of which is that there is no standardized definition of what a subspecies is (see Section 3.3.1).

This focus on species suggests that many systematists believe that species are somehow more important or legitimate than taxa of other (especially higher) nomenclatural levels. This point of view is held by at least some conservation biologists, as shown by this quote from Haig et al. (2006: 1585): "Species are generally recognized as the fundamental units of taxonomy." For this reason, the potential validity of the species level needs to be analyzed more thoroughly than that of higher taxa.

As we will see, even in this case, we are far from a consensus about the definition of this (specific) level, for various reasons. The most important one is that many species concepts coexist; Lherminier and Solignac (2005) listed 146 attempts at defining the species, many of which correspond to a different concept! The definition that garners the most support is probably the so-called biological species concept (a reproductive community; abbreviated **BSC** from now on), but even among proponents of this concept (such as Avise and Liu 2011: 709), there are disagreements over whether it should be evaluated on gene flows that occur naturally or capacity to obtain viable hybrids in captivity. Even if we agree to consider only naturally occurring gene flow, this criterion is difficult to use, as indicated by Mishler and Donoghue (1982: 497):

> Variation in morphology, ecology, and breeding is enormous and complex; there are discontinuities of varying degree in each of these factors and the discontinuities are often not congruent. There may often be **roughly continuous reduction** in the degree of **cohesion** due to **gene flow** as more inclusive groups of organisms are considered. The acquisition of reproductive isolating mechanisms appears in many cases to be fortuitous and such isolation is neither the cause of morphological or ecological divergence nor is it necessary for divergence to occur.

Congruent statements were made by other authors (such as Maddison 1997), and Mishler (1999) recently reasserted these conclusions.

Mishler and Donoghue's (1982) claims have received additional support from recent studies, notably on the classical case of cichlid teleost species flocks in the East African Great Lakes. These studies showed the fleeting existence of reproductive communities, which can be split by ecological differences and visual recognition systems or lumped by hybridization, which occurred in the last decades when water quality decreased, rendering it murky and hampering the visually-driven mate recognition systems (Spinney 2010). Other studies showed that introgressive hybridization played a major role in their impressive evolutionary radiation, especially in the last 200,000 years (Salzburger 2018: 709). Many intermediates exist between cichlid species, and some of them form ring species (in which the most divergent populations are reproductively isolated from each other, but they are connected by a geographic ring of intermediate populations that interbreed, thus forming a continuum). Partly because of these phenomena and because of recent divergences (some barely a few hundred to a few thousand years ago), species delimitation in cichlids is difficult, even when invoking multiple species concepts, as explained by Salzburger (2018: 707):

> The **classic species concepts provide little guidance for species delineation** in cichlids. The most widely used definition in biology for the category species, the **biological species concept**[49,50], is **not very practical** when applied to cichlids and **often fails** on the basis that **reproductive isolation is usually incomplete between sister taxa**. In fact, many East African cichlid species are intercrossable[51,52], even when belonging to distinct phylogenetic lineages and being derived f**rom different adaptive radiations**[53,54], and cichlids do interbreed in the wild as evidenced by occasionally observed hybrid specimens[55] as well as molecular analyses demonstrating substantial levels of gene flow between species[41,56,57]. Delineating cichlid species by means of **genetic markers is problematic** too. DNA barcoding, a widely used method for identifying species on the basis of the mitochondrial COX1 gene[58], performs poorly when applied to cichlids[59]. This is not surprising, given the high levels of DNA sequence similarity in cichlids (for example, the average genomewide sequence divergence between Lake Malawi cichlids is only $0.1–0.25\%$)[27] as well as mitochondrial haplotype sharing between species[42,43]. Defining species according to the **phylogenetic species concept is equally problematic**, as there is no a priori level of genetic distinctiveness above which two sister taxa should be considered different species and because reciprocal monophyly of sister taxa does not help in deciding whether these are populations of the same species or different species. Grouping individuals (or lineages thereof) into species according to their shared ecology—as suggested in the **ecological species concept**—is difficult in cichlids, as there is substantial niche and resource overlap and, hence, little competitive exclusion (whereby two species cannot stably coexist in the same ecological niche) between some species[17,60,61]. In practice, to facilitate the expedient naming of distinct taxonomic units in cichlids,

species are seen as clusters of individuals that are morphologically and ecologically similar and distinct from other such clusters (that is, the **vernacular species concept**).[62]

However, the claim that intercrossability decreases gradually as lineages diverge is not new. A pioneer of genetics and of studies on the speciation process wrote (Carson 1975: 84):

> On the other hand, reproductive isolation is fraught with difficulties. The principal one is to decide exactly where the species boundary can be drawn between two gene pool communities. Drawing this line increases in difficulty as one compares and studies closely related species, especially in plants. The multitude of difficult questions that come up tend to blur the biological species concept and to throw its general usefulness into some doubt. For example, because of the weak isolating mechanisms observed in many plants, reproductive isolation does not provide a very useful criterion for the delimiting of species.

Such a gradual decrease in intercrossability is not fully coherent with the hypothesis that species arise through relatively brief events that differ from normal, background evolution. An early variant of this hypothesis, formulated by Mayr (1954: 170), involved genetic revolutions in small peripheral populations isolated from the bulk of the species, which are now known as "peripheral isolates." Mayr (1954) postulated that genetic integration within gene complexes prevented important genetic changes most of the time, but that these constraints could be partly lifted in peripheral isolates, particularly through a founder effect if few individuals contributed to the peripheral isolate, and if these individuals represented only part of the genetic variability of the main population. Rapid evolution in a small population (the peripheral isolate) on a relatively small area would also explain the relative rarity of transitional forms (between species) in the fossil record (Mayr 1954: 179). The notion of genetic revolutions thus provided a theoretical justification for the punctuated equilibrium model proposed by Eldredge and Gould (1972: 114) to explain evolutionary patterns in the fossil record. These points are neatly summarized by this quote from Eldredge and Gould (1972: 114):

> In this view, the importance of peripheral isolates lies in their small size and the alien environment beyond the species border that they inhabit—for only here are selective pressures strong enough and the inertia of large numbers sufficiently reduced to produce the "genetic revolution" (Mayr 1963: 533) that overcomes homeostasis.

The model of punctuated equilibria, in its more recent form of a speciational model, has nevertheless received some support from body size evolution patterns in Cenozoic mammals (Bokma et al. 2016). A more recent, perhaps slightly more realistic version of the genetic revolution concept involves only small sets of genes (perhaps in the order of 10) that form "coadapted, internally balanced gene complexes," whereas the rest of the genome can vary much more freely (Carson 1975). Significant change within these integrated complexes would be possible during flush-crash founder cycles. In this model, the flush phase would be characterized by a relaxation of natural selection during a great population increase, during which individuals with genomes conferring lower fitness would be able to reproduce (whereas they would have been eliminated in normal circumstances). In the following crash phase, only few individuals would survive, and survival might depend more on chance than on fitness, which would result in a strong founder effect (Carson 1975: 89). Such limited genetic revolutions could result, among others, from chromosomal inversions, which had been observed earlier by Dobzhansky (1944). Dubois (1982: 41) even suggested that more than 90% of the speciation events involve chromosomal rearrangements. However, even such speciation models have limited empirical support (Lherminier and Solignac 2005: 535). For instance, among the East African cichlids, Salzburger (2018: 711) reported:

> At the level of entire chromosomes, there is little variation among the East African cichlids, with chromosome numbers ranging from 2n = 40 to 2n = 46 according to karyotyping; most species have 2n = 44 (refs[98,99]) (Table 3.1). Differences in chromosome number, thus, do not seem to have an important role in the origin or maintenance of cichlid species.

This quote does not exclude more minor chromosomal rearrangements (such as inversions), but there is certainly no support from cichlids that the speciation process frequently involves chromosomal rearrangements. The speciation process is still poorly understood and defies simple ideas. For instance, it does not simply depend on an overall degree of genetic distance.

The idea of genetic revolutions would make species more natural because the speciation process associated with such revolutions would be very fast (at least at the geological time scale) and would mean that for most of their existence, populations of a given species would be rather stable (phenotypically and genetically) and would be separated from other species by a reproductive barrier. Lherminier and Solignac (2005: 533) argued that this represents a different form of typological thinking about species because between genetic revolutions, species would be stabilized by a strong functional integration of the genome that would prevent significant change.

Modern genetics has shown that the speciation process is far more complex than implied by the concept of genetic revolutions and that it depends also on the geographic context, especially in isolation (Feder et al. 2014). There are also taxonomic differences in the speciation process. Among embryophytes, autopolyploid speciation (rather than speciation linked to chromosome rearrangements) is expected to prevail in the next few centuries because of climate change (Gao 2019). Among elopomorph teleosts, simple chromosomal rearrangements may have played an important role in the speciation process as argued by Sousa et al. (2021), but these conclusions are based chiefly on four species. By contrast, Barby et al. (2018) found a surprising karyotypic stability in six species of notopterid teleosts forming two main clades that diverged around 100 Ma. Thus, it would be premature to extrapolate the role of chromosomal rearrangements in the speciation process among teleosts, let alone to larger clades of metazoans. Among insects, chromosomal rearrangements may often be linked with the speciation process, but here too, the number of well-documented cases is very low, and the presence of much karyotypic polymorphism within *Euchroma gigantea* (*Coleoptera: Buprestidae*) suggests a complex relationship, if any, between speciation and karyotype (Xavier et al. 2018), unless *Euchroma gigantea* actually encompasses several entities that should be considered species. Similar claims could be made about *Drosophila* (Reis et al. 2018). The scant data available so far (compared to the more than 1,500,000 currently recognized species) do not allow validating the species concept based on presumed chromosomal rearrangements. Modern genetics showed that most chromosomic rearrangements are neutral and without notable effects. Some are eliminated because they are disadvantageous, but geographic isolation can facilitate the persistence of neutral or advantageous rearrangements, and if reproductive isolation is reached over time, they may come to characterize a new species, but they need not be causally related to the speciation process (Lherminier and Solignac 2005: 536).

The BSC is inapplicable outside sexually reproducing organisms, so this completely excludes two domains of life (*Bacteria* and *Archaea*) and some eukaryotes. Indeed, under the *Prokaryotic Code* (Parker et al. 2019), thresholds in measures of genetic divergence are used to delimit species (Oren 2011: 438) and, if applied to eukaryotes, the same thresholds would yield much more inclusive taxa (see Chapter 2). The exclusive use of this genetic distance criterion to rank taxa had been discussed and rejected by at least some zoologists long ago (Dubois 1982). The BSC is also inapplicable to extinct taxa. Last but not least, even though the BSC is supported by many systematists in principle, it has been applied to a fairly small minority of cases. Haig et al. (2006: 1585) even stated that "Nevertheless, there is little evidence outside of *Drosophila* (e.g., Ayala et al. 1974) that this criterion [reproductive isolation] has been routinely employed," and that most "plant taxa" had been recognized based on morphological criteria.

Under the phenetic species concept, Levin (1979) proposed that each species was a cluster of organisms in Euclidean space separated from other such clusters by gaps of similar sizes. However, reality does not validate this concept; phenotypic clusters are instead nested within each other with gaps of continuously varying size (Mishler 1999). In addition to this, application of this criterion is difficult in some fields, like paleontology, in which we often have few, fragmentary specimens of a given putative species. But even when the fossil record is fairly rich, delimiting species under the phenetic species concept (arguably the easiest one to apply in paleontology) can be tricky, partly because fossils are often distorted. This is illustrated by the fact that 21 of Robert Broom's *Dicynodon* holotypes were subsequently argued to belong to the species *Oudenodon baini* (Wyllie 2003: 3)!

The various concepts do not agree on monophyly; neither the BSC nor the phenetic species concept require it, but under other concepts, monophyly is required (Lherminier and Solignac 2005; Mishler 1999). To sum up, the search for a universal criterion to rank taxa at the species level has been in vain so far. We can only admire the wisdom of early evolutionists such as Darwin who considered that species "were simply less transient varieties and left the species problem at that" (Mishler and Wilkins 2018: 2).

### 3.3.3 Genera

The genus is less commonly considered to have an objective basis than species. However, Linnaeus considered that genera were the results of creation, and Darwin "never entirely abandoned" the idea that they were real (Wilkins 2018: 158). Dubois (1982, 1988) suggested using the reproductive criterion (already omnipresent in the BSC) to define genera. This may be the only recent proposal to provide an objective basis of the genus as a taxonomic level (if we exclude proposals that deal with taxa of all levels with a same criterion, as for geological age; see next section). Note that Dubois distinguishes nomenclatural levels, which are relative and temporarily assigned to taxa, from taxonomic levels, which are based on properties of taxa. Dubois et al. (2021: 21, 24) argued that only species and genus taxonomic levels can be defined objectively; for higher levels, they knew of no suitable defining criteria.

Dubois' (1982, 1988) proposal is to recognize as a genus the largest clade composed of species that can give viable adult hybrids, whether or not they are fertile. This criterion would be used for inclusion only; within each genus, some pairs of species might not be able to yield such hybrids. Experiments would need to be used because the lack of hybrids between some species in nature would not be sufficient ground to put these species into different genera. The rationale is that, according to Dubois (1982, 1988), species capable of hybridizing (at least in the laboratory) are genetically compatible and that this is a global measure of genetic similarity concerning both structural and regulatory genes, and that this might be more biologically significant than immunological (or even sequence) similarity.

This suggestion has not been followed by many systematists, as far as I know. Perhaps one of the main reasons, which was recognized by Dubois (1982: 30), is that this would result in radical changes to the nomenclature of some taxa, such as birds. In some avian taxa, the number of intergeneric hybrids is high, so use of this criterion would result in many genera becoming junior subjective synonyms. Dubois (1982: 32) suggested that birds have been over-split compared with other vertebrate taxa for two reasons: first, they have been more intensively studied than many other taxa (mammals excepted), and second, plumage characters have been over-emphasized, and these appear to evolve fast. This may well be, but trying to uniformize the ranking of taxa using hybridization potential or any other objective criterion would surely result in substantial nomenclatural changes. Michener (1963: 155) had already argued that:

> the idea that forms that hybridize have to belong to the same genus would create **endless nomenclatorial uncertainties** because who can tell when or under what novel experimental conditions the right pair of gametes might meet to produce a hybrid between species previously placed in separate genera?

Among birds, the hybridization patterns and their taxonomic and nomenclatural implications have been intensively investigated in *Anatidae* (Johnsgard 1960; Gonzalez et al. 2009). The compilation of taxonomic occurrences of about 400 interspecific hybrid combinations shows a pervasive pattern of intergeneric and even inter-tribal hybrids in *Anatidae* (Johnsgard 1960: table 3.1). According to Johnsgard (1960: 28), this "vindicates the submerging [synonymy] of the previously upheld subfamilies *Cygninae* and *Anserinae*," a nomenclatural decision that was subsequently upheld (Livezey 1997; Gonzalez et al. 2009). Furthermore, inter-tribal hybrids are found in most tribes, with the exception of *Dendrocygnini*, which is most distantly related to other anserids and may have appeared in the Paleogene, between 35 Ma and 65 Ma ago; the other major taxa that can hybridize started diversifying about 18–41 Ma ago (Gonzalez et al. 2009). A small proportion of the hybrids (4.6%, or 22 out of 479) originate from parental species that belong to two subfamilies (*Anserinae* and *Anatinae*), according to the nomenclature used by Gonzalez et al. (2009: 315). Thus, applying Dubois' (1982) suggestion to this clade would result in synonymizing many genera and most tribes of *Anatidae*, and would even result in considering the clade composed of *Anatinae* and *Anserinae* as a single genus.

Use of the hybridization potential criterion to delimit genera would not affect the nomenclature of all large taxa evenly; mammals would be less affected, according to Dubois (1982). This reflects to an extent the speed at which reproductive isolation is achieved in the various clades. Dubois (1982: table 3.2) reported that in placental mammals, the average age of divergence between species that can produce viable adult hybrids is about 2–3 Ma, but for birds, it is 20–23 Ma, and for anurans, about 21 Ma. Obviously, using this criterion would yield mammalian genera that are much younger than avian and anuran genera, and this would conflict with using geological age as a ranking criterion (see Section 3.3.4.1). A similar phenomenon would occur in other taxa. For instance, Mishler and Donoghue (1982: 498) mentioned that if potential to interbreed were adopted to delimit taxa, "the family Orchidaceae, with approximately 20,000 species at present (covering a great range of variation), might be lumped into just a few species because horticulturalists have produced so many bi- and pluri-generic hybrids." Dubois (1982: 34, 1988: 31) argued that such transitory nomenclatural instability that would accompany adoption of his genus concept would be compensated by greater subsequent stability, and he has still supported such proposals recently (e.g., Dubois et al. 2021: 21) while recognizing that systematists keep using various genus concepts (Dubois et al. 2021: 85). Nomenclatural stability could be achieved for the names currently designating genera (arguably more easily and with much less transitory instability, but without regard to rank) with the *PhyloCode* (see Chapter 4), but Dubois has been a vocal opponent of that code (Dubois 2005a, 2005b).

Given the arguments presented here, the relative instability of delimitation of many genera is not surprising. There is no rescuing the genus by evoking a given number of included species. Even the most enthusiastic proponents of an objective status for the genus recognize that the number of species included in individual genera can be highly variable (Dubois 1982: 42). As an extreme example, more than 2,000 species are currently recognized in the scarab genus *Onthophagus* (Minelli and Fusco 2013: 307).

All these arguments concern the objective nature of the genus as a nomenclatural category, not the objective nature of the taxa that have been ranked as genera. There is now a fairly broad consensus (though not unanimity) that higher taxa (above the species level) should be clades and, as such, it is not the reality of these taxa that is questioned, but the objective nature of Linnaean categories, like species, genus, family or any other such category.

### 3.3.4 Higher Taxa

#### 3.3.4.1 Geological Time of Origin

Hennig (1966: 154) recognized that if Linnaean categories were to be useful in science, they had to have an objective basis: "If systematics is to be a science it must bow to the self-evident requirement that objects to which the same label is given must be comparable in some way." Hennig (1950, 1966) initially suggested ranking taxa according to their age of origin, for a good logical reason: "the absolute rank order cannot be independent of the age of the group, since in the phylogenetic system the coordination and subordination of groups is **by definition** set by their relative age of origin" (Hennig 1966: 160). He first established a ranking scheme for arthropods that consisted in six subdivisions of the geological timescale (Table 3.2) that seemed to work reasonably well. This approach of using "time bands" to rank taxa inspired other authors to adopt it, under the name of "time banding," despite its obvious problems (see the following). Indeed, Hennig (1966: 184) realized that the number of categories that could be recognized using the time of origin was far greater than the six periods into which he had subdivided the timescale. He also noted that the rank of mammalian taxa could not be determined using the same equivalence between age and rank as insects (Hennig 1966: 187). Even among arthropods, there were problems with this approach. Thus, Hennig (1966: 190) noted that:

> The re-evaluation would be even more drastic for some groups. Certain of the so-called genera of the Ostracoda (Crustacea: *Bythocypris, Bairdia, Macrocypris, Pontocypris, Cytherella*), for example, apparently are known with certainty as far back as the Ordovician. This means that each of these "genera" would have to receive a higher rank (one from the "class stage") than the entire group of land vertebrates (Tetrapoda), which probably arose in the Upper Devonian.

However, this did not lead him to immediately drop the suggestion to use geological age of origin to rank taxa. Hennig (1966: 191) explained that:

> The suggested compromise is to designate different time scales for different animal groups that would make it possible to retain the present absolute ranking of most subgroups. For example, for mammals (and probably also for birds) the time portion V (Fig. 58) [from Oligocene to Upper Cretaceous; see Table 3.2]

could be assigned to the "ordinal stage" and the time portion IV [Triassic to Lower Cretaceous] to the "class stage." The present rank hierarchy of groups of mammals would scarcely be changed. For insects, on the other hand, the time portion III [Mississippian to Permian] could be assigned to the "ordinal stage" and the time portion II [Cambrian to Devonian] to the "class stage." With the help of a **conversion chart** we could determine that the "orders" of mammals and birds cannot be compared with the "orders of insects, but perhaps with the "tribe stage."

The requirement of a conversion chart was not a very elegant solution, and Hennig must have sensed it. Thus, Hennig (1969, 1981) quickly abandoned this recommendation and suggested dropping absolute ranks (Linnaean categories). He explained it thus:

> I have refrained from giving any categorical rank ("order," "suborder," etc.) to groups of higher rank. I have done this because I have found that the fundamental questions of phylogenetic systematics so often become bound up with the **subsidiary question of the rank of each group**, and I wanted to avoid this kind of **unfruitful debate**.

**(Hennig 1981: xviii)**

Dropping Linnaean categories was a wise choice because Griffiths (1973) showed that Hennig's initial suggestion ("each terminal twig in the phylogenetic tree of the fossil group would be not only a species, but at the same time a representative of a higher category, such as the 'ordinal stage'") raised problems. For instance, each taxon present in the Triassic would represent an order, and if these became extinct in the Triassic, each of these orders might include potentially a single species, thus generating much redundancy. Redundancy is generated by RN

in asymmetrical parts of the Tree of Life, but redundancy is admitted to be a problem even by proponents of RN (such as Dubois et al. 2021: 27). This point was ignored in most subsequent suggestions to use geological age of origin of taxa to rank them, probably because these works focused on extant taxa (for instance, Avise and Johns 1999; Avise and Mitchell 2007). Griffiths (1973: 340) also advocated dropping Linnaean categories, among other reasons because "these names were intended as terms of Aristotelian (essentialist) logic and are therefore inappropriate to modern classifications." Griffiths (1976: 170) further suggested that if taxa had to be ranked into age classes, the most obvious way to do so "would be to use names connoting the geological periods; for instance, Paleotaxon, Eotaxon, Oligotaxon, Miotaxon, and Pliotaxon for classes of taxa originating during each epoch of the Tertiary." This suggestion did not generate much enthusiasm.

The systematic community appears to have selectively assimilated Hennig's thoughts on this topic. Thus, geological age of origin was subsequently suggested as the preferred criterion to rank taxa by Sibley and Ahlquist (1990) and Avise and Johns (1999). The latter study credited Hennig (1966) with the idea of using geological age of origin of taxa to rank them and shared with Hennig his concern that current rank allocations of taxa are meaningless for comparative biology. Avise and Johns (1999) proposed a detailed protocol for rank allocation, and even raised the possibility to use alternatives to the classical Linnaean categories, such as a simple single-letter designation (in their example, they used the letters D–Q) for ranks. These proposals were partly prompted by progress in molecular dating. However, Avise and Johns (1999) neither cited Hennig (1969, 1981) nor mentioned that he recommended dropping Linnaean categories altogether. Even though they explained that the details of implementation should be worked out by the systematic community, they suggested a specific set of time bands to illustrate their approach (Avise and Johns 1999: 7361):

> A scaling of taxonomic ranks to conventional geological windows is appealing because the latter are well known, are fortuitously about equal in number to conventional Linnaean ranks (Table 2), and often are associated with important evolutionary events such as mass extinctions (e.g., the Permian and Cretaceous) and adaptive radiations (perhaps in the Cambrian).

This proposed set of temporal bands is reproduced here as Table 3.3 because it was proposed again, with only cosmetic modifications, by subsequent studies.

A problem with this time banding approach is that extinct taxa represented by ancient fossils could not belong to species or genera. Thus, what is now known as *Tyrannosaurus rex* would be a suborder in this scheme (Table 3.3), but paleontologists need to be able to erect new species! This problem had been detected long ago and Farris (1976: 279) thus advocated using taxon duration rather than age for ranking purposes. However, this creates new problems because, as pointed out by Lücking (2019: 219), two sister-taxa of unequal duration could be ranked differently, whereas Hennig (1966) emphasized that sister-taxa should be of the same rank.

**TABLE 3.2**

Relationship between age of origin of taxa and their absolute rank according to Hennig (1966). The first and second column (on the left) follow Hennig (1966: fig. 58); the two columns to the right follow Hennig's (1966) text (pp. 185–187), but were not drawn as such by Hennig.

| Number of the time period | Geological age of the taxon | Corresponding rank for insects (and, by extension, other arthropods) | Corresponding rank for mammals (and, by extension, other vertebrates) |
|---|---|---|---|
| VI | Present Miocene | Genus? | ? |
| V | Oligocene Upper Cretaceous | Tribe | Order |
| IV | Lower Cretaceous Triassic | Family | Class |
| III | Permian Mississippian | Order | ? |
| II | Devonian Cambrian | Class | ? |
| I | Precambrian | Phylum | ? |

**TABLE 3.3**

Temporal banding proposed by Avise and Johns (1999: table 3.1) to rank taxa (format slightly modified). Specific ranking was excluded from this scheme because Avise and Johns (1999) proposed to use the biological isolation criterion.

| Taxonomic rank | Proposed designation | Geological episode | Temporal band | |
|---|---|---|---|---|
| Domain | (A) | Archaean | 2.5–3.6 | Bya |
| Kingdom | (B) | Proterozoic | 0.55–2.5 | Bya |
| Phylum | (C) | Cambrian | 500–550 | Mya |
| Subphylum | (D) | Ordovician | 440–500 | Mya |
| Superclass | (E) | Silurian | 410–440 | Mya |
| Class | (F) | Devonian | 350–410 | Mya |
| Subclass | (G) | Carboniferous | 290–350 | Mya |
| Cohort | (H) | Permian | 250–290 | Mya |
| Superorder | (I) | Triassic | 205–250 | Mya |
| Order | (J) | Jurassic | 145–205 | Mya |
| Suborder | (K) | Cretaceous | 65–145 | Mya |
| Superfamily | (L) | Paleocene | 56–65 | Mya |
| Family | (M) | Eocene | 33–56 | Mya |
| Subfamily | (N) | Oligocene | 24–33 | Mya |
| Tribe | (O) | Miocene | 5–24 | Mya |
| Genus | (P) | Pliocene | 2–5 | Mya |
| Subgenus | (Q) | Pleistocene | 0–2 | Mya |
| Species | — | — | — | |

More recently, Avise and Mitchell (2007) proposed alternative "time banding" approach using "timeclips" (illustrated with the same time bands as Avise and Johns 1999). Their new proposal differs substantially from previous ones. They suggested that:

> The first letter in the code, printed in uppercase font, designates the clade's nodal origin: A in the Recent epoch; B, Pleistocene; C, Pliocene; and so on consecutively back to the Archaean (R) . . . This first letter is followed by a colon and then by two lowercase letters mnemonically abbreviating the geological episode. We further suggest that the timeclip be printed in **bold**, bracketed as [. . . .], and connected directly to the clade's taxon name as either a prefix or a suffix (e.g., **[D:mi]** Hominoidea; or *Drosophila* **[F:eo]**). For publication or other formal purposes, the authority for the temporal estimate could be appended to the timeclip (e.g., Drosophila [F:eo Johndoe 2000]) with the source cited in references.

Avise and Mitchell's (2007) proposal is not aimed at standardizing Linnaean categories; the current rank allocations could remain unaltered, and the timeclips would provide objective data that would allow to compare taxa. Thus, Avise and Mitchell (2007) proposed this approach to provide "a practical way to retain the familiar Linnaean system and simultaneously promote the incorporation of new phylogenetic discoveries from molecular biology, paleontology, or other relevant evolutionary disciplines." This approach suggests that it is desirable to retain Linnaean categories even though they contain no objective information about taxa. This is, in any

case, the preference of Avise and Mitchell (2007), who stated that "Although we prefer the retention of ranking hierarchies in biological classifications, our current proposal could also be implemented in rank-free systems such as PhyloCode." Avise (2008) added that:

> Serendipitously, there are 17 supraspecific ranks in modern versions of the Linnaean hierarchy (29) and also 17 primary subdivisions in the traditional geological time scale (30) [from Archaean to Pleistocene], thus affording the possibility of a perfect one-to-one allocation of taxonomic rank to geological episode (Figure 3.2).

In a slightly more recent study on this topic, Avise and Liu (2011: 713) suggested that:

> For the first time since Linnaeus, conceptual frameworks as well as empirical approaches now exist to stabilize and universally standardize taxonomic assignments across all forms of life [through time banding]. The open question is: will systematists finally choose to adopt some such system?

However, one may wonder if such long-lasting temporal subdivisions (Table 3.3) would be sufficient. The Archaean lasted for more than 1 billion years. Closer to us, the Miocene, with its "mere" 18 Ma (approximately) of duration, would encompass ages of 5 to 23 Ma, during which much taxonomic diversification took place, notably among mammals and birds.

Naomi (2014) also proposed a system using time banding to add information to taxon names. Under this proposal, taxa are inherently unranked, but endings indicate the temporal band in which they originated. This proposal is fairly complex (perhaps too much to be convenient to most practicing systematists) because it includes rules for both a standard set of temporal bands, and sets of non-standard temporal bands, which can be used in the (plausibly rather common) cases when application of a standard set of temporal bands results in disruptive name changes. However, under Naomi's (2014) proposal, three kinds of clades are distinguished: holoclades (equivalent to branch-based taxa under PN), synclade (node-based clades under PN) and basic clades. This latter kind is problematic; it is defined as a "simple branch," which suggests a single evolutionary lineage. However, this is not so simple because Naomi (2014: 9) indicates that such "simple branches" may have side branches, which may be either extinct species or "very small clades," and the latter may comprise several species. Given the absence of an exact threshold of clade size, or even a way to measure it, the distinction between basic clades (also called "simple branches") and the other two kinds of clades (holoclades and synclades) is subjective. This particular time banding proposal could cause further nomenclatural instability because changes in nodal ages or topology can change the names of "basic clades," and even change the status of these "basic clades" into holoclades, or vice versa (Naomi 2014: figure 3.7). Naomi (2014: 10) admitted that taxa of a same band are not equivalent because they do not have exactly the same time of origin; he then argued that they could nevertheless be considered "semi-comparable" for convenience (Naomi 2014: 11).

The recurring problematic choice of time bands was discussed by Zachos (2011: 733) in the context of using time banding to reassess the rank of taxa, but these objections also stand in the context of using time banding in comparative biology. He stated:

> Temporal banding would, initially, provide the otherwise vacuous Linnean categories with some kind of information (time since origin), although there is a major weakness, too: in past adaptive radiations, the splitting of major extant lineages occurred quite quickly so that the temporal bands would have to be very narrow. Otherwise, nested clades would be assigned the same rank so that one or more, say, orders of mammals would be part of another order. This would destroy the encaptic character of classification! If, however, the temporal bands are very narrow, we will need many of them, and hence many categories.

Nothing precludes the adoption of narrower time bands. However, it seems simpler to simply use nodal ages in the context of comparative analyses which incorporate timetrees (such as phylogenetic independent contrasts) than to lump such ages into discrete categories, a process that necessarily discards data. More basically, as stated by Zachos (2011: 733), "Comparing an early-Miocene taxon with a late-Miocene one makes less sense than comparing the latter with an early-Pliocene taxon if time is the yardstick by which we decide what is comparable and what is not!" Also, as recognized by Zachos, "Inevitably, we would still have the same rank for taxa of very different age, although now researchers should neglect the rank and look at the time-clip instead. So, at best, the ranks are superfluous. Why then stick to them at all?"

Despite the reservations evoked here, several studies (listed by Lücking 2019: 205) applied temporal banding to rank various taxa, such as mammals, birds and lichens. For instance, Jønsson et al. (2016), like Naomi (2014), developed a "least disruption" approach to attribute ranks to taxa in a given clade by making as few changes as possible to ensure consistency between rank allocations and time of origin. Jønsson et al. (2016) applied this approach to corvid families and genera and found that the optimal (maximal) age thresholds were 11.79 Ma for genera and 21.62 Ma for families. They reported that 51 genera out of 125 and 22 families out of 30 were unchanged (presumably in delimitation) by their time banding procedure. Cai et al. (2019) used Jønsson et al.'s (2016) strict temporal banding to rank babbler (*Aves: Passeriformes*) taxa into six families and 50 genera, and reported that "94% of the families and 60% of the genera were unchanged from the current taxonomic list."

Strict temporal banding was criticized by proponents of RN (such as Dubois et al. 2021: 22). Lücking (2019) pointed out that the "least disruption" approach was also problematic:

> It makes no sense to postulate an objective method for ranking, when in the end that approach is arbitrarily adjusted to a [sic.] each group in a way that it best fits preexisting classifications that were **considered subjective in the first place**. Also, if

clade-specific adjustment is accepted, **where would one draw the limit** between clade-based temporal bands? Should each kingdom have its own bands? Or each phylum or class within each kingdom? Each family or genus within each order? If that is the envisioned practice, then the approach could essentially be **watered down to no adjustments at all**.

> If temporal banding is used as an absolute criterion to determine ranks, it must use **the same bands for the same ranks across all organisms**, requiring a broad consensus across microbiologists, zoologists, mycologists, and botanists. Such a compromise is much less likely to materialize than the efforts towards a unified nomenclatural *BioCode* for all organisms (Greuter et al. 2011).

These are indeed serious objections to the "least disruption" approach to time banding (for an assessment of the likely fate of the draft *BioCode*, see Chapter 2). Lücking (2019) also argued that temporal banding should be used only in combination with phenotypic disparity. Lücking (2019: 241) argued:

> Temporal banding is a powerful tool to assess rank-based classifications in addition to phenotypic disparity. Both approaches can be executed quantitatively and should be combined in a balanced way to define ranks in a way to both maximize information content and provide taxonomic and nomenclatural stability. At the genus and family level, two ranks that have important practical implications, temporal banding should be employed very conservatively and adjustments in ranks should only be proposed when the timeline identifies a clear outlier.

The goal of not disrupting current rank allocations is coherent with nomenclatural stability, but the approach advocated by Lücking (2019), by combining two criteria (geological age of origin and phenotypic disparity), reduces information content because it is not possible to know from such a nomenclature to what extent it reflects each of these two criteria for any taxon.

Lücking (2019: 225) also objected to strict time banding to rank taxa because of its automatic character:

> There is an obvious trend to render classifications "objective" to the point that they can be carried out mechanistically by non-experts: if strict temporal banding would be broadly adopted, **anyone could run a molecular clock** on practically any taxon with published sequence data, determine temporal bands, and then **mess around with classifications** [nomenclatures] using largely automated algorithms. In contrast, the meaningful evaluation of phenotypic disparity requires a level of expertise in a group, first and foremost in order to properly recognize and code traits. The latter is a virtue, not an impediment!

There are two problems with this quote. Trivially, performing molecular dating is not such a simple task that anyone could do it! But, more importantly, the fact that ranking taxa could be done automatically through algorithms is not an argument against this procedure. After all, ranking taxa is arguably far from the most important task awaiting systematists. Indeed, automatization could benefit systematics, given that much of the extant biodiversity is still unknown and that progress in our knowledge of the Tree of Life is accelerating, partly due to the increasing flow of new molecular sequences. These factors, coupled with the low (and unfortunately dwindling) number of systematists, make automated processes in systematics critically important. This prompted the development of an "automated phylogenetic taxonomy" that used a set of algorithms (collectively called "*mor*") to automatically collect molecular sequences of genes (the nuclear-encoded large subunit ribosomal DNA, or nuc-lsu rDNA for short), build phylogenies from these sequences and label some nodes that represent important taxa (Hibbett et al. 2005). This initiative was restricted to mushroom-forming fungi, but it was a great success as long as this project kept running.

From a practical point of view, strict temporal banding is also hampered by uncertainties about the age of taxa. Lücking (2019: 227) reported the example of the "kingdom Fungi," for which times of origin ranging from 660 Ma to 2.15 Ga had been proposed. But such problems occur at all timescales, even in the Neogene, as shown by recent works on Old World camelids that have been previously evoked; the age of the crown-clade that includes the Dromedary and the Bactrian camel varied about 8-fold, from 1 Ma to 8 Ma (Geraads et al. 2020). This is only a practical problem that can potentially be resolved by future technical progress, but in the meantime, this approach may not be the most convenient.

Most of the works reviewed here had assumed that Linnaean categories did not match well with the geological time of origin of taxa, but was this assumption warranted, and how bad was the mismatch? To determine this, Avise and Liu (2011) looked at the geological age of origin of many orders, families and genera of tetrapod vertebrates and decapod crustaceans. Their survey confirmed that taxa of a given rank differed widely in geological age of origin. Thus, the order Decapoda is about 437 Ma old, whereas orders of mammals and birds are only about 81.2 and 85.5 Ma, according to their survey (and these numbers reflect molecular estimates; the paleontological consensus would yield a younger age for avian and mammalian orders). For families, the discrepancies were equally spectacular, with ages of about 201 Ma for decapods, but only 37 Ma for mammals. The pattern for genera is not too different, with an age of 60 Ma for decapods, but only 9.6 Ma for mammals. These results appear to show an anthropocentric bias because the attention that systematists have given to taxa seems to decrease with phylogenetic distance to Man; hence, taxa of a given rank are increasingly older as they are distant from us. This would explain that mammalian taxa of a given rank have the most recent age, followed closely by other amniotes (especially birds, which, as endotherms, seem to generate more interest than ectothermic reptiles), amphibians and, finally, decapod crustaceans. This can be exemplified for genera, where the mean ages reported by Avise and Liu (2011: table 3.1) are 9.6 Ma for mammals, 27.7 Ma for birds, 31.5 Ma for ectothermic reptiles, 37.3 Ma for amphibians and 60.2 Ma for decapod crustaceans, but the numbers for families and genera yield the same sequence of increasing order across taxa.

Even among closely related taxa, ages of origin at a given rank differ widely. Thus, Avise and Liu (2011) reported that families of mammals range in age from 7.3 to 80.5 Ma, and for decapod crustaceans, families may be from 94 to 430 Ma old. Genera also display spectacular age differences; in mammals, they may be from 0.1 to 40 Ma old, and for decapods, from 16.8 to 135.1 Ma old. Thus, Avise and Liu (2011: 711) stated that "for each vertebrate group, we conclude that clades of a given taxonomic rank can be associated with a wide range of evolutionary ages, meaning that there is little consistency with respect to time of origin."

This long analysis of temporal banding shows that it is not a practical solution to rank taxa or, arguably, even to add information to taxon names (ranked or not) and that dropping absolute ranks altogether might be simpler and more coherent with evolutionary theory.

### 3.3.4.2 Other Ranking Criteria and Conclusion on Ranking Practices

Pheneticists suggested using similarity coefficients to assign ranks (Sneath and Sokal 1962: 859). However, this was proposed explicitly for phenetic groups, which Sneath and Sokal (1962) proposed to call "phenoms" to distinguish them from taxa recognized by other systematists (such as genera, tribes and families under RN), because the latter "have evolutionary, nomenclatural and other connotations." Of course, ranking based on similarity coefficients would be difficult to extend beyond a single study because such ranking would have to rest on the same set of characters to be meaningful. Consequently, phenoms were never used widely by systematists.

More recently, the degree of molecular divergence, which is linked with evolutionary time (at least under a hypothesis of a molecular clock), has also been envisioned to rank taxa. However, this measure proved just as problematic, as was shown long ago. Thus, Wallace et al. (1973: 11) concluded that "It appears likely that the genomes of frogs within a single genus can differ at least as much as do the genomes of mammals in different orders," and similar problems were raised by more recent studies (such as Avise and Liu 2011: 709). Wallace et al. (1973) realized that this raised a dilemma: if anuran taxa were elevated in rank to those equivalents of those of mammals showing comparable molecular divergence in the albumin, this would cause much nomenclatural instability. But not doing it means that anuran genera are really not comparable with mammalian genera. Such examples are numerous, but such distances have been used by bacteriologists to delimit taxa, possibly because they were less encumbered by taxonomic tradition (see Section 2.4.5). Avise and Liu (2011: 712) concluded:

> Thus, evolutionary dates of origin can now be added to the **pantheon of biological variables** for which nested ranks in current Linnean taxonomies are **highly inconsistent**. Other such variables include such diverse features as numbers of species per taxon, magnitudes of phenotypic divergence (e.g. in morphology, behaviour, physiology, etc.), ecological roles, speciation rates, geographic ranges, genetic distances in proteins or nucleic acids, calibrations for

molecular clocks, and just about any other biological variable that you might think of. In other words, even two-and-a-half centuries after Linnaeus, systematists have yet to standardize (or even develop criteria for standardizing) the nested Linnean categories (ranks) that serve as the organizational and nomenclatural foundation for essentially all areas of biology.

Even the species, which many people view as more "real" than higher taxa (wrongly, in my opinion), is problematic in this respect. Boyd (1999: 180) had already noticed that "controversies about the species level seem to revolve around whether certain groups of similar populations should be grouped into the same subspecies, species, or genus." Darwin had already concluded that the nature of variations displayed by individuals, populations, races, subspecies and species made ranking a subjective exercise (Wilkins 2018: 165). This was already in the first edition of the *Origin*, in which Darwin (1859: 48) stated:

> Many years ago, when comparing, and seeing others compare, the birds from the separate islands of the Galapagos Archipelago, both one with another, and with those from the American mainland, I was much struck how **entirely vague and arbitrary** is the distinction between species and varieties.

Still today, given the multiple (more than 140) species definitions that have been used in biology (Lherminier and Solignac 2005), there is no objective, consensual way of determining if a group of organisms forms a species.

Folk taxonomies may not have absolute ranks (this remains a controversial point; see Chapter 1), possibly because they include names of taxa which human populations know reasonably well. Raven et al. (1971) thus argued:

> It would therefore be meaningless to ask a Tzeltal speaker what properties the taxa indicated by his set of generic names have in common. Only when the classificatory system is extended to hundreds of thousands of poorly known organisms do we begin to ask for a "definition" of genera and species. Confronted with this difficulty, the human mind is all too ready to accept spurious generalities such as the "biological species concept."

Raven et al. (1971) argued that the development of printing and the possibility to produce widely distributed, lasting inventories of biodiversity led to an increase in the number of recognized taxa and, consequently, of taxa of intermediate ranks. The development of Linnaean categories was thus in a way a historical accident. A relative hierarchy of taxa was clearly needed, but absolute categories above the genus level, which were introduced by Magnol (1689) and Tournefort (1694), developed further by Linnaeus (see Section 1.3) and subsequently adopted by most systematists, do not appear to be justified by comparisons with folk taxonomies, and even less by what we now know about evolution. Indeed, reliance on Linnaean categories has long been considered artificial and a weakness of systematics by other scientists (Lecointre and Le Guyader 2016: 32), philosophers and historians of science (Winsor 2006: 156).

A survey of the ranking criteria that have actually been used by systematists confirms the subjective nature of such rankings. Smith (1994: 92–93) listed the five main criteria that had been used to rank taxa. These are (quoting only the titles of each item):

1. High categorial rank as a topological consequence of a group achieving considerable [phenotypic and taxonomic] diversity;
2. High categorial rank as a topological consequence of perceived morphological distinctiveness;
3. High categorial rank as a result of sister-group relationships [with a high-ranking taxon];
4. High taxonomic rank given to a paraphyletic ancestral group after abstraction of a number of well-defined monophyletic groups; and
5. High categorial rank given because of ignorance [about taxonomic affinities].

Given this heterogeneous set of criteria, it is clear that the search for a posteriori ontological justification for equivalency of taxa of a given rank (as these ranks have actually been allocated to them) is off to a very bad start! As Minelli (2000: 345) concluded: "Absolute ranks are a myth. A dangerous myth, indeed." In fact, even some of the most enthusiastic proponents of RN (some of whom are simultaneously the most vocal opponents of PN) admit that the search for objective criteria to attribute absolute ranks to higher taxa is futile (e.g., Dubois et al. 2019: 18–19). Thus, Dubois et al. (2021: 5) stated:

> we stress the fact that **nomenclatural ranks do not have biological definitions or meanings** and that they should never be used in an 'absolute' way (e.g., to express degrees of genetic or phenetic divergence between taxa or hypothesised ages of cladogeneses) but in a 'relative' way: two taxa which are considered phylogenetically as sister-taxa should always be attributed to the same nomenclatural rank, but **taxa bearing the same rank in different 'clades' are by no means 'equivalent'**, as the number of ranks depends largely on the number of terminal taxa (species) and on the degree of phylogenetic resolution of the tree.

Similar statements can be found in Dubois et al. (2021: 21) and elsewhere. This conclusion is also confirmed by the history of systematics. Pavlinov (2021: 15) argued that RN was initially "based on essentialist ontology, which is irrelevant now."

## 3.4 The Consequences

### 3.4.1 Nomenclatural Instability

An inevitable consequence of our inability to find an objective basis for absolute ranks (Linnaean categories) is nomenclatural instability. This is epitomized by the eternal struggle between "splitters" and "lumpers" (systematists who tend to split or lump more taxa together under a given name than the norm), which occurs at all nomenclatural levels (species, genera, families and so on). Lamarck had already struggled with this problem with nominal species (Lherminier and Solignac 2005: 38). The occurrence of such instability in rank-based nomenclature is fairly uncontroversial as even its most vocal adepts have admitted it (for instance, Dubois 1982: 11; Löbl 2015) and documented many empirical examples. A theoretical example exploring the extent of potential instability in a simple four-taxon case, was previously presented (**Figure 3.10**).

Empirical examples of the inability of rank-based nomenclature to provide stability abound. A rather familiar one was provided in a talk given by Mike Keesey (also on behalf of his co-author Rutger Jansma) at the third meeting of the International Society for Phylogenetic Nomenclature that convened in Halifax (Canada) in July 2008. Keesey (cited in Laurin and Bryant 2009: 336) pointed out that:

> various taxon names typified by *Homo* or *Homo sapiens* (from *Homo* to *Hominoidea*) in RN have been associated with between two (*Hominoidea*) and six (*Hominidae*) nested clades. This instability in delimitation results partly from phylogenetic uncertainty, but **mostly from personal preferences** and the recent shift towards monophyletic taxa. In contrast, vernacular names, such as "apes," "lesser apes," "great apes," and "African great apes" tend to consistently apply to one clade, when used for a clade. Given the **ambiguous meaning of formal names**, and the **greater precision of vernacular names**, he recommends coining new formal names derived from vernacular names in PN, rather than converting currently-used formal names.

This example is striking in that the principles of rank-based nomenclature as implemented in the *Zoological Code* provided less nomenclatural stability for the name *Hominidae* than the (unregulated) vernacular language did for the equivalent term ("hominids").

Another good empirical example of unnecessary instability generated by RN was provided by de Queiroz and Gauthier (1992). Among squamates, the taxa *Agamidae* and *Chamaelonidae* have long been thought to be closely related and to form the taxon *Acrodonta* (**Figure 3.13**). In the early 1990s, some phylogenies suggested that *Chamaeleonidae* was nested within *Agamidae*. Under RN, if we want to maintain monophyly of the taxa (which is not required by the *Zoological Code* but is now an established taxonomic practice), one of these names must be discarded because otherwise, we would have a family within a family, which is contrary to basic principles of RN. Given the topology of the tree, it would make sense to retain *Agamidae* for the clade (if one of the family names, rather than *Acrodonta*, is to be used) because it is associated with ancestor 1, which is the ancestor of the whole clade. Unfortunately, this solution is forbidden by the *Zoological Code* because *Chamaeleonidae* was erected before *Agamidae*, so it has priority. Thus, *Chamaeleonidae*, which was initially associated with ancestor 2, now becomes associated with the much older ancestor 1 and the name becomes attached to a much more inclusive clade (**Figure 3.13**). The clade that was formerly called *Chamaeleonidae* must now be renamed (most simply, *Chamaeleoninae*, a

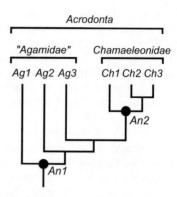

**FIGURE 3.13** Phylogeny and nomenclature of the squamate clade *Acrodonta*. The taxon Chamaleonidae was suggested to be nested within *Agamidae* (both considered families under RN). To avoid having a family nested within a family, if we wish for taxa to be monophyletic, under RN, the oldest of the two (*Chamaeleonidae*) has priority, but this results in a rather large change in contents.

*Source:* Redrawn from Queiroz and Gauthier (1992: figure 3).

subfamily), even though neither its content nor its phylogeny have changed. This illustrates that under RN, association of a taxon name with a rank is more important than its association with a clade or ancestor or with the included lower-ranking taxa. Fortunately, subsequent research suggests that both taxa (*Agamidae* and *Chamaeleonidae*) are monophyletic (Okajima and Kumazawa 2010), so phylogenetic progress does not, in this case, create these unfortunate nomenclatural conclusions.

However, nomenclatural instability can be created by other sources in RN. Let's return to the example of the taxa *Acrodonta, Agamidae* and *Chamaeleonidae* under the more favorable situation in which all these taxa are monophyletic, as discussed by de Queiroz (2012). Their rank could change (**Figure 3.14**), either because of a personal preference or to accommodate newly discovered taxa, or because a greater resolution of the squamate tree pushes a systematist to reallocate some names to less inclusive taxa. The taxa *Agamidae* and *Chamaeleonidae* would become *Agaminae* and *Chamaeleoninae* as they are reranked from family to subfamily level, which is only moderately disrupting. However, to reflect the change from suborder to family rank, the taxon *Acrodonta* must be renamed *Chamaeleonidae* because this is the oldest families that exists in this clade. This exemplifies the deleterious nomenclatural effect that absolute ranks (Linnaean categories) play in RN; this is because in this system, taxon names retain their association with the original ranks, rather than with the original taxa. As explained by de Queiroz (2012: 139):

> This example illustrates that under rank-based nomenclature, taxon names are more closely associated with categorical ranks than they are with taxa. When the ranks of taxa are changed, taxon names retain their associations with the original ranks, rather than retaining their associations with the original taxa. This situation implies that the rank-based system effectively treats the rank of a taxon as though it is more important to the concept of

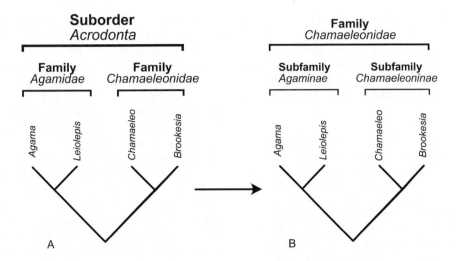

**FIGURE 3.14** Nomenclatural instability under rank-based nomenclature caused by changes in rank.

*Source:* Redrawn from de Queiroz (2012: figure 1).

that taxon than are ideas about properties such as composition, diagnostic characters, or phylogenetic relationships. It is therefore inconsistent with the widespread opinion among biologists that the rank of a taxon is less significant than are those other properties.

These problems affect names that have standardized rank endings; in zoology, this affects only the names of the family group. Thus, under RN, zoological names at other ranks are applied in a manner reminiscent of Linnaeus and other 18th century systematists (de Queiroz 2012: 139), which implies a closer link with taxonomic composition and diagnostic characters. Note that this inappropriate emphasis on ranks is not restricted to zoological names. For instance, Article 1 of the Preamble of the *Cultivated Plant Code* (Brickell et al. 2016) states: "The purpose of giving a name to a taxon is to supply a means of referring to it and to indicate **to which category it is assigned**, rather than to indicate its **characters** or **history**." This indicates that the *Cultivated Plant Code* deliberately ties names more closely to categories than to characters or evolutionary history. These examples show that PN better preserves the association between names and taxa than RN when the phylogeny changes or when rank changes, both of which frequently occur, as shown by the history of systematics (for instance, Bryant 1996; Brochu 1999; Laurin and Smithson 2020).

The problem of name changes caused by rank changes was previously illustrated using a zoological example, but note that this source of nomenclatural instability linked with ranks is even greater in botany because the *Botanical Code* (which regulates names of fungi, in addition to photosynthetic organisms, such as alga, and even cyanobacteria) has ranked taxa with mandatory endings extending up to the phylum (see Section 2.4.3 "*Botanical Code*"). Similarly, this problem affects taxa of all ranks, from classes to subtribes of prokaryotic organisms (see Section 2.4.5 "*Prokaryotic Code*") and the ranks from subgenus to realm (which is above kingdom) for names of virus taxa (see Section 2.4.6 "*Code for Viruses*").

Nomenclatural instability generated by RN has far-reaching negative impacts, even in the apparently distant field of biochronology. Thus, some biozones are based on the stratigraphic distribution of a nominal genus, notably those of the Permo-Triassic continental strata of the Karoo Basin in South Africa (Rubidge et al. 2013). One of these used to be called the *Dicynodon* Assemblage Zone (AZ for here on). A recent phylogenetic analysis recently showed that *Dicynodon*, as previously delimited, was in fact polyphyletic and that many of the 16 valid species previously attributed to *Dicynodon* have a different appearance date in the fossil record (Kammerer et al. 2011: figure 162). Thus, this zone has been redefined and renamed the *Daptocephalus* Assemblage Zone (Viglietti et al. 2016). Obviously, taxonomic and nomenclatural progress can lead to improvements in biochronology, as shown by this example.

As we have seen, dealing with ranks, including when attempting to apply the principle of priority, raises a number of problems. The codes have been developed and clarified over time to deal as best as they can with synonymy and homonymy in the context of rank-based nomenclature, but, as a result, they have become long and complex. So complex, in fact, that even the ICZN has sometimes failed to apply its own rules! Bock (1994: 7) documented a case (Opinion 1068, by Melville 1977) in which a family-group name was conserved at one nomenclatural level (*Leptosomatidae* Filipjev 1916) but was rejected at another (lower) level (*Leptosomatini* Filipjev 1916). Yet, the code then in force, like the current one, stipulated that "A name established for a taxon at any rank in the family group is deemed to have been simultaneously established for nominal taxa at all other ranks in the family group" (Article 36.1, ICZN 1999). More generally, Bock (1994: 8) stated that:

> Family-group nomenclature is as difficult for the International Commission on Zoological Nomenclature as it is for individual zoologists, including ornithologists, judging from the problems experienced by the ICZN in dealing with avian family-group names over the past four decades.

If the commission that drafts and updates a code fails to respect it (either by ignorance or deliberately), how can the rest of the systematic community be expected to abide by it?

This complexity of RN is not problematic only for systematists. Hibbett et al. (2005: 665–667) explained this in their description of an "automated phylogenetic taxonomy" (and nomenclature):

> The *mor* system demonstrates that the core elements of phylogenetic taxonomy can be automated. This system also demonstrates an **advantage of rank-free taxonomy** [nomenclature] relative to traditional taxonomy, which is that taxa in a rank-free system are amenable to algorithmic interpretation. Of course, the groups delimited by *mor* could be classified using Linnaean categories. However, that would negate another advantage of rank-free classification, which is that the names of taxa are stable in the face of rearrangements in tree topologies. This last feature is critical for automated taxonomic systems, because trees may change from week to week, resulting in arrangements that **violate the hierarchy of Linnaean ran**ks. Conceivably, algorithms could be devised that would detect conflicts between the hierarchy of Linnaean ranks and the nested relationships among clades, but automated solutions to such conflicts are probably not a realistic possibility (as shown by the *Coprinus* example). If taxonomy is to be fully automated, it will be necessary to adopt a system of **rank-free classification**.

For these reasons, the automated algorithms (*mor*) presented by Hibbett et al. (2005) used phylogenetic nomenclature to automatically label nodes with names of large fungal clades.

### 3.4.2 Misuse of Ranks in Conservation Biology and Comparative Biology

A further consequence of the subjectivity of ranking taxa under RN is that the ranks thus attributed to taxa convey disinformation. As mentioned in the preface of the *PhyloCode*:

> The existence of ranks encourages researchers to use taxonomies inappropriately, treating taxa at the same rank **as though they were comparable in some biologically meaningful way**—for example, when they count genera or families to study past and present patterns of biological diversity.

> **(Cantino and de Queiroz 2020: xii)**

This is especially problematic in conservation biology, which often relies on taxon counts at a given rank to assess biodiversity. This, in turn, can create peer pressure not to synonymize species-level taxa that do not match our species concept. For instance, it is now reasonably well-established that the brown bear (*Ursus arctos*) is paraphyletic because some brown bears are closer to the polar bear (*Ursus maritimus*) than to other brown bears (Ashrafzadeh et al. 2016), and both taxa have hybridized in the last 100,000 years,

plausibly more than once and in more than one location (Edwards et al. 2011). Yet, both species are retained, presumably because declaring the polar bear a mere subspecies or a race would hamper conservation efforts, given that conservation emphasizes threatened species (rather than nominal taxa of other ranks). Note that the IUCN has a list of threatened species (www.iucn.org/resources/conservation-tools/iucn-red-list-threatened-species), but no list of threatened genera or subspecies.

This, however, is only a small part of the problem of what Bertrand et al. (2006) called "**taxonomic surrogacy**," which consists in counting supraspecific taxa of a given rank (often genera or families) as a substitute of the number of species to assess biodiversity. This approach rests on the assumption that counts of higher taxa can be reliable indicators of the number of species, and since it is easier to count higher taxa than species (partly because identifying a specimen to family or genus level is easier than to species level), this could save time and money. Leaving aside for now the problems linked with the species concept (see Chapter 6), this approach raises many problems because it is based on the false premise that taxa of a given rank are somehow comparable, an assumption that predates the rise of "tree thinking" in biology (O'Hara 1988, 1997), and which is known to be wrong even among the most enthusiastic proponents of RN (such as Dubois et al. 2021: 10; see also Section 3.3.4.2). To exemplify this, O'Hara (1997: 325) cited an example from Raikow (1986), which "concerns the traditional orders of birds, the largest of which is the Passeriformes which by itself contains about half of all bird species, with the other 30 or so traditional orders containing all the rest." This may seem surprising, but Raikow (1986: 258) concluded that we asked the wrong question:

> Why, then, are there so many kinds of passerine birds? Probably not because they possess any key innovation of potential historical significance; none is evident among their synapomorphies. The "success" of the group, in terms of number of species, is more likely an artifact of its classificatory history [like the rank of order being attributed to it] than a consequence of any particular biological feature.

This was probably not the best example to illustrate the pitfalls of pre-"tree thinking" evolutionary biology because passerine birds appear to be remarkably diversified. This, however, is not because one order out of about 30 includes half of the bird species; rather, it is because the taxon Passeriformes does not seem to have originated earlier than most other avian taxa currently ranked as orders (Prum et al. 2015), so this implies an unusually fast diversification rate for passeriforms. Raikow (1986: 257) hinted at this problem when he mentioned that mousebirds (Coliiformes) formed "a group roughly as old as the passerines but with only six extant species." However, the great variability in the number of included taxa in taxa of higher ranks is indeed a serious problem for taxonomic surrogacy. For families, the number of included nominal species ranges from one to about 60,000, the highest figures being among insects (Lücking 2019: 202).

In conservation biology, taxonomic surrogacy has been shown to work poorly. This has been illustrated by studies aimed at establishing a network of biological reserves in which each higher taxon is represented at least once. Thus, summarizing an earlier study, Bertrand et al. (2006: 151) concluded:

> The authors have used data from a wide range of South African plants and animals (mammals, birds, plants, butterflies, termites, antlions, scarab beetles and buprestid beetles) to compare the complementary sets resulting from the use of genera and families with the networks representing the species. With the removal of termites and antlions which were either poorly surveyed or represented by few species, **the average overlap was 27.7% for genera and species-based sets and 4.5% for families and species**. Thus this study provided support against the congruence in complementarity across taxonomic levels.

These are hardly encouraging results, but taxon surrogacy has had little more success in the use of indicator taxa to monitor global biodiversity. Bertrand et al. (2006: 156) stated:

> Based on the assumption that the richness of a particular taxon, or set of taxa, can mirror global species richness, numerous indicator taxa have been proposed . . . However, this surrogacy approach has been severely criticised based on studies which, for a wide range of biotas, exhibits little correspondence in taxon richness.

Taxonomic surrogacy has frequently been used in comparative and evolutionary biology. For instance, many studies on the evolution of biodiversity over time have counted taxa of a given rank, often families (for instance, Benton 1985, 1989; Sahney and Benton 2008) or genera (e.g., Day et al. 2015), rather than a more objective measure of biodiversity, such as evolutionary lineages (although in paleontology, this probably results partly of the difficulty in identifying some specimens to the species level, given that the species-level systematics of many nominal genera is in a state of flux). Taxonomic surrogacy has also been used to study correlations between extinction and origination rates, fractal patterns in biodiversity or evidence for external causes of extinctions (Bertrand et al. 2006: 150).

Griffiths (1974: 118) had commented long ago that:

> What biologists should be talking about in historical comparisons is not genera and families, but groups of Paleocene origin, groups of Cretaceous origin, groups of Jurassic origin etc. (to whatever degree of precision is required for the purpose in hand). As our understanding of the history of organisms increases the Linnaean category terms should be gradually abandoned.

This visionary statement was echoed long after by Hedges and Kumar (2009: 16), who pointed out the great benefits that could be reaped from a timetree of Life:

> The immense value of having a robust Timetree of Life—for all fields of science—cannot be overstated. It will provide a means for estimating rates of change for almost anything biological—for example, morphological structures, behaviors, genes, proteins, non-coding regions of genomes—in any group of organisms. In that sense it will catalyze a Renaissance in comparative biology. For paleontologists, geologists, geochemists, and climatologists, it will provide a biological timeline for comparison, prediction, and synchronization with Earth history. In turn this will help formulate better hypotheses for how the biosphere has evolved on Earth and provide insights into evolutionary mechanisms in the Universe.

By contrast, the deleterious effects of taxonomic surrogacy are numerous. To take a paleontological example, Brocklehurst et al. (2013: 475) noted that:

> Across the Kungurian/Roadian boundary [which is also the early/middle Permian boundary], an interval of extinction is visible, and diversity falls from 26 synapsid species in the late Kungurian to 16 in the early Roadian (Figure 3.1A). Interestingly the number of genera falls by only one (Figure 3.1B) [from 15 to 14], possibly due to the large number of polyspecific genera from the Kungurian.

This discrepancy in extinction rate (38% at the specific level vs 6.7% at the generic level) illustrates the pitfalls of using higher taxon counts to assess evolution of biodiversity over time. This confirms previous studies (cited in Bertrand et al. 2006: 151) that found taxon surrogacy to poorly predict specific richness. Yet, most paleontological studies on large-scale biodiversity evolution used counts of genera (for instance, Alroy et al. 2001; Foote 2003, 2007; Day et al. 2015), families (e.g., Raup and Sepkoski 1982; Benton 1985, 1989; Erwin et al. 1987; Sepkoski 1993; Sahney and Benton 2008; Sahney et al. 2010) or more rarely (especially in older studies), even orders (for instance, Sepkoski 1978; Archibald and Deutschman 2001; see review in Smith and Patterson 1988). This situation is rather puzzling given that the problems with taxon surrogacy and the lack of equivalence of taxa of a given rank have long been known. As recently pointed out by Minelli (2000: 340), Michel Adanson (1727–1806), one of the best pre-evolutionary systematists, already alerted contemporary systematists about this problem, for instance by pointing out that no systematist had provided a definition of "genus," except for artificial ones (Adanson 1763: cv). But two and a half centuries later, many systematists continue counting genera!

## 3.5 Not Enough Ranks, or Too Many?

Under binominal nomenclature, redundancy is unavoidable because each new species must be inserted into a genus. If it happens to be the sister-group of a clade that includes at least two genera (or at least one taxon of a higher rank), this requires erection of a redundant (monotypic) genus (one that includes a single species). But at least some rank-based codes encourage more redundancy. For instance, the *Zoological Code* states (ICZN 1999: Appendix B4) that "An author establishing a new nominal taxon should clearly state the higher (more inclusive) taxa (such as family, order, class) to which the taxon is assigned." This completes Article 11.9.3 of the same code, which states (in conformity with the principle of binominal nomenclature as implemented in the current RN codes): "A species-group name must be published in unambiguous combination with a generic name." Thus, a newly discovered nominal species that is the sister-group of a clade ranked as a class not only requires erection of a redundant genus, but also of a new family, order and class—an undesirable property of RN (Farris 1976).

Progress in phylogenetics and in our knowledge of biodiversity has also revealed an additional weakness in RN as currently implemented in the main RN codes. Namely, the number of ranks used in most studies is not nearly enough ranks to rank all known taxa. This problem was already faced by Linnaeus, who is commonly considered to have used no more than six formal, named ranks, from variety (in plants) to kingdom (see Section 1.3). This created problems, notably for the application of his rank-based nomenclature, as pointed out by Papavero et al. (2001: 28), and, in fact, Linneaus may well have used far more ranks than normally admitted (Dubois 2007a). Thus, Linnaeus (1758) placed the many species that he recognized in his genus *Curculio* (*Insecta, Coleoptera*) two (not formally ranked) taxa called "*Brevirostres*" and "*Longirostres*," and these include three and two lower-ranking (but not formally ranked and unnamed) taxa, at least according to the interpretation of Dubois (2007a), which he admits is tentative. The alternative is that at least some of these putative taxa of unspecified rank, most of which are unnamed, are simply successive levels of an identification key (Dubois 2007a: 83). Nevertheless, if Dubois' (2007a) interpretation that these are genuine taxa is correct, Linnaeus actually used 17 ranks, although 11 of them were not formally named and half of the taxa at these ranks were unnamed. Many other examples of taxa of intermediate (and unnamed) ranks are documented in Linnaeus' writings. One is the nominal genus *Phalaena* (no longer valid) that Linnaeus (1758: 496) erected for moths (*Lepidoptera*), and for which he listed seven subgenera, some of which are subdivided into three additional levels between subgeneric and specific (Dubois 2007a: 99). Taxa of at least two of these three additional levels cannot be available because the *Zoological Code* (Article 6.2) does not allow for ranks between the subgenus and the aggregate of species. On at least one occasion, namely in his genus *Papilio*, he recognized six "phalanges," but did not mention this rank elsewhere (Dubois 2007a: 86).

Even though the rank-based codes allow use of many more ranks than those used by Linnaeus (even under Dubois' 2007a interpretation), this number is still limiting. A recent example is documented in a controversy about the nomenclature of the speciose anuran taxon *Rana*, typically ranked as a genus (Dubois 2007b). This taxon is a good example to illustrate the limitations imposed by the *Zoological Code* because it is very speciose. Indeed, despite the old nominal genus *Rana* being partly dismantled over many years (and according to

the nomenclature recognized by some authors) by raising the rank of taxa formerly recognized as subgenera (such as *Pelophylax*; Dubois and Ohler 1994) to genera, Amphibiaweb (https://amphibiaweb.org/lists/Ranidae.shtml) still recognized (on June 19, 2022) 107 species in the nominal genus *Rana*. Of course, more species are attributed to this genus according to other nomenclatures—for instance, if *Pelophylax* is considered to be a subgenus. Thus, Hillis and Wilcox (2005: 299) considered that the genus *Rana* included about 250 species. Considering the hierarchical relationships between taxa reflected in their figure 2 (a simplified tree with several polytomies of various sizes), they named only well-supported clades, leaving many weakly supported clades unnamed. Despite this, they considered recognize up to five levels of taxa between the nominal genus *Rana* and nominal species included therein, and they considered all taxa at these intermediate levels as subgenera. They justified this by quoting Article 10.4 of the *Zoological Code*, which states that:

> A uninominal name proposed for a genus-group division of a genus, even if proposed for a secondary (or further) subdivision, is deemed to be a subgeneric name even if the division is denoted by a term such as "section" or "division."

However, Dubois (2007b: 323) correctly pointed out that in the genus-series, only two ranks are available. As clarified in Article 42.1 of the *Zoological Code*, the genus-series "encompasses all nominal taxa at the ranks of genus and subgenus." The only additional rank available between genus and species is the aggregate of species (Article 6.2). As a result of this problem (and others reviewed by Dubois 2007b), many of the names proposed by Hillis and Wilcox (2005) are not valid under the *Zoological Code*. For instance, within the genus *Rana*, Hillis and Wilcox (2005) recognized a subgenus *Novirana*, which includes a subgenus *Sierrana*, which includes a subgenus *Pantherana*, which includes a subgenus *Stertirana*, which includes a subgenus *Lacusirana*. According to the *Zoological Code*, no more than one of these can be valid. Of course, one level could be named (differently) by recognizing an aggregate of species, but this would still leave three names without a rank (hence, invalid under the *Zoological Code*, unless these taxa are considered paraphyletic). Thus, even with the only moderate degree of phylogenetic resolution recognized by Hillis and Wilcox (2005: fig. 2), only a fairly small proportion of the clades can be named using the *Zoological Code*. The problem is even more acute with *Drosophila*, given that over 1,600 species are already recognized, and that it is estimated that only about 75% of drosophilid biodiversity is known (O'Grady and DeSalle 2018: 18).

This high number of species within a nominal genus is problematic under RN as implemented in the *Zoological Code* because the latter only allows two ranks between genus and species: the subgenus, in the genus series (Article 42.1), and the aggregates of species (Article 6.2). Names at both of these ranks are subject to the principle of priority (Article 23.3.3). In the family series, the *Zoological Code* is more flexible; Article 35.1 of the *Zoological Code* states that "The family group encompasses all nominal taxa at the ranks of superfamily,

family, subfamily, tribe, subtribe, and any other rank below superfamily and above genus that may be desired." Thus, if we count these ranks from the three series subjected to the rule of priority (family, genus and species), the *Zoological Code* minimally allows 11 ranks (in the species series, aggregates of subspecies and subspecies are allowed). If we combine the use of the prefixed "super," "sub" and "infra" with the ranks family and tribe to obtain more ranks in the family series without inventing exotic ranks (such as supertribes, which are allowed, but not mentioned in the code), this total increases to 14 ranks.

It has been suggested that taking into consideration higher-ranking series (not submitted to the rule of priority), about 36 ranks can be used (Simpson 1961; Laurin 2005). However, this did not consider the possibility to recognize additional ranks that have not frequently been used in the family-series in the past. This is possible under the *Zoological Code* by virtue of Article 35.1, which does not even specify how these additional ranks should be named, but this is not necessarily possible under other RN codes. The ranks traditionally recognized by zoologists above the family series include the order, class, phylum and kingdom series. More recently (Woese et al. 1990), the domain series has been recognized for the taxa *Bacteria, Archaea* and *Eucarya* (for eukaryotes), which are still thought to be the basalmost divisions of known biodiversity. Given that the paper that suggested recognizing domains above the kingdom series (Woese et al. 1990) was cited 7,775 times according to Google Scholar (searched June 19, 2022), it was clearly well-received by the systematic community. None of these levels (above the family series) are tightly regulated by the *Zoological Code* (for instance, priority does not apply), but this differs in other RN codes. Nevertheless, if we also consider the old tradition of using the prefixes "super," "sub" and "infra," we get about 20 ranks from the five higher-ranking series (above the family series). When combined with the up to 11 to 14 ranks subjected to the rule of priority under the *Zoological Code*, this still only amounts to only 31 to 34 ranks, so Simpson's (1961) estimate of 36 ranks already appears generous (but feasible by generating additional ranks in the family series). However, to accurately rank the known biodiversity (by assigning sister-taxa the same rank) would presumably require well over a hundred ranks, possibly thousands (precise counts are difficult to obtain because they depend on the reference phylogeny and require considering all life forms simultaneously). Note that flexibility in the *Zoological Code* in the family series, and the very scant regulation of that code above this in the class series, allows use of many more ranks (not subject to priority), and this flexibility has been exploited by Dubois et al. (2021: 33) to recognize 25 ranks between class and genus, and Dubois (2006) proposed a system that could produce up to 209 ranks. The high number of ranks (allowed above the genus series level), can be illustrated by examples at various levels, in addition to the example of the taxon *Rana* given previously.

We may start with the example of the taxon *Chordata*, which may be reasonably familiar to most readers given that we are chordates. *Chordata* is typically ranked as a phylum, and *Vertebrata* as a subphylum. The most inclusive superclass of vertebrates (which is the most favorable to RN by minimizing the number of intermediate ranks between it and the taxon

*Vertebrata*) is *Osteichthyes*. We probably know only a small proportion of the extinct taxa that would require additional ranks between the subphylum *Vertebrata* and the superclass *Osteichthyes*, but what we do know already creates serious challenges. To demonstrate this, we may use a simplified phylogeny based on a recent review of the literature (**Figure 3.15**). This simple tree shows that at least ten ranks are required for taxa more inclusive than *Osteichthyes* but less inclusive than *Vertebrata*. However, among the ranks that can be generated by the classically recognized series and prefixes in zoological nomenclature, only the rank of infraphylum is available. What should be done with the nine other taxa?

At a higher taxonomic level, it has long been obvious that more ranks would be required. Already, when Woese et al. (1990) proposed to recognize the new rank "Domain" above the Kingdom series, his Domain *Eukarya* included six terminal taxa, one of which was "animals," which is typically ranked as a Kingdom. This highly simplified tree (Woese et al. 1990: fig. 1) already exhausted the ranks available, assuming that we would recognize sub- and infradomains to accommodate this hierarchy. However, recent works have greatly increased resolution of the relevant taxa and show many nested clades between the Domain *Eukarya* and the Kingdom *Animalia* (or *Metazoa*). Thus, using the phylogeny of Burki et al. (2020: fig. 1), despite a low resolution among the main eukaryotic taxa, there are at least three nested taxa (and probably many more, once the tree is properly resolved) between the root of *Eukarya* and *Opisthokonta*. Considering the more resolved eukaryotic phylogeny of Torruella et al. (2015: fig. 1), which does not include all main eukaryotic clades, there are at least four nested taxa between the root of *Eukarya* and *Opisthokonta*, and five more from the root of *Opisthokonta* to *Metazoa*. Thus, even if we raised *Eukarya* to the informal rank of Super-Domain, to rank all nine intermediate taxa implied by the phylogeny of Torruella et al. (2015: fig. 1) by combining the series Domain and Kingdom and the prefixes "super," "sub" and "infra," only eight ranks would be available and *Metazoa* would rank below the level of Infra-Kingdom, which would contradict its usual Kingdom ranking. This may be why Torruella et al. (2015) and Burki et al. (2020) recognize the informal rank of "supergroup." This does not solve all problems; Burki et al. (2020: 51) recognized that:

> Decisions about which major groupings are considered supergroups have always been arbitrary, but the increasing absence of distinguishing biological features makes this more apparent. Paradoxically, the improved resolution of the tree makes the problem worse, not better. . . . Moreover, there is a blurry line between orphan lineages, which often have just a few known species, and the least speciose supergroups. If a diversity-poor orphan is shown to be evolutionarily unrelated to all supergroups, does that make it a new supergroup?

Thus, we see that the *Zoological Code* does not provide enough ranks for taxa of at least three levels: between taxa ranked as genera and species (see previous *Rana* example), between phylum and classes (see previous Vertebrata example)

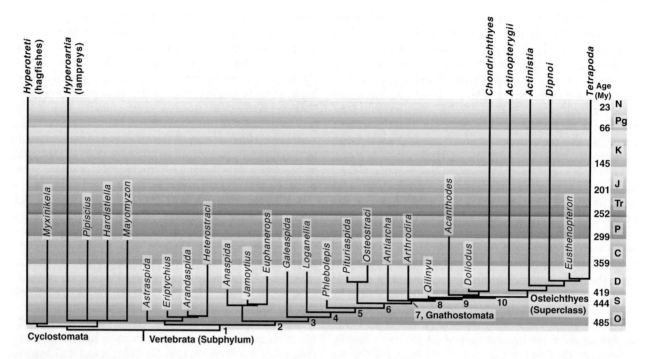

**FIGURE 3.15**  Vertebrate phylogeny, showing that at least ten ranks would be required to accommodate the taxa between the subphylum *Vertebrata* and the superclass *Osteichthyes*. Note that only major clades are represented in this simplified tree; if all known extinct taxa were shown, even more ranks would be required.

*Source:* Modified from Laurin (2021: figure 14.1). Colors of geological stages (in the e-book version; not named for lack of space) follow the Commission for the Geological Map of the World (CGMW), located in Paris (France). Abbreviations of geological periods: C, Carboniferous; D, Devonian; J, Jurassic; K, Cretaceous; N, Neogene; O, Ordovician; P, Permian; Pg, Paleogene; S, Silurian; and Tr, Triassic.

and above the kingdom level (see *Eukarya* example in the previous paragraphs). Other RN codes might behave slightly differently, but they also generate this problem (see Chapter 2). It could also be argued that the more ranks we recognize, the harder it is to rank taxa (also see Section 3.3), and the more rank reshuffling may be required as the tree changes.

Because of these limitations imposed by the initially low number of available ranks, various solutions have been proposed to increase the number of ranks. All current rank-based codes allow far more ranks than were ever used by Linnaeus (see Chapter 2), but this is still far from sufficient. Thus, the expression "species in statu nascendi" has been coined for taxa intermediate between subspecies that can still hybridize and reproductively isolated species. This expression was probably first proposed by Fægri (1937: 410), who proposed the following (not entirely clear) definition: "lines of evolution in the moment of dividing, but they have not divided as yet, and as we cannot to-day predict how the results will be when the division is accomplished, we might just as well resign." This expression was used in numerous studies (370 according to a Google Scholar search carried out on May 8, 2022). The term "Artenkreis" (meaning "species circles" or "species rings") had been proposed earlier by Rensch (1928) for closely related, imperfectly reproductively isolated taxa, and this term also had substantial success, judging by the 819 papers that used this term according to Google Scholar (search carried out

on May 8, 2022). Mayr (1931: 2) proposed to replace this term by "superspecies," which clarifies that contrary to the "*species in statu nascendi*," "superspecies" are above the species level. Superspecies are composed of semispecies, although this latter term has subsequently been used for *species in statu nascendi* (Minelli 2022a).

More radical solutions to increase the number of ranks more substantially have also been made. Thus, Farris (1976) proposed additional prefixes and suggested that prefixes could be juxtaposed to produce a potentially unlimited number of ranks. This system would use a decimal ranking system (ranks would be rational numbers). The primary rank would be the most important and be analogous to an integer; each prefix, starting from the last (immediately preceding the main rank name), would denote increasingly small decimals (Table 3.4). For instance, a gigasuperorder would receive a rank of 5.14 because orders have a rank of 5, the prefix super had a modifier value of 1 and giga a value of 4. This rank would be greater than that of a supersuborder, which has a value of 4.91 in this system because the prefix sub has a modifier value of -1 (applied to the first decimal, this subtracts 0.1 from 5, which represents the order). As the number of ranks required increases, this ingenious system would become a bit cumbersome; would it be convenient to compare gigasupersubclasses with megagigamicropicophyla? This may be one reason why very few systematists have used it, as far as I know.

## TABLE 3.4

Primary ranks and prefixes to modify ranks proposed by Farris (1976: tables 1, 2).

| Primary ranks | | Prefixes | |
|---|---|---|---|
| **Name** | **Numerical rank** | **Prefix** | **Modifier value** |
| Kingdom | 9 | Giga | +4 |
| Phylum | 8 | Mega | +3 |
| Class | 7 | Hyper | +2 |
| Cohort | 6 | Super | +1 |
| Order | 5 | (none) | 0 |
| Family | 4 | Sub | −1 |
| Tribe | 3 | Infra | −2 |
| Genus | 2 | Micro | −3 |
| Species | 1 | Pico | −4 |

Farris' (1976) scheme allows the creation of many additional ranks, but how many would be required to rank all named taxa? And what about the more numerous still unnamed and undiscovered taxa? We can try to compute the number of ranks required to describe the currently known taxa, which can conservatively be estimated as a minimum of 2 million, given that between 1.3 and 1.8 million species are currently recognized (Lücking 2019: 205) and that if the actual Tree of Life (TOL) were strictly dichotomic, this would imply the existence of about as many supraspecific taxa. The actual number of required ranks depends on the shape of the TOL. If it were strictly symmetrical (**Figure 3.16A**), for $N_{tx}$ terminal taxa, the number of required ranks would be $Log_2 N_{tx}+1$ (Laurin 2005: 77). Thus, for 2 million terminal taxa (species or subspecies), 22 ranks would be required, which is not problematic. However, if the tree were completely asymmetrical (**Figure 3.16B**) for $N_{tx}$ terminal taxa, the number of required ranks would be $N_{tx}$, which would clearly not be convenient. Of course, the actual shape of the TOL is unknown, but it is clearly not close to either extreme, so the number of ranks required for all known taxa is impossible to determine accurately, but it is clearly great, plausibly in the order of 1,000 to 50,000 (Laurin 2005). This suggests that if we wanted to rank all taxa, solutions like the use of multiple prefixes for a given taxon (e.g., Farris 1976) would be required.

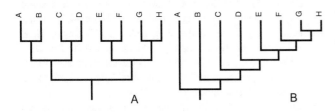

**FIGURE 3.16** Perfectly symmetrical (A) and asymmetrical (B) phylogenies.

## 3.6 Conclusion: Ranks Must Go

Given all the caveats mentioned here, it is no surprise that Zachos (2011: 733) concluded:

> The Linnean hierarchy was introduced in a pre-evolutionary paradigm, and there is no necessity to stick desperately to it. Systematics and taxonomy in an evolutionary paradigm have to come up with reproducible results and testable conclusions. Linnean ranks are useful for neither! These ranks appear to be of no benefit other than that we are used to them.

Zachos (2011) failed to mention that Linnaean categories are retained because they are required to use rank-based nomenclature, but now that PN is a viable alternative, it may be time to drop these categories.

## 3.7 Diagnoses Do Not Delimit Taxa

While Linnaean ranks fail to delimit taxa, some readers might think that the diagnosis serves this purpose. However, diagnoses neither define nor delimit taxa precisely, partly because they include several character states, only some of which are derived. As Sereno (2005: 596) stated, "The traditional 'differential diagnosis' amounts to a grab bag of symplesiomorphies and synapomorphies that may, or may not, be present in most group members." Another reason that diagnoses do not stabilize taxon delimitation is that they can be revised as frequently as desired (typically, each study on a taxon includes a revised diagnosis). Indeed, their purpose is a little different: they act as very basic (binary) identification keys allowing a systematist to decide if a specimen should be attributed to a taxon, or if a low-ranking taxon should be assigned to a higher-ranking taxon. In this respect, the glossary of the *Zoological Code* is misleading in defining "definition" as "A statement in words that purports to give those characters which, in combination, uniquely distinguish a taxon," which clearly refers to a diagnosis. Diagnoses describe taxa, but they do not define them. This leaves taxa undelimited under RN.

## REFERENCES

Adanson, M. 1763. Familles des plantes. 1ère partie. Vincent, Paris, cccxxv + 189 pp.

Alroy, J. 1999. The fossil record of North American mammals: Evidence for a Paleocene evolutionary radiation. Systematic Biology 48:107–118.

Alroy, J., C. R. Marshall, R. K. Bambach, K. Bezusko, M. Foote, F. T. Fürsich, T. A. Hansen, S. M. Holland, L. C. Ivany, D. Jablonski, D. K. Jacobs, D. C. Jones, M. A. Kosnik, S. Lidgard, S. Low, A. I. Miller, P. M. Novack-Gottshall, T. D. Olszewski, M. E. Patzkowsky, D. M. Raup, K. Roy,

J. J. Sepkoski, Jr., M. G. Sommers, P. J. Wagner, and A. Webber. 2001. Effects of sampling standardization on estimates of Phanerozoic marine diversification. Proceedings of the National Academy of Sciences of the United States of America 98:6261–6266.

American Ornithologists' Union. 1886. The code of nomenclature and check-list of North American birds adopted by the American Ornithologists' Union. American Ornithologists' Union, New York, viii + 392 pp.

Arai, M. 2014. Aptian/Albian (Early Cretaceous) paleogeography of the South Atlantic: A paleontological perspective. Brazilian Journal of Geology 44:339–350.

Archibald, J. D. 1999. Divergence times of eutherian mammals. Science 285:2031a.

Archibald, J. D. and D. H. Deutschman. 2001. Quantitative analysis of the timing of the origin and diversification of extant placental orders. Journal of Mammalian Evolution 8:107–124.

Arcila, D., R. A. Pyron, J. C. Tyler, G. Ortí, and R. Betancur-R. 2015. An evaluation of fossil tip-dating versus node-age calibrations in tetraodontiform fishes (Teleostei: Percomorphaceae). Molecular Phylogenetics and Evolution 82:131–145.

Arens, N. C. and I. D. West. 2008. Press-pulse: A general theory of mass extinction? Paleobiology 34:456–471.

Ascarrunz, E., J.-C. Rage, P. Legreneur, and M. Laurin. 2016. *Triadobatrachus massinoti*, the earliest known lissamphibian (Vertebrata: Tetrapoda) re-examined by μCT-Scan, and the evolution of trunk length in batrachians. Contributions to Zoology 85:201–234.

Ascarrunz, E., M. R. Sánchez-Villagra, R. Betancur-R, and M. Laurin. 2019. On trends and patterns in macroevolution: Williston's law and the branchiostegal series of extant and extinct osteichthyans. BMC Evolutionary Biology 19:117.

Ashrafzadeh, M. R., M. Kaboli, and M. R. Naghavi. 2016. Mitochondrial DNA analysis of Iranian brown bears (*Ursus arctos*) reveals new phylogeographic lineage. Mammalian Biology 81:1–9.

Avise, J. C. 2008. Three ambitious (and rather unorthodox) assignments for the field of biodiversity genetics. Proceedings of the National Academy of Sciences of the United States of America 105:11564–11570.

Avise, J. C. and G. C. Johns. 1999. Proposal for a standardized temporal scheme of biological classification for extant species. Proceedings of the National Academy of Sciences of the United States of America 96:7358–7363.

Avise, J. C. and J.-X. Liu. 2011. On the temporal inconsistencies of Linnean taxonomic ranks. Biological Journal of the Linnean Society 102:707–714.

Avise, J. C. and D. Mitchell. 2007. Time to standardize taxonomies. Systematic Biology 56:130–133.

Ayala, F. J., M. L. Tracey, D. Hedgecock, and R. C. Richmond. 1974. Genetic differentiation during the speciation process in *Drosophila*. Evolution 28:576–592.

Ayres, D. L., M. P. Cummings, G. Baele, A. E. Darling, P. O. Lewis, D. L. Swofford, J. P. Huelsenbeck, P. Lemey, A. Rambaut, and M. A. Suchard. 2019. BEAGLE 3: Improved performance, scaling, and usability for a high-performance computing library for statistical phylogenetics. Systematic Biology 68:1052–1061.

Barby, F. F., P. Ráb, S. Lavoué, T. Ezaz, L. A. C. Bertollo, A. Kilian, S. R. Maruyama, E. Aguiar de Oliveira, R. F. Artoni, and M. H. Santos. 2018. From chromosomes to genome: Insights into the evolutionary relationships and biogeography of Old World knifefishes (Notopteridae; Osteoglossiformes). Genes 9:306.

Battistuzzi, F. U. and S. B. Hedges. 2009. Eubacteria; pp. 106–115 in S. B. Hedges and S. Kumar, The timetree of life. Oxford University Press, New York.

Benoit, J., F. Abdala, P. R. Manger, and B. S. Rubidge. 2016. The sixth sense in mammalian forerunners: Variability of the parietal foramen and the evolution of the pineal eye in South African Permo-Triassic eutheriodont therapsids. Acta Palaeontologica Polonica 61:777–789.

Benton, M. J. 1985. Mass extinction among non-marine tetrapods. Nature 316:811–814.

Benton, M. J. 1989. Mass extinctions among tetrapods and the quality of the fossil record. Philosophical Transactions of the Royal Society of London, Series B 325:369–386.

Benton, M. J. 1999. Early origins of modern birds and mammals: Molecules vs. morphology. BioEssays 21:1043–1051.

Benton, M. J. 2005. Vertebrate palaeontology. Blackwell, Oxford, 455 pp.

Benton, M. J., A. M. Dunhill, G. T. Lloyd, and F. G. Marx. 2011. Assessing the quality of the fossil record: Insights from vertebrates. Geological Society, London, Special Publications 358:63–94.

Bertrand, Y., F. Pleijel, and G. W. Rouse. 2006. Taxonomic surrogacy in biodiversity assessments, and the meaning of Linnaean ranks. Systematics and Biodiversity 4:149–159.

Betts, J., R. P. Young, C. Hilton-Taylor, M. Hoffmann, J. P. Rodríguez, S. N. Stuart, and E. Milner-Gulland. 2020. A framework for evaluating the impact of the IUCN red list of threatened species. Conservation Biology 34:632–643.

Bock, W. J. 1994. History and nomenclature of avian family-group names. Bulletin of the American Museum of Natural History (USA) 222:1–281.

Boev, Z. and N. Spassov. 2009. First record of ostriches (Aves, Struthioniformes, Struthionidae) from the late Miocene of Bulgaria with taxonomic and zoogeographic discussion. Geodiversitas 31:493–507.

Bokma, F., M. Godinot, O. Maridet, S. Ladevèze, L. Costeur, F. Solé, E. Gheerbrant, S. Peigné, F. Jacques, and M. Laurin. 2016. Testing for Depéret's rule (body size increase) in mammals using combined extinct and extant data. Systematic Biology 65:98–108.

Bookstein, F. L. 1994. Can biometrical shape be a homologous character?; pp. 197–227 in B. K. Hall (ed.), Homology: The hierarchial basis of comparative biology. Academic Press, London.

Boyd, R. 1999. Homeostasis, species, and higher taxa; pp. 141–185 in R. A. Wilson (ed.), Species: New interdisciplinary essays. MIT Press, Cambridge, MA.

Bremer, K. 1988. The limits of amino acid sequence data in angiosperm phylogenetic reconstruction. Evolution 42:795–803.

Brickell, C. D., C. Alexander, J. J. Cubey, J. C. David, M. H. A. Hoffman, A. C. Leslie, V. Malécot, and X. Jin. 2016. International code of nomenclature for cultivated plants. International Society for Horticultural Science, Katwijk aan Zee, The Netherlands, xvii + 190 pp.

Brochu, C. A. 1999. Phylogenetics, taxonomy, and historical biogeography of Alligatoroidea. Journal of Vertebrate Paleontology 19:9–100.

Brocklehurst, N. 2017. Rates of morphological evolution in Captorhinidae: An adaptive radiation of Permian herbivores. PeerJ 5:e3200.

Brocklehurst, N. 2018. An examination of the impact of Olson's extinction on tetrapods from Texas. PeerJ 6:e4767.

Brocklehurst, N., C. F. Kammerer, and J. Fröbisch. 2013. The early evolution of synapsids, and the influence of sampling on their fossil record. Paleobiology 39:470–490.

Bromham, L., M. J. Phillips, and D. Penny. 1999. Growing up with dinosaurs: Molecular dates and the mammalian radiation. Trends in Ecology and Evolution 14:113–118.

Bryant, H. N. 1996. Explicitness, stability, and universality in the phylogenetic definition and usage of taxon names: A case study of the phylogenetic taxonomy of the Carnivora (Mammalia). Systematic Biology 45:174–189.

Bryant, H. N. 2001. Character polarity and the rooting of clado-grams; pp. 319–337 in G. P. Wagner (ed.), The character concept in evolutionary biology. Academic Press, London.

Buckley, T. R., P. Arensburger, C. Simon, and G. K. Chambers. 2002. Combined data, Bayesian phylogenetics, and the origin of the New Zealand cicada genera. Systematic Biology 51:4–18.

Buffrénil, V. de, A. Canoville, S. E. Evans, and M. Laurin. 2015. Histological study of karaurids, the oldest known (stem) urodeles. Historical Biology 27:109–114.

Burki, F., A. J. Roger, M. W. Brown, and A. G. Simpson. 2020. The new tree of eukaryotes. Trends in Ecology & Evolution 35:43–55.

Cai, T., A. Cibois, P. Alström, R. G. Moyle, J. D. Kennedy, S. Shao, R. Zhang, M. Irestedt, P. G. Ericson, and M. Gelang. 2019. Near-complete phylogeny and taxonomic revision of the world's babblers (Aves: Passeriformes). Molecular Phylogenetics and Evolution 130:346–356.

Cain, A. J. 1959. The post-Linnaean development of taxonomy. Proceedings of the Linnean Society of London 170:234–244.

Cantino, P. D. and K. de Queiroz. 2020. International code of phylogenetic nomenclature (PhyloCode): A phylogenetic code of biological nomenclature. CRC Press, Boca Raton, Florida, xl + 149 pp.

Cantino, P. D., J. A. Doyle, S. W. Graham, W. S. Judd, R. Olmstead, D. E. Soltis, P. S. Soltis, and M. J. Donoghue. 2007. Towards a phylogenetic nomenclature of Tracheophyta. Taxon 56:E1–E44.

Carroll, R. L. 1997. Patterns and processes of vertebrate evolution. Cambridge University Press, Cambridge, 448 pp.

Carson, H. L. 1975. The genetics of speciation at the diploid level. The American Naturalist 109:83–92.

Chippindale, P. T., R. M. Bonett, A. S. Baldwin, and J. J. Wiens. 2004. Phylogenetic evidence for a major reversal of life-history evolution in Plethodontid salamanders. Evolution 58:2809–2822.

Cooper, A. and D. Penny. 1997. Mass survival of birds across the Cretaceous-Tertiary boundary: Molecular evidence. Science 275:1109–1113.

Cracraft, J. 1983. Species concepts and speciation analysis; pp. 159–187 in R. F. Johnston (ed.), Current ornithology. Springer, New York, NY.

Cui, P., R. Ji, F. Ding, D. Qi, H. Gao, H. Meng, J. Yu, S. Hu, and H. Zhang. 2007. A complete mitochondrial genome sequence of the wild two-humped camel (*Camelus bactrianus ferus*): An evolutionary history of Camelidae. BMC Genomics 8:241.

Darwin, C. 1859. On the origin of species by means of natural selection or the preservation of favoured races in the struggle for life. John Murray, London, 502 pp.

Davies, T. W., M. A. Bell, A. Goswami, and T. J. Halliday. 2017. Completeness of the eutherian mammal fossil record and implications for reconstructing mammal evolution through the Cretaceous/Paleogene mass extinction. Paleobiology 43:521–536.

Davín, A. A., E. Tannier, T. A. Williams, B. Boussau, V. Daubin, and G. J. Szöllősi. 2018. Gene transfers can date the tree of life. Nature Ecology & Evolution 2:904–909.

Day, M. O., J. Ramezani, S. A. Bowring, P. M. Sadler, D. H. Erwin, F. Abdala, and B. S. Rubidge. 2015. When and how did the terrestrial mid-Permian mass extinction occur? Evidence from the tetrapod record of the Karoo Basin, South Africa. Proceedings of the Royal Society B: Biological Sciences 282:20150834.

de Queiroz, K. 1985. The ontogenetic method for determining character polarity and its relevance to phylogenetic systematics. Systematic Zoology 34:280–299.

de Queiroz, K. 2012. Biological nomenclature from Linnaeus to the PhyloCode. Bibliotheca Herpetologica 9:135–145.

de Queiroz, K. and J. Gauthier. 1992. Phylogenetic taxonomy. Annual Review of Ecology and Systematics 23:449–480.

Didier, G., M. Fau, and M. Laurin. 2017. Likelihood of tree topologies with fossils and diversification rate estimation. Systematic Biology 66:964–987.

Didier, G. and M. Laurin. 2020. Exact distribution of divergence times from fossil ages and tree topologies. Systematic Biology 69:1068–1087.

Didier, G. and M. Laurin. 2021. Distributions of extinction times from fossil ages and tree topologies: The example of some mid-Permian synapsid extinctions. PeerJ 9:e12577.

Dobzhansky, T. 1944. Chromosomal races in *Drosophila pseudoobscura* and *Drosophila persimilis*; pp. 47–144 in T. Dobzhansky and C. Epling (eds.), Contributions to the genetics, taxonomy, and ecology of *Drosophila pseudoobscura* and its relatives. Carnegie Institution of Washington, Washington, DC.

Dornburg, A., J. M. Beaulieu, J. C. Oliver, and T. J. Near. 2011. Integrating fossil preservation biases in the selection of calibrations for molecular divergence time estimation. Systematic Biology 60:519–527.

Drummond, A. J., S. Y. W. Ho, M. J. Phillips, and A. Rambaut. 2006. Relaxed phylogenetics and dating with confidence. PLoS Biology 4:e88.

Dubois, A. 1981. Quelques réflexions sur la notion de genre en zoologie. Bulletin de la Société Zoologique de France 106:503–513.

Dubois, A. 1982. Les notions de genre, sous-genre et groupe d'espèces en zoologie à la lumière de la systématique évolutive. Monitore Zoologico Italiano-Italian Journal of Zoology 16:9–65.

Dubois, A. 1988. The genus in zoology: A contribution to the theory of evolutionary systematics. Mémoires du Muséum National d'Histoire Naturelle 140:1–122.

Dubois, A. 2005a. Proposed rules for the incorporation of nomina of higher-ranked zoological taxa in the international code of zoological nomenclature. 1: Some general questions, concepts and terms of biological nomenclature. Zoosystema 27:365–426.

Dubois, A. 2005b. Proposals for the incorporation of nomina of higher-ranked taxa into the code. Bulletin of Zoological Nomenclature 62:200–209.

Dubois, A. 2006. Proposed rules for the incorporation of nomina of higher-ranked zoological taxa in the International code of zoological nomenclature. 2: The proposed rules and their rationale. Zoosystema 28:165–258.

Dubois, A. 2007a. Nomina zoologica linnaeana. Zootaxa 1668:81–106.

Dubois, A. 2007b. Naming taxa from cladograms: A cautionary tale. Molecular Phylogenetics and Evolution 42:317–330.

Dubois, A., A. M. Bauer, L. M. Ceríaco, F. Dusoulier, T. Frétey, I. Löbl, O. Lorvelec, A. Ohler, R. Stopiglia, and E. Aescht. 2019. The Linz *Zoocode* project: A set of new proposals regarding the terminology, the principles and rules of zoological nomenclature: First report of activities (2014–2019). Bionomina 17:1–111.

Dubois, A. and A. Ohler. 1994. Frogs of the subgenus *Pelophylax* (Amphibia, Anura, genus *Rana*): A catalogue of available and valid scientific names, with comments on name-bearing types, complete synonymies, proposed common names, and maps showing all type localities. Zoologica Poloniae 39:139–204.

Dubois, A., A. Ohler, and R. Pyron. 2021. New concepts and methods for phylogenetic taxonomy and nomenclature in zoology, exemplified by a new ranked cladonomy of recent amphibians (Lissamphibia). Megataxa 5:1–738.

Edwards, C. J., M. A. Suchard, P. Lemey, J. J. Welch, I. Barnes, T. L. Fulton, R. Barnett, T. C. O'Connell, P. Coxon, and N. Monaghan. 2011. Ancient hybridization and an Irish origin for the modern polar bear matriline. Current Biology 21:1251–1258.

Egerton, F. N. 2012. History of ecological sciences, part 41: Victorian naturalists in Amazonia—Wallace, Bates, Spruce. Bulletin of the Ecological Society of America 93:35–60.

Eldredge, N. and S. J. Gould. 1972. Punctuated equilibria: An alternative to phyletic gradualism; pp. 82–115 in T. J. M. Schopf (ed.), Models in paleobiology. Freeman, Cooper & Company, San Francisco.

Eriksson, T. 1998. AutoDecay ver. 4.0 (program distributed by the author). https://github.com/TorstenEriksson/AutoDecay

Erwin, D. H., J. W. Valentine, and J. J. Sepkoski. 1987. A comparative study of diversification events: The early Paleozoic versus the Mesozoic. Evolution 41:1177–1186.

Evans, S. E., A. R. Milner, and F. Mussett. 1988. The earliest known salamanders (Amphibia, Caudata): A record from the Middle Jurassic of England. Geobios 21:539–552.

Fægri, K. 1937. Some fundamental problems of taxonomy and phylogenetics. The Botanical Review 3:400–423.

Faith, D. P. 1992. Conservation evaluation and phylogenetic diversity. Biological Conservation 61:1–10.

Farris, J. S. 1976. Phylogenetic classification of fossils with recent species. Systematic Zoology 25:271–282.

Farris, J. S. 1983. The logical basis of phylogenetic analysis; pp. 7–36 in N. Platnick and V. Funk (eds.), Advances in cladistics, vol. 2. Columbia University Press, New York, NY.

Feder, J. L., P. Nosil, and S. M. Flaxman. 2014. Assessing when chromosomal rearrangements affect the dynamics of speciation: Implications from computer simulations. Frontiers in Genetics 5:295.

Felsenstein, J. 1985. Confidence limits on phylogenies: An approach using the bootstrap. Evolution 39:783–791.

Felsenstein, J. 1993. PHYLIP (phylogeny inference package), version 3.5 c. www.dbbm.fiocruz.br/molbiol/main.html

Felsenstein, J. 2001. The troubled growth of statistical phylogenetics. Systematic Biology 50:465–467.

Foote, M. 2001. Inferring temporal patterns of preservation, origination, and extinction from taxonomic survivorship analysis. Paleobiology 27:602–630.

Foote, M. 2003. Origination and extinction through the Phanerozoic: A new approach. The Journal of Geology 111:125–148.

Foote, M. 2007. Extinction and quiescence in marine animal genera. Paleobiology 33:261–272.

Gao, J. 2019. Dominant plant speciation types: A commentary on: "Plant speciation in the age of climate change". Annals of Botany 124:iv–vi.

Gaudry, A. 1866. Considérations générales sur les animaux fossiles de Pikermi. F. Savy, Paris, 68 pp.

Gavryushkina, A., D. Welch, T. Stadler, and A. J. Drummond. 2014. Bayesian inference of sampled ancestor trees for epidemiology and fossil calibration. PLoS Computational Biology 10:e1003919.

Gavryushkina, A. and C. Zhang. 2020. Total-evidence dating and the fossilized birth–death model; pp. 175–193 in S. Y. Ho (ed.), The molecular evolutionary clock. Springer Nature Switzerland, Cham, Switzerland.

Geoffroy Saint-Hilaire, I. 1859. Histoire naturelle générale des règnes organiques. Masson, Paris, 523 pp.

Geraads, D., G. Didier, A. Barr, D. Reed, and M. Laurin. 2020. The fossil record of camelids demonstrates a late divergence between Bactrian camel and dromedary. Acta Palaeontologica Polonica 65:251–260.

Goloboff, P. A., C. I. Mattoni, and A. S. Quinteros. 2006. Continuous characters analyzed as such. Cladistics 22:589–601.

Goloboff, P. A., C. I. Mattoni, and A. S. Quinteros. 2008. TNT, a free program for phylogenetic analysis. Cladistics 24:774–786.

Goloboff, P. A., M. Pittman, D. Pol, and X. Xu. 2019. Morphological data sets fit a common mechanism much more poorly than DNA sequences and call into question the Mkv model. Systematic Biology 68:494–504.

Goloboff, P. A., A. Torres, and J. S. Arias. 2018. Weighted parsimony outperforms other methods of phylogenetic inference under models appropriate for morphology. Cladistics 34:407–437.

Gonzalez, J., H. Düttmann, and M. Wink. 2009. Phylogenetic relationships based on two mitochondrial genes and hybridization patterns in Anatidae. Journal of Zoology 279:310–318.

Graur, D. and W. Martin. 2004. Reading the entrails of chickens: Molecular timescales of evolution and the illusion of precision. Trends in Genetics 20:80–86.

Greuter, W., G. Garrity, D. L. Hawskworth, R. Jahn, P. M. Kirk, S. Knapp, J. McNeill, E. Michel, D. J. Patterson, R. L. Pyle, and B. J. Tindall. 2011. Draft BioCode (2011): Principles and rules regulating the naming of organisms. New draft, revised in November 2010. Bionomina 3:26–44.

Griffiths, G. C. D. 1973. Some fundamental problems in biological classification. Systematic Zoology 22:338–343.

Griffiths, G. C. D. 1974. On the foundations of biological systematics. Acta Biotheoretica 23:85–131.

Griffiths, G. C. D. 1976. The future of Linnaean nomenclature. Systematic Biology 25:168–173.

Guindon, S. 2020. Rates and rocks: Strengths and weaknesses of molecular dating methods. Frontiers in Genetics 11:526.

Haeckel, E. 1866. Generelle Morphologie der Organismen. Reimer, Berlin, 1036 pp.

Haig, S. M., E. A. Beever, S. M. Chambers, H. M. Draheim, B. D. Dugger, S. Dunham, E. Elliott-Smith, J. B. Fontaine, D. C. Kesler, and B. J. Knaus. 2006. Taxonomic considerations in listing subspecies under the US Endangered Species Act. Conservation Biology 20:1584–1594.

Hall, B. K., ed. 2007. Fins into limbs: Evolution, development and transformation. University of Chicago Press, Chicago, 433 pp.

Halliday, T. J., P. Upchurch, and A. Goswami. 2017. Resolving the relationships of Paleocene placental mammals. Biological Reviews 92: 521–550.

Hanken, J. 1999. Larvae in amphibian development and evolution; pp. 61–108 in B. K. Hall and M. H. Wake (eds.), The origin and evolution of larval forms. Academic Press, London.

Hassanin, A., F. Delsuc, A. Ropiquet, C. Hammer, B. J. V. Vuuren, C. Matthee, M. Ruiz-Garcia, F. Catzeflis, V. Areskoug, T. T. Nguyen, and A. Couloux. 2012. Pattern and timing of diversification of Cetartiodactyla (Mammalia, Laurasiatheria), as revealed by a comprehensive analysis of mitochondrial genomes. Comptes rendus Biologies 335:32–50.

Hauser, D. L. and W. Presch. 1991. The effect of ordered characters on phylogenetic reconstruction. Cladistics 7:243–265.

Heath, T. A., J. P. Huelsenbeck, and T. Stadler. 2014. The fossilized birth-death process for coherent calibration of divergence-time estimates. Proceedings of the National Academy of Sciences 111:E2957–E2966.

Hedges, S. B. and S. Kumar. 2009. Discovering the timetree of life; pp. 3–18 in S. B. Hedges and S. Kumar (eds.), The timetree of life. Oxford University Press, New York.

Heintzman, P. D., G. D. Zazula, J. A. Cahill, A. V. Reyes, R. D. MacPhee, and B. Shapiro. 2015. Genomic data from extinct North American *Camelops* revise camel evolutionary history. Molecular Biology and Evolution 32:2433–2440.

Hennig, W. 1950. Grundzuge einer Theorie der phylogenetischen Systematik. Deutscher Zentralverlag, Berlin, 370 pp.

Hennig, W. 1966. Phylogenetic systematics. University of Illinois Press, Urbana, Chicago, London, 263 pp.

Hennig, W. 1969. Die Stammesgeschichte der Insekten. Kramer, Frankfurt am Main, 436 pp.

Hennig, W. 1981. Insect phylogeny. John Wiley & Sons, Chichester, xi + 514 pp.

Hibbett, D. S., H. R. Nilsson, M. Snyder, M. Fonseca, J. Costanzo, and M. Shonfeld. 2005. Automated phylogenetic taxonomy: An example in the homobasidiomycetes (mushroom-forming Fungi). Systematic Biology 54:660–667.

Hillis, D. M. and T. P. Wilcox. 2005. Phylogeny of the new world true frogs (*Rana*). Molecular Phylogenetics and Evolution 34:299–314.

Ho, S. Y. W., M. J. Phillips, A. Cooper, and A. J. Drummond. 2005. Time dependency of molecular rate estimates and systematic overestimation of recent divergence times. Molecular Biology and Evolution 22:1561–1568.

Höhna, S., T. Stadler, F. Ronquist, and T. Britton. 2011. Inferring speciation and extinction rates under different sampling schemes. Molecular Biology and Evolution 28:2577–2589.

Hugall, A. F., R. Foster, and M. S. Y. Lee. 2007. Calibration choice, rate smoothing, and the pattern of tetrapod diversification according to the long nuclear gene RAG-1. Systematic Biology 56:543–563.

ICZN. 1999. International code of zoological nomenclature. The International Trust for Zoological Nomenclature, London, 306 pp. www.iczn.org/the-code/the-international-code-of-zoological-nomenclature/the-code-online/

Irisarri, I., D. S. Mauro, F. Abascal, A. Ohler, M. Vences, and R. Zardoya. 2012. The origin of modern frogs (Neobatrachia) was accompanied by acceleration in mitochondrial and nuclear substitution rates. BMC Genomics 13:1–19.

Janvier, P. 1996. Early vertebrates. Oxford University Press, Oxford, 393 pp.

Jayaswal, V., J. Robinson, and L. Jermiin. 2007. Estimation of phylogeny and invariant sites under the general Markov model of nucleotide sequence evolution. Systematic Biology 56:155–162.

Jenkins, F. A., Jr., D. M. Walsh, and R. L. Carroll. 2007. Anatomy of *Eocaecilia micropodia*, a limbed caecilian of the Early Jurassic. Bulletin of the Museum of Comparative Zoology 158:285–365.

Johnsgard, P. A. 1960. Hybridization in the Anatidae and its taxonomic implications. The Condor 62:25–33.

Jønsson, K. A., P.-H. Fabre, J. D. Kennedy, B. G. Holt, M. K. Borregaard, C. Rahbek, and J. Fjeldså. 2016. A supermatrix phylogeny of corvoid passerine birds (Aves: Corvides). Molecular Phylogenetics and Evolution 94:87–94.

Joyce, W. G., J. F. Parham, T. R. Lyson, R. C. M. Warnock, and P. C. J. Donoghue. 2013. A divergence dating analysis of turtles using fossil calibrations: an example of best practices. Journal of Paleontology 87:612–634.

Kammerer, C. F., K. D. Angielczyk, and J. Fröbisch. 2011. A comprehensive taxonomic revision of *Dicynodon* (Therapsida, Anomodontia) and its implications for dicynodont phylogeny, biogeography, and biostratigraphy. Journal of Vertebrate Paleontology 31:1–158.

King, B. 2019. Which morphological characters are influential in a Bayesian phylogenetic analysis? Examples from the earliest osteichthyans. Biology Letters 15:20190288.

Laurin, M. 1998. A reevaluation of the origin of pentadactyly. Evolution 52:1476–1482.

Laurin, M. 2005. The advantages of phylogenetic nomenclature over Linnean nomenclature; pp. 67–97 in A. Minelli, G. Ortalli, and G. Sanga (eds.), Animal names. Instituto Veneto di Scienze, Lettere ed Arti, Venice.

Laurin, M. 2008. The splendid isolation of biological nomenclature. Zoologica Scripta 37:223–233.

Laurin, M. 2010a. How vertebrates left the water. University of California Press, Berkeley, xv + 199 pp.

Laurin, M. 2010b. The subjective nature of Linnaean categories and its impact in evolutionary biology and biodiversity studies. Contributions to Zoology 79:131–146.

Laurin, M. 2012. Recent progress in paleontological methods for dating the tree of life. Frontiers in Genetics 3:1–16.

Laurin, M. 2021. Introduction to the phylogenetic diversity of skeletal tissues; pp. 291–293 in V. de Buffrénil, A. J. de Ricqlès, L. Zylberberg, K. Padian, M. Laurin, and A. Quilhac (eds.), Vertebrate skeletal histology and paleohistology. CRC Press, Boca Raton.

Laurin, M. and H. N. Bryant. 2009. Third meeting of the International Society for Phylogenetic Nomenclature: A report. Zoologica Scripta 38:333–337.

Laurin, M. and T. R. Smithson. 2020. Anthracosauria; pp. 751–754 in K. de Queiroz, P. D. Cantino, and J. A. Gauthier (eds.), Phylonyms: A companion to the PhyloCode. CRC Press, Boca Raton, Florida.

Lecointre, G. and H. Le Guyader. 2016. Classification phylogénétique du vivant: tome 1. Belin, Paris, 4th ed., 584 pp.

Legendre, S. 1987. Les immigrations de la "Grande Coupure" sont-elles contemporaines en Europe occidentale. In International Symposium on Mammalian Biostratigraphy and Paleoecology of the European Paleogene, edited by N. Schmidt-Kittler. Münchner Geowissenschaftliche Abhandlungen, Reihe A-Geologie und Paläontologie, 141–148.

Lemierre, A., A. Folie, S. Bailon, N. Robin, and M. Laurin. 2021. From toad to frog, a CT-based reconsideration of *Bufo servatus*, an Eocene anuran mummy from Quercy (France). Journal of Vertebrate Paleontology 41:e1989694.

Levin, D. A. 1979. The nature of plant species. Science 204:381–384.

Lewis, P. O. 2001. A likelihood approach to estimating phylogeny from discrete morphological character data. Systematic Biology 50:913–925.

Lherminier, P. and M. Solignac. 2005. De l'espèce. Syllepse, Paris, XI + 694 pp.

Linnaeus, C. 1735. Systema Naturae. Haak, Lugduni Batavorum, 1st ed., 12 pp.

Linnaeus, C. 1758. Systema naturae. 10th ed. Holmiae (Laurentii Salvii), Stockholm, 824 pp.

Livezey, B. C. 1997. A phylogenetic classification of waterfowl (Aves: Anseriformes), including selected fossil species. Annals of Carnegie Museum 66:457–496.

Löbl, I. 2015. Stability under the international code of zoological nomenclature: A bag of problems affecting nomenclature and taxonomy. Bionomina 9:35–40.

Lu, P. J., M. Yogo, and C. R. Marshall. 2006. Phanerozoic marine biodiversity dynamics in light of the incompleteness of the fossil record. Proceedings of the National Academy of Sciences 103:2736–2739.

Lücking, R. 2019. Stop the abuse of time! Strict temporal banding is not the future of rank-based classifications in fungi (including lichens) and other organisms. Critical Reviews in Plant Sciences 38:199–253.

Luckow, M. and R. A. Pimentel. 1985. An empirical comparison of numerical Wagner computer programs. Cladistics 1:47–66.

Maddison, W. P. 1997. Gene trees in species trees. Systematic Biology 46:523–536.

Maddison, W. P., M. J. Donoghue, and D. R. Maddison. 1984. Outgroup analysis and parsimony. Systematic Zoology 33:83–103.

Magallón, S. 2020. Principles of molecular dating; pp. 67–81 in S. Y. Ho (ed.), The molecular evolutionary clock. Springer Nature Switzerland, Cham, Switzerland.

Magnol, P. 1689. Prodromus historiae generalis plantarum in quo familiae plantarum per tabulas disponuntur. Pech, Montpellier, 79 pp.

Marjanović, D. 2021. The making of calibration sausage exemplified by recalibrating the transcriptomic timetree of jawed vertebrates. Frontiers in Genetics 12:535.

Marjanović, D. and M. Laurin. 2007. Fossils, molecules, divergence times, and the origin of lissamphibians. Systematic Biology 56:369–388.

Marjanović, D. and M. Laurin. 2008. Assessing confidence intervals for stratigraphic ranges of higher taxa: The case of Lissamphibia. Acta Palaeontologica Polonica 53:413–432.

Marjanović, D. and M. Laurin. 2014. An updated paleontological timetree of lissamphibians, with comments on the anatomy of Jurassic crown-group salamanders (Urodela). Historical Biology 26:535–550.

Marshall, C. R. 1994. Confidence intervals on stratigraphic ranges: Partial relaxation of the assumption of randomly distributed fossil horizons. Paleobiology 20:459–469.

Marshall, C. R. 1997. Confidence intervals on stratigraphic ranges with nonrandom distributions of fossil horizons. Paleobiology 23:165–173.

Marshall, C. R. 2008. A simple method for bracketing absolute divergence times on molecular phylogenies using multiple fossil calibration points. The American Naturalist 171:726–742.

Marshall, C. R. 2019. Using the fossil record to evaluate timetree timescales. Frontiers in Genetics 10:1049.

Mayr, E. 1931. Birds collected during the Whitney South Sea Expedition: 12, notes on *Halcyon chloris* and some of its subspecies. American Museum Novitates 469:1–10.

Mayr, E. 1954. Change of genetic environment and evolution; pp. 157–180 in J. S. Huxley (ed.), Evolution as a process. Allen & Unwin, London.

Mayr, E. 1969. Principles of systematic zoology. MacGraw-Hill, New York, 428 pp.

Melville, R. V. 1977. Leptosomatidae in Aves and Nematoda: Resolution of homonymy arising from similarity in the names of the type-genera. Bulletin of Zoological Nomenclature 33:159–161.

Michener, C. D. 1963. Some future developments in taxonomy. Systematic Zoology 12:151–172.

Mickevich, M. F. 1982. Transformation series analysis. Systematic Zoology 31:461–478.

Minelli, A. 1998. Molecules, developmental modules, and phenotypes: A combinatorial approach to homology. Molecular Phylogenetics and Evolution 9:340–347.

Minelli, A. 2000. The ranks and the names of species and higher taxa, or a dangerous inertia of the language of natural history; pp. 339–351 in M. T. Ghiselin and A. E. Leviton (eds.), Cultures and institutions of natural history: Essays in the history and philosophy of science. California Academy of Sciences, San Francisco.

Minelli, A. 2016. Tracing homologies in an ever-changing world. Rivista di estetica 56:40–55.

Minelli, A. 2022a. Species; in B. Hjørland and C. Gnoli (eds.), ISKO Encyclopedia of Knowledge Organization (IEKO). www.isko.org/cyclo/species

Minelli, A. 2022b. The species before and after Linnaeus: Tension between disciplinary nomadism and conservative nomenclature; pp. 191–226 in J. Wilkins, I. Pavlinov, and F. Zachos (eds.), Species problems and beyond: Contemporary issues in philosophy and practice. CRC Press, Boca Raton, FL.

Minelli, A. and G. Fusco. 2013. Homology; pp. 289–322 in K. Kampourakis (ed.), The philosophy of biology: A companion for educators. Springer, Dordrecht.

Mishler, B. D. 1999. Getting rid of species?; pp. 307–315 in R. Wilson (ed.), Species: New interdisciplinary essays. MIT Press, Cambridge, MA.

Mishler, B. D. and M. J. Donoghue. 1982. Species concepts: A case for pluralism. Systematic Zoology 31:491–503.

Mishler, B. D. and J. S. Wilkins. 2018. The hunting of the SNaRC: A snarky solution to the species problem. Philosophy, Theory, and Practice in Biology 10:1–18.

Montgelard, C., F. M. Catzeflis, and E. Douzery. 1997. Phylogenetic relationships of artiodactyls and cetaceans as deduced from the comparison of cytochrome b and 12S rRNA mitochondrial sequences. Molecular Biology and Evolution 14:550–559.

Müller, K. 2004. PRAP: Computation of Bremer support for large data sets. Molecular Phylogenetics and Evolution 31:780–782.

Naomi, S.-I. 2014. Proposal of an integrated framework of biological taxonomy: A phylogenetic taxonomy, with the method of using names with standard endings in clade nomenclature. Bionomina 7:1–44.

O'Grady, P. M. and R. DeSalle. 2018. Phylogeny of the genus *Drosophila*. Genetics 209:1–25.

O'Hara, R. J. 1988. Homage to Clio, or, toward an historical philosophy for evolutionary biology. Systematic Zoology 37:142–155.

O'Hara, R. J. 1997. Population thinking and tree thinking in systematics. Zoologica Scripta 26:323–329.

Okajima, Y. and Y. Kumazawa. 2010. Mitochondrial genomes of acrodont lizards: Timing of gene rearrangements and phylogenetic and biogeographic implications. BMC Evolutionary Biology 10:1–15.

O'Leary, M. A., J. I. Bloch, J. J. Flynn, T. J. Gaudin, A. Giallombardo, N. P. Giannini, S. L. Goldberg, B. P. Kraatz, Z.-X. Luo, J. Meng, X. Ni, M. J. Novacek, F. A. Perini, Z. S. Randall, G. W. Rougier, E. J. Sargis, M. T. Silcox, N. B. Simmons, M. Spaulding, P. M. Velazco, M. Weksler, J. R. Wible, and A. L. Cirranello. 2013. The placental mammal ancestor and the post: K-Pg radiation of placentals. Science 339:662–667.

Oren, A. 2011. How to name new genera and species of prokaryotes?; pp. 437–463 in F. Rainey and A. Oren (eds.), Taxonomy of prokaryotes. Elsevier, Amsterdam.

Papavero, N., J. Llorente-Bousquets, and J. M. Abe. 2001. Proposal of a new system of nomenclature for phylogenetic systematics. Arquivos de Zoologia Museu de Zoologia da Univeristade de São Paulo 36:1–145.

Parker, C. T., B. J. Tindall, and G. M. Garrity. 2019. International code of nomenclature of prokaryotes: Prokaryotic code (2008 revision). International Journal of Systematic and Evolutionary Microbiology 69:S1–S111. https://doi.org/10.1099/ijsem.0.000778

Pavlinov, I. Y. 2021. Taxonomic nomenclature: What's in a name: Theory and history. CRC Press, Boca Raton, 276 pp.

Phillips, M. J. and C. Fruciano. 2018. The soft explosive model of placental mammal evolution. BMC Evolutionary Biology 18:104.

Phillips, M. J., G. C. Gibb, E. A. Crimp, and D. Penny. 2010. Tinamous and moa flock together: Mitochondrial genome sequence analysis reveals independent losses of flight among ratites. Systematic Biology 59:90–107.

Pierotti, R. 2020. Historical links between ethnobiology and evolution: Conflicts and possible resolutions. Studies in History and Philosophy of Science Part C: Studies in History and Philosophy of Biological and Biomedical Sciences 81:101277.

Platnick, N. I. 1987. An empirical comparison of microcomputer parsimony programs. Cladistics 3:121–144.

Posada, D. 2008. jModelTest: Phylogenetic model averaging. Molecular Biology and Evolution 25:1253–1256.

Pough, F. H., R. M. Andrews, M. L. Crump, A. H. Savitzky, K. D. Wells, and M. C. Brandley. 2018. Herpetology. Sinauer Associates, Sunderland, MA, xv + 591 pp.

Prum, R. O., J. S. Berv, A. Dornburg, D. J. Field, J. P. Townsend, E. M. Lemmon, and A. R. Lemmon. 2015. A comprehensive phylogeny of birds (Aves) using targeted next-generation DNA sequencing. Nature 526:569–573.

Pyron, R. A. 2011. Divergence-time estimation using fossils as terminal taxa and the origins of Lissamphibia. Systematic Biology 60:466–481.

Raikow, R. J. 1986. Why are there so many kinds of passerine birds? Systematic Zoology 35:255–259.

Raup, D. M. and J. J. Sepkoski. 1982. Mass extinctions in the marine fossil record. Science 215:1501–1503.

Raven, P. H., B. Berlin, and D. E. Breedlove. 1971. The origins of taxonomy. Science 174:1210–1213.

Reis, M., C. P. Vieira, R. Lata, N. Posnien, and J. Vieira. 2018. Origin and consequences of chromosomal inversions in the *virilis* group of *Drosophila*. Genome Biology and Evolution 10:3152–3166.

Rensch, B. 1928. Grenzfälle von Rasse und Art. Journal für Ornithologie 76:222–231.

Rineau, V., A. Grand, R. Zaragüeta, and M. Laurin. 2015. Experimental systematics: Sensitivity of cladistic methods to polarization and character ordering schemes. Contributions to Zoology 84:129–148.

Rineau, V., R. Zaragüeta, I. Bagils, and M. Laurin. 2018. Impact of errors on cladistic inference: Simulation-based comparison between parsimony and three-taxon analysis. Contributions to Zoology 87:25–40.

Roelants, K., D. J. Gower, M. Wilkinson, S. P. Loader, S. D. Biju, K. Guillaume, L. Moriau, and F. Bossuyt. 2007. Global patterns of diversification in the history of modern amphibians. Proceedings of the National Academy of Sciences of the United States of America 104:887–892.

Romer, A. S. 1966. Vertebrate paleontology. University of Chicago Press, Chicago, 468 pp.

Ronquist, F., S. Klopfstein, L. Vilhelmsen, S. Schulmeister, D. L. Murray, and A. Rasnitsyn. 2012b. A total-evidence approach to dating with fossils, applied to the early radiation of the Hymenoptera. Systematic Biology 61:973–999.

Ronquist, F., M. Teslenko, P. van der Mark, D. L. Ayres, A. Darling, S. Höhna, B. Larget, L. Liu, M. A. Suchard, and J. P. Huelsenbeck. 2012a. MrBayes 3.2: Efficient Bayesian phylogenetic inference and model choice across a large model space. Systematic Biology 61:539–542.

Roveri, M., R. Flecker, W. Krijgsman, J. Lofi, S. Lugli, V. Manzi, F. J. Sierro, A. Bertini, A. Camerlenghi, and G. De Lange. 2014. The Messinian salinity crisis: Past and future of a great challenge for marine sciences. Marine Geology 352:25–58.

Rubidge, B. S., D. H. Erwin, J. Ramezani, S. A. Bowring, and W. J. de Klerk. 2013. High-precision temporal calibration of Late Permian vertebrate biostratigraphy: U-Pb zircon constraints from the Karoo Supergroup, South Africa. Geology 41:363–366.

Ruta, M., J. C. Cisneros, T. Liebrecht, L. A. Tsuji, and J. Müller. 2011. Amniotes through major biological crises: Faunal turnover among parareptiles and the end-Permian mass extinction. Palaeontology 54:1117–1137.

Sahney, S. and M. J. Benton. 2008. Recovery from the most profound mass extinction of all time. Proceedings of the Royal Society B: Biological Sciences 275:759–765.

Sahney, S., M. J. Benton, and H. J. Falcon-Lang. 2010. Rainforest collapse triggered Carboniferous tetrapod diversification in Euramerica. Geology 38:1079–1082.

Salzburger, W. 2018. Understanding explosive diversification through cichlid fish genomics. Nature Reviews Genetics 19:705–717.

Sanderson, M. J. 2002. Estimating absolute rates of molecular evolution and divergence times: A penalized likelihood approach. Molecular Biology and Evolution 19:101–109.

San Mauro, D., M. Vences, M. Alcobendas, R. Zardoya, and A. Meyer. 2005. Initial diversification of living amphibians predated the breakup of Pangaea. The American Naturalist 165:590–599.

Sauquet, H. 2013. A practical guide to molecular dating. Comptes Rendus Palevol 12:355–367.

Sepkoski, J. J., Jr. 1978. A kinetic model of phanerozoic taxonomic diversity I: Analysis of marine orders. Paleobiology 4:223–251.

Sepkoski, J. J., Jr. 1993. Ten years in the library: New data confirm paleontological patterns. Paleobiology 19:43–51.

Sereno, P. C. 2005. The logical basis of phylogenetic taxonomy. Systematic Biology 54:595–619.

Shaffer, H. B. and M. L. McKnight. 1996. The polytypic species revisited: Genetic differentiation and molecular phylogenetics of the tiger salamander *Ambystoma tigrinum* (Amphibia: Caudata) complex. Evolution 50:417–433.

Shen, X.-X., D. Liang, M.-Y. Chen, R.-L. Mao, D. B. Wake, and P. Zhang. 2016. Enlarged multilocus data set provides surprisingly younger time of origin for the plethodontidae, the largest family of salamanders. Systematic Biology 65:66–81.

Sibley, C. G. and J. E. Ahlquist. 1990. Phylogeny and classification of birds: A study in molecular evolution. Yale University Press, New Haven, xxiii + 976 pp.

Sidor, C. A., D. A. Vilhena, K. D. Angielczyk, A. K. Huttenlocker, S. J. Nesbitt, B. R. Peecook, S. Steyer, R. M. H. Smith, and L. A. Tsuji. 2013. Provincialization of terrestrial faunas following the end-Permian mass extinction. Proceedings of the National Academy of Sciences of the United States of America 110:8129–8133.

Silvestro, D., J. Schnitzler, L. H. Liow, A. Antonelli, and N. Salamin. 2014. Bayesian estimation of speciation and extinction from incomplete fossil occurrence data. Systematic Biology 63:349–367.

Simpson, G. G. 1961. Principles of animal taxonomy. Columbia University Press, New York, 247 pp.

Slowinski, J. B. 1993. "Unordered" versus "Ordered" characters. Systematic Biology 42:155–165.

Smith, A. B. 1994. Systematics and the fossil record: Documenting evolutionary patterns. Blackwell Science, Oxford, viii + 223 pp.

Smith, A. B. and P. M. Barrett. 2012. Modelling the past: New generation approaches to understanding biological patterns in the fossil record. Biology Letters 8:112–114.

Smith, A. B. and C. Patterson. 1988. The influence of taxonomic method on the perception of patterns of evolution; pp. 127–216 in M. K. Hecht and B. Wallace (eds.), Evolutionary biology, vol. 23. Plenum Press, New York.

Smith, H. F., W. Parker, S. H. Kotzé, and M. Laurin. 2017. Morphological evolution of the mammalian cecum and cecal appendix. Comptes Rendus Palevol 16:39–57.

Sneath, P. H. and R. R. Sokal. 1962. Numerical taxonomy. Nature 193:855–860.

Sousa, R. P. C., G. C. Silva-Oliveira, I. O. Furo, A. B. de Oliveira-Filho, C. D. B. de Brito, L. Rabelo, A. Guimarães-Costa, E. H. C. de Oliveira, and M. Vallinoto. 2021. The role of the chromosomal rearrangements in the evolution and speciation of Elopiformes fishes (Teleostei; Elopomorpha). Zoologischer Anzeiger 290:40–48.

Spinney, L. 2010. Dreampond revisited: A once-threatened population of African fish is now providing a view of evolution in action: Laura Spinney asks what Lake Victoria cichlids have revealed about speciation. Nature 466:174–176.

Springer, M. S., C. A. Emerling, R. W. Meredith, J. E. Janečka, E. Eizirik, and W. J. Murphy. 2017. Waking the undead: Implications of a soft explosive model for the timing of placental mammal diversification. Molecular Phylogenetics and Evolution 106:86–102.

Springer, M. S., N. Foley, P. Brady, J. Gatesy, and W. Murphy. 2019. Evolutionary models for the diversification of placental mammals across the KPg boundary. Frontiers in Genetics 10:1241.

Stadler, T. 2010. Sampling-through-time in birth–death trees. Journal of Theoretical Biology 267:396–404.

Stadler, T., A. Gavryushkina, R. C. Warnock, A. J. Drummond, and T. A. Heath. 2018. The fossilized birth-death model for the analysis of stratigraphic range data under different speciation modes. Journal of Theoretical Biology 447:41–55.

Stadler, T. and Z. Yang. 2013. Dating phylogenies with sequentially sampled tips. Systematic Biology 62:674–688.

Sterli, J., D. Pol, and M. Laurin. 2013. Incorporating phylogenetic uncertainty on phylogeny-based paleontological dating and the timing of turtle diversification. Cladistics 29:233–246.

Stevens, P. F. 1980. Evolutionary polarity of character states. Annual Review of Ecology and Systematics 11:333–358.

Strauss, D. and P. M. Sadler. 1989. Classical confidence intervals and Bayesian probability estimates for ends of local taxon ranges. Mathematical Geology 21:411–427.

Su, Z., Z. Wang, F. López-Giráldez, and J. P. Townsend. 2014. The impact of incorporating molecular evolutionary model into predictions of phylogenetic signal and noise. Frontiers in Genetics 2:1–12.

Swofford, D. L. 2002. PAUP* phylogenetic analysis using parsimony (*and other methods). Version 4.0a, build 167 (updated on Feb. 1, 2020).

Swofford, D. L. and W. P. Maddison. 1987. Reconstructing ancestral character states under Wagner parsimony. Mathematical Biosciences 87:199–229.

Tabor, C. R., C. G. Bardeen, B. L. Otto-Bliesner, R. R. Garcia, and O. B. Toon. 2020. Causes and climatic consequences of the impact winter at the Cretaceous-Paleogene boundary. Geophysical Research Letters 47:e60121.

Tassy, P. 2011. Trees before and after Darwin. Journal of Zoological Systematics and Evolutionary Research 49:89–101.

Torruella, G., A. De Mendoza, X. Grau-Bové, M. Antó, M. A. Chaplin, J. Del Campo, L. Eme, G. Pérez-Cordón, C. M. Whipps, and K. M. Nichols. 2015. Phylogenomics reveals convergent evolution of lifestyles in close relatives of animals and fungi. Current Biology 25:2404–2410.

Tournefort, J. P. de. 1694. Elemens de botanique ou methode pour connoitre les plantes. L'Imprimerie Royale, Paris, 379 pp.

Vance, E. 2017. The axolotl paradox: An iconic salamander species, celebrated and studied around the world, is racing towards extinction. Nature 551:286–289.

Vermeij, G. J. 1999. A serious matter with character-taxon matrices. Paleobiology 25:431–433.

Viglietti, P. A., R. B. Benson, R. M. Smith, J. Botha, C. F. Kammerer, Z. Skosan, E. Butler, A. Crean, B. Eloff, and S. Kaal. 2021. Evidence from South Africa for a protracted end-Permian extinction on land. Proceedings of the National Academy of Sciences 118:e2017045118.

Viglietti, P. A., R. M. Smith, K. D. Angielczyk, C. F. Kammerer, J. Fröbisch, and B. S. Rubidge. 2016. The *Daptocephalus* assemblage zone (Lopingian), South Africa: A proposed biostratigraphy based on a new compilation of stratigraphic ranges. Journal of African Earth Sciences 113:153–164.

Wake, M. H. and J. Hanken. 1982. Development of the skull of *Dermophis mexicanus* (Amphibia: Gymnophiona), with comments on skull kinesis and amphibian relationships. Journal of Morphology 173:203–223.

Wallace, D. G., M.-C. King, and A. C. Wilson. 1973. Albumin differences among ranid fogs: Taxonomic and phylogenetic implications. Systematic Zoology 22:1–13.

Warnock, R. C., T. A. Heath, and T. Stadler. 2020. Assessing the impact of incomplete species sampling on estimates of speciation and extinction rates. Paleobiology 46:137–157.

Warnock, R. C. M., Z. Yang, and P. C. J. Donoghue. 2012. Exploring uncertainty in the calibration of the molecular clock. Biology Letters 8:156–159.

Wiens, J. J. 2001. Character analysis in morphological phylogenetics: Problems and solutions. Systematic Biology 50:689–699.

Wignall, P. B., Y. Sun, D. P. Bond, G. Izon, R. J. Newton, S. Védrine, M. Widdowson, J. R. Ali, X. Lai, H. Jiang, H. Cope, and S. H. Bottrell. 2009. Volcanism, mass extinction, and carbon isotope fluctuations in the Middle Permian of China. Science 324:1179–1182.

Wilkins, J. S. 2018. Species: The evolution of the idea. CRC Press, Boca Raton, xxxviii + 389 pp.

Winsor, M. P. 2006. The creation of the essentialism story: An exercise in metahistory. History and Philosophy of the Life Sciences 28:149–174.

Woese, C. R., O. Kandler, and M. L. Wheelis. 1990. Towards a natural system of organisms: Proposal for the domains Archaea, Bacteria, and Eucarya. Proceedings of the National Academy of Sciences 87:4576–4579.

Wu, H., X. Guang, M. B. Al-Fageeh, J. Cao, S. Pan, H. Zhou, L. Zhang, M. H. Abutarboush, Y. Xing, Z. Xie, and J. Wang. 2014. Camelid genomes reveal evolution and adaptation to desert environments. Nature Communications 5:5188.

Wyllie, A. 2003. A review of Robert Broom's therapsid holotypes: Have they survived the test of time? Palaeontologia Africana 39:1–19.

Xavier, C., R. V. S. Soares, I. C. Amorim, D. C. Cabral-de-Mello, and R. d. C. de Moura. 2018. Insights into the karyotype evolution and speciation of the beetle *Euchroma gigantea* (Coleoptera: Buprestidae). Chromosome Research 26:163–178.

Zachos, F. E. 2011. Linnean ranks, temporal banding, and time-clipping: Why not slaughter the sacred cow? Biological Journal of the Linnean Society 103:732–734.

Zhang, C., T. Stadler, S. Klopfstein, T. A. Heath, and F. Ronquist. 2016. Total-evidence dating under the fossilized birth–death process. Systematic Biology 65:228–249.

Zuckerkandl, E. and L. Pauling. 1965. Evolutionary divergence and convergence in proteins; pp. 97–166 in V. Bryson and H. J. Vogel (eds.), Evolving genes and proteins. Academic Press, New York.

# 4

## Phylogenetic Nomenclature

### 4.1 History of Phylogenetic Nomenclature

We saw in Chapter 1 that the idea of evolution and the Tree of Life metaphor became widely accepted in the second half of the 19th century, but, despite this, phylogeny did not become important under most rank-based codes (Chapter 2). Rank-based nomenclature (RN from here on) raised problems when our knowledge of the phylogeny improved, as discussed in Chapter 3, and this led to much nomenclatural confusion. This problem became especially acute in the last few decades, in which progress in phylogenetics was fast. As summarized by Sereno et al. (2005: 1):

> Over the last 25 years, there has been a dramatic increase in resolution of branch points in the tree of life driven principally by three factors: an increasing volume of phylogenetic research, widespread use of computer-assisted analysis (Swofford et al. 1996; Felsenstein 2004), and the influx of molecular data (Hillis et al. 1996).

The problems raised by RN in this new context are twofold. They include the lack of explicit delimitation of taxa, which has resulted in some names, such as *Mammalia*, being applied to multiple nested clades (Rowe and Gauthier 1992), a problem that is especially acute in paleontology because the fossil record documents transitions between many obvious phenotypic gaps in the extant fauna that had been used to delimit taxa (Michener 1963: 153). The second problem involves the numerous unnecessary name changes required by the application of the rule of priority within given ranks, as occurred in some squamate clades (de Queiroz and Gauthier 1990). This situation is unfortunate because the rank-based codes were designed precisely to prevent this, as eloquently put by Strickland et al. (1842: 2):

> So long as naturalists differ in the views which they are disposed to take of the natural affinities of animals there will always be diversities of classification, and the only way to arrive at the true system of nature is to allow perfect liberty to systematists in this respect. But the evil complained of is of a different character. It consists in this, that when naturalists are agreed as to the characters and limits of an individual group or species, they still disagree in the **appellations** by which they distinguish it. A genus is often designated by **three or four, and a species by twice that number of precisely equivalent synonyms**; and in the absence of any rule on the subject, the naturalist is wholly at a loss what nomenclature to adopt.
>
> **(Bold font used for emphasis is mine, unless noted otherwise.)**

This quote clearly states that one of the necessary conditions to have a "true system of nature" (a good taxonomy or natural classification) is to have a consensus on the affinities (what we now call phylogeny, which was conceptualized differently at the time). Another requirement is a nomenclatural system that will provide unequivocal allocation of names to groups. Rank-based nomenclature cannot really do that simply because it was not designed with phylogeny in mind. Rather, it is based on types, which exist, and absolute ranks (Linnaean categories), which do not exist in nature. This artificial system has long been considered inadequate and outdated, and not only among proponents of phylogenetic nomenclature (PN from here on), as shown by this quote from Papavero et al. (2001: 4–5):

> the Linnaean categorial ranks were not introduced to indicate sistergroup relationships, and they were originally not linked with the idea of organismic evolution. They were introduced to serve as classification of the organisms on the basis of Aristotelian logic (Griffiths 1974: 118, 1976). Now that we know that (and how) the biotic diversity is underlain by a natural system, categorial ranking in the form used by Linné contradicts our knowledge about the structure of the living world and **must be abandoned**.

Systematists thus began conceiving a new nomenclatural system that would integrate the phylogeny from the ground up; this new system is thus probably associated with the rise of "tree thinking" in systematics (O'Hara 1997). This new nomenclature would be designed to name clades and would not rely on Linnaean categories. Of course, this could be achieved through various nomenclatural conventions. In the following, we will consider the developments initiated by Hennig that culminated in the development of PN as implemented in the *PhyloCode* (Cantino and de Queiroz 2020). Only a very brief summary of an alternative system proposed by Papavero et al. (2001) is required here. This proposal differs strongly from PN by striving to produce as few new names as possible and by naming all taxa on a phylogeny (Papavero et al. 2001: 45). However, this system has at least three important drawbacks: first, it requires using numbers in taxon names to indicate hierarchy; second, and more importantly, topological changes could result in a cascade of name changes; third, to name a large clade, the names and date at which all included taxa were erected are required. These practical problems, which are especially acute for large clades, might partly explain why this system has not been widely adopted by the systematic community, as shown by the fact that as of this writing (June 12, 2022), it has only 24 Google Scholar citations.

Hennig (1966, 1969, 1981) pioneered the developments of PN by emphasizing crown and total groups that are delimited solely by the phylogeny, rather than ranks. This also represents a shift in taxon delimitation from phenotypic clusters to using tree topology and major divergences, like those between speciose extant clades.

Hennig's views on taxonomy and nomenclature took time to be accepted in the systematic community; indeed, these views remain to an extent controversial today. Early supporters include Griffiths (1974), who argued that:

> As our understanding of the history of organisms increases the **Linnaean category terms should be gradually abandoned**. But this abandonment of the terms will still leave certain formal problems of nomenclature, since there is a series of Linnaean category concepts (clustered about the family level) which have been built into group names by the use of suffices.

However, even many followers of Hennig, such as Nelson and Platnick (1981), continued to view taxa as classes on the arguably erroneous premise that they have defining properties (Ghiselin 1984: 106). This view suggests that diagnoses can somehow define taxa, at least to the extent that they are based on derived character states. However, this ignores the fact that the presence of more than one derived state in a diagnosis would sooner or later lead to ambiguous delimitation and that any character can be lost. Thus, if the taxon *Tetrapod* (to pick an example evoked by Ghiselin 1984) were defined by the presence of a limb with digits, snakes and gymnophionans (among others) should be excluded, but no systematist has proposed this recently.

On the contrary, Ghiselin (1984) pointed out that from an evolutionary perspective, it makes much more sense to view taxa as individuals—as entities that can be defined only by their history, not by their intrinsic properties (unless you consider ancestry such a property; see Chapter 6 for a more thorough treatment of this topic). Such views suggest that rank-based nomenclature is inappropriate and that taxa should be viewed as parts of the Tree of Life—ideas that would subsequently be central to the development of PN. Griffiths (1976: 170) thus concluded that "the Linnaean category sequence has become an empty formalism which has lost its theoretical foundation." Such opinions, shared by many others, led some systematists to develop PN.

This conceptual shift required new vocabulary, such as the expressions "stem group," "crown goup" (or "crown clade") and "total group" (or "total clade"). A crown clade is delimited by extant taxa; for instance, *Amniota* was defined thus by Laurin and Reisz (2020b): "The smallest crown clade containing *Homo sapiens* Linnaeus 1758 (*Synapsida*), *Testudo graeca* Linnaeus 1758 (*Testudines*), and *Crocodylus* (originally *Lacerta*) *niloticus* Laurenti 1768 (*Diapsida*)." This definition clearly refers to the smallest clade that includes all extant amniotes. Of course, it includes many extinct taxa as well; *Amniota* has a rich fossil record. But rather than seeking when exactly the amniotic egg or other key characters of amniotes

appeared and delimiting the taxon on this basis, the basalmost divergence between extant representatives (synapsids and sauropsids, according to most recent phylogenies) delimits the taxon.

The **stem group** is paraphyletic and consists of all extinct taxa that are more closely related to their crown group than to other crown groups. For instance, among amniotes, *Mammalia* has also been defined as a crown group (Rowe 2020b), and its stem group includes all extinct amniotes that are outside this crown group and that are closer to *Mammalia* than to other mutually exclusive crown clades (**Figure 4.1**). This stem group includes Permo-Carboniferous taxa such as *Ophiacodon*, *Edaphosaurus* and *Dimetrodon*, which look very different from mammals (**Figure 4.2**), but also mid- or late Permian therapsids such as dinocephalians, gorgonopsians and therocephalians (the latter extended into the Triassic; Angielczyk and Kammerer 2018), Triassic cynodonts such as *Thrinaxodon* and *Cynognathus*, and even Jurassic cynodonts, such as *Bienotherium* and *Diarthrognathus* (Ruta et al. 2013; Angielczyk and Kammerer 2018), which resemble mammals much more closely. In this example, the crown clade closest to *Mammalia* is *Reptilia*.

A total group is composed of a crown clade and its stem group. The total group of mammals has usually been called *Synapsida*, although this name, which was coined well before the advent of cladistics, was initially applied to the stem group only (see Section 4.4.3.1).

The lineage that is composed of the **direct ancestors** of the crown group can be called the "**stem lineage**," although Meier and Richter (1992) proposed to call it the ancestral lineage. In most cases, its characteristics can only be inferred through nodal inferences (using parsimony, maximum likelihood or Bayesian inference, for instance). Some ancestors must be

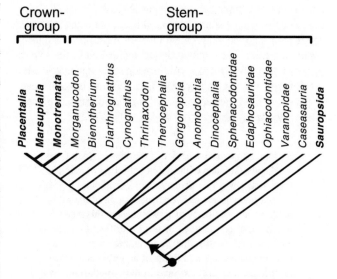

**FIGURE 4.1** Synapsid phylogenetic tree (emphasizing taxa discussed in the text) to show the crown-clade (*Mammalia*) and the stem-group of synapsids. The total group is composed of the crown plus the stem. *Sauropsida* includes a crown-clade (*Reptilia*; not shown) and some stem-taxa, at least under some topologies. Extant taxa are in bold type. The crown-group is shown with thicker lines.

**FIGURE 4.2** Fleshed-in reconstruction of the synapsids *Ophiacodon mirus* (late Carboniferous and early Permian; upper left), *Edaphosaurus pogonias* (early Permian; upper right), the dinocephalian *Titanophoneus* (middle Permian; lower left) and the therocephalian *Pristerognathus* (middle Permian; lower right). All drawing by ДиБгд (DiBgd), public domain, accessible at https://commons.wikimedia.org/wiki/File:Ophiacodon_mirus.jpg, https://commons.wikimedia.org/wiki/File:Edaphosaurus_pogonias.jpg, https://upload.wikimedia.org/wikipedia/commons/4/4d/Doliosauriscus1DB.jpg and https://en.wikipedia.org/wiki/Therocephalia#/media/File:Pristeroognathus_DB.jpg.

documented in the fossil record (Foote 1996), but Hennig (1966) first pointed out that they were difficult to identify. The stem lineage should not be confused with the stem group. Among synapsids, the stem group is abundantly represented in the fossil record (**Figure 4.2**), but few, if any, of these fossils represent direct ancestors of mammals; thus, *Ophiacodon* and *Edaphosaurus* belong to the mammalian stem group, but not to its stem lineage, as shown by the various autapomorphies of these taxa (such as the very long dorsal neural spines with lateral knobs of edaphosaurids).

The first phylogenetic definitions were published in a monograph on saurischian dinosaurs, ironically in the introductory part of lengthy diagnoses (Gauthier 1986). Subsequently, the basic principles of PN were exposed in a series of papers by de Queiroz and Gauthier (1990, 1992, 1994). Further developments were published shortly thereafter by many authors; a non-exhaustive list sorted in chronological order of publication includes Rowe and Gauthier (1992), Bryant (1994), Sundberg and Pleijel (1994), Schander and Thollesson (1995), Lee (1996a, 1996b, 1998), Wyss and Meng (1996), Brochu (1997), Cantino et al. (1997), Kron (1997), Baum et al. (1998), Cantino (1998), Eriksson et al. (1998), Härlin (1998), Hibbett

and Donoghue (1998), Moore (1998), Schander (1998a, 1998b), Sereno (1998, 1999), Mishler (1999) and Pleijel (1999).

Not all of these papers need to be summarized here, but one can illustrate the kind of progress that was being made. For instance, early phylogenetic definitions were often formulated using explicit references at a common ancestor and its descendants. Thus, Gauthier et al. (1988: 106) defined *Tetrapoda* as "the most recent common ancestor of extant Lissamphibia and Amniota, and all of its descendants." Some systematists objected that Hennig and his followers had emphasized the difficulty in identifying such ancestors. Lee (1998) pointed out that explicit reference to such ancestors was unnecessary in PN. Node-based taxa can be defined in the form "the least-inclusive clade that includes [the taxa] B and C." Branch-based taxa (which Lee 1998 called "stem-based") can be defined as "the most-inclusive clade that includes [taxa] B (or C) but not [taxon] A." Finally, Lee (1998) suggested that apomorphy-based taxa can be diagnosed as "the clade diagnosed by trait X." This last type of definition needs to be supplemented by ". . . . synapomorphic with taxon A" to be complete, to allow unambiguous identification, given the possibility of convergence. This is easily demonstrated by an example. The definition "the clade diagnosed by endothermy" could refer to either birds (or a more inclusive clade given that many Mesozoic dinosaurs, and possibly even other Permo-Triassic archosaurs, appear to have displayed high metabolic rates) or mammals (or a more inclusive clade of synapsids). However, "the clade diagnosed by endothermy synapomorphic with *Homo sapiens*" unambiguously refers to a synapsid clade that includes mammals; birds and other Mesozoic archosaurs are clearly ruled out.

PN was applied to many taxa in the 1990s; this includes Paleozoic amniotes (Laurin 1991; Laurin and Reisz 1995), Mesozoic dinosaurs (Holtz 1996), crocodilians (Brochu 1999), gastropod mollusks (Roth 1996), basidiomycete fungi (Swann et al. 1999) and angiosperms (Judd et al. 1993, 1994; Bremer 2000; Alverson et al. 1999).

Work on the *PhyloCode* started only in 1997, when M. Donoghue, P. Cantino and K. de Queiroz decided to organize a workshop (which convened in August 1998) for this purpose (Cantino and de Queiroz 2020: xiv). The initial development of the *PhyloCode* was thus simultaneous with further developments and debates in PN that continue to this day (in 2023) and will presumably persist in the future.

The first public version of the *PhyloCode* was posted on the Internet in April 2000, and systematists were encouraged to send suggestions to improve it. These comments were forwarded to the *PhyloCode* Advisory Group (which supervised the development of the *PhyloCode* from 2000 to 2004) for consideration. Publication of the *PhyloCode* triggered numerous comments, some of which were critical and often misleading (for instance, Benton 2000; Nixon and Carpenter 2000; Stuessy 2000, just to mention those published in 2000), whereas others were supportive (for instance, Bremer 2000; Brochu and Sumrall 2001; Ereshefsky 2001, to mention only some of the earliest ones that were not published by the co-authors of the *PhyloCode*).

A second workshop on PN, organized by M. Donoghue, J. Gauthier, P. Cantino and K. de Queiroz, convened at Yale University in July 2002. It was attended by 20 systematists from five countries. In addition to discussing some proposed changes (several of which were adopted), some fundamental issues were tackled. For instance, it was decided that species names would be regulated by a distinct code, rather than the *PhyloCode* for clades, to avoid delaying the code for clades.

The First International Phylogenetic Nomenclature Meeting convened in July 2004 at the Muséum National d'Histoire Naturelle in Paris. This meeting was attended by 70 systematists from 11 countries (Laurin and Cantino 2004). In addition to featuring 36 talks, it inaugurated the International Society for Phylogenetic Nomenclature (ISPN from now on), which has been overseeing the development and promotion of the *PhyloCode* since then. One of its committees, the Committee on Phylogenetic Nomenclature, initially composed of nine elected members, took over responsibility for approving changes to the code in 2005.

The ISPN convened a second time, in 2006 at Yale University (New Haven). Various general nomenclatural issues were discussed, and this allowed to determine how much support various proposals had among the ISPN membership (Laurin and Cantino 2007). Thus, a proposal to allow "unrestricted emendations" of phylogenetic definitions to be published by any

systematist, without involvement of the CPN (Committee on Phylogenetic Nomenclature), was well-received. This mechanism allows association between a name and a clade to be retained when the phylogeny changes, as may occur, especially if the initial definition was poorly formulated. This process is arguably more flexible than the appeals for conservation that have to be approved by the ICZN for zoological names. The plans that had been made at the first ISPN meeting to work on a code for species distinct from a code for clades was abandoned. Under that plan, the species names would consist of the specific epithet, the name of the author who erected the species, the year of publication of the work in which that species was erected, along with the page number where the species is erected, following suggestions by Lanham (1965) that minimize taxonomic information contained species names, as explained in Section 6.2.2.2. The reasons to abandon plans for a "species code" were explained by Laurin and Cantino (2007: 113):

> First, species names would be different under rank-based and phylogenetic nomenclature (e.g. "*Homo sapiens*" vs. "*sapiens* Linnaeus 1758"). Second, if the definitions of species names under the *PhyloCode* and the rank-based codes would not differ fundamentally (i.e. the species that includes a particular type specimen), then the utility of publishing and registering every converted species name—a very time-consuming endeavor—is highly questionable. Third, the introduction of species into the *PhyloCode* might be interpreted as introducing a rank (Mishler 1999), which might be considered inconsistent with the independence from ranks in the rest of the code. Finally, differences in the handling of types by the *ICZN* and the *International Code of Botanical Nomenclature (ICBN)* complicate the development of a code dealing with species names of all life forms.

Instead, it was decided that the regulation of species names would be left to the RN codes but that the *PhyloCode* would incorporate conventions to communicate on the status (monophyletic or not) of the genus name (now Recommendation 21.3A), which would be considered to be a prenomen (the first part of the species name) rather than a genuine taxon name. This approach was described in detail by Dayrat et al. (2008) and incorporated into the *PhyloCode* as Article 21. Note that interpreting the first part of a binominal name as a praenomen (this slightly different spelling is used in earlier papers) had already been suggested long ago by Griffiths (1974: 120), and this proposal had been endorsed by Papavero et al. (2001: 45).

The third (and still most recent) meeting of the ISPN convened in Dalhousie University in Halifax, Canada, in July 2008. It focused on how to complete in a timely fashion development of *RegNum* (the ISPN-supported online registration database for names and definitions established under the *PhyloCode*) and *Phylonyms*, the "*Companion Volume*" of the *PhyloCode*, which would include the first set of names established under that code (Laurin and Bryant 2009). At this third meeting, the editors of *Phylonyms* announced that they had set up an editorial board to speed up work on that monograph.

Despite the formation of an editorial board, *Phylonyms* (de Queiroz et al. 2020) and the *PhyloCode* (Cantino and de Queiroz 2020) were published only in 2020, many years behind the initial schedule. Nevertheless, these two publications marked the advent of regulated PN, and as such, mark a turning point in the history of biological nomenclature. The next event in this field was the publication (in August 2022) of the first paper (Johnson et al. 2022) published by the *Bulletin of Phylogenetic Nomenclature*, which facilitates publication of nomenclatural acts in conformity with the *PhyloCode*.

## 4.2 Basic Principles

### 4.2.1 Three Kinds of Definitions

PN allows three basic kinds of definitions (each with many possible variants): minimum (formerly node-based), maximum (formerly branch-based) and apomorphy-based (**Figure 4.3**). Under the *PhyloCode* (Cantino and de Queiroz 2020), the specimens or characters that play a key role in these definitions (and which are analogous with types under the rank-based codes) are called specifiers. In most cases, these are specimens, but given that specimen numbers are not familiar to most readers, the *PhyloCode* allows using species as specifiers, with the explicit rule that when a species is used as a specifier, its type is the actual specifier, so ultimately, all definitions established under the *PhyloCode* have specimens as specifiers.

Minimum (formerly node-based) definitions can be formulated thus: the smallest clade that includes specimens (or species) A and B (or any number of specifiers). Maximum (formerly branch-based) definitions can be formulated in this way: the largest clade that includes specimen (or species) A

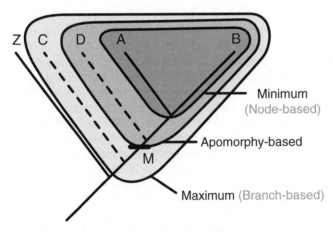

**FIGURE 4.3** The three basic kinds of phylogenetic taxon name definitions: minimum (formerly node-), maximum (formerly branch-) and apomorphy-based. In this hypothetical example, initially (when the definitions are formulated), only taxa A, B and Z are known. If three names are defined, each using a different kind of definitions, the three nested taxa are redundant but not synonymous because their definitions are of different kinds. Note that if taxa C and D are subsequently discovered, the three nested taxa defined as previously indicated are no longer redundant, and assigning C and D to these taxa does not require subjective decisions, contrary to what would happen under rank-based nomenclature.

(and eventually, B, C and so on) but not X (and eventually, Y, Z and so on). Because of distinctions that were emphasized by Martin et al. (2010) between two kinds of trees (or rather, tree representations), node-based and branch-based definitions were replaced in the current version of the *PhyloCode* by the equivalent expressions "minimum-clade" and "maximum-clade" definitions, respectively. However, given that the bulk of the literature on PN has used the older names, both sets of names are used in this book.

Apomorphy-based definitions differ from branch- and node-based definitions in using both specimens and a character as specifiers. They can be formulated thus: the smallest clade that possesses the apomorphy M synapomorphic with A (eventually, B, C and so on). In the case of apomorphy-based definitions, only the nominal species included in the definition implicitly refers to a specimen; the apomorphy does not. The presence of at least one specifier that is a species or a specimen ensures that if the apomorphy has appeared in other clades through convergence, the definition refers only to the intended clade (see also Section 4.1). This fixes one of the problems with apomorphy-based definitions that was raised by Bryant (1994: 128), when he pointed out that defining the name *Tetrapoda* as "the first vertebrate to possess digits and all its descendants" worked only if digits appeared only once in vertebrate evolution. This seems to be the case (Clack 2002; Laurin 2010), but at some points, some embryologists and paleontologists suggested tetrapod diphyly (for instance, Holmgren 1933; Jarvik 1963, 1986).

### 4.2.2 The Problematic Apomorphy-Based Definitions

It could be argued that apomorphy-based definitions are more difficult to apply objectively than other kinds of definitions, as argued by Bryant (1994) and Sereno (1998, 2005). This is because there is often more than one possible (or optimal) reconstruction of character evolution on a given tree (Bryant 1994: 128). Also, each systematist has his own concept of a given character (Sereno 1998, 2005), so delimitation under apomorphy-based definitions is inherently less precise than under other kinds of phylogenetic definitions. Evolution often proceeds by small steps, so a continuum of structures may have existed at some point, or the evolutionary leaps may have been much smaller than we think. When looking only at extant taxa, we often see conspicuous phenotypic gaps, but the fossil record documents many intermediates that fill these gaps. Furthermore, what may seem at first sight a simple character may be conceptualized as a complex of characters. Both of these considerations may lead to considerable problems in applying apomorphy-based definitions. This can be illustrated through two examples, the origins of the pentadactyl limb and of the feather.

Let's start with the tetrapod limb. When looking at extant taxa only, discriminating between vertebrate fins and digited limbs is not a problem because a substantial phenotypic gap separates these structures. The paired fins of extant dipnoans (our closest extant finned relatives) share several similarities with the fins of other Paleozoic sarcopterygians, but differ sharply from the earliest limbs with digits (Laurin 2010; **Figure 4.4**). Most

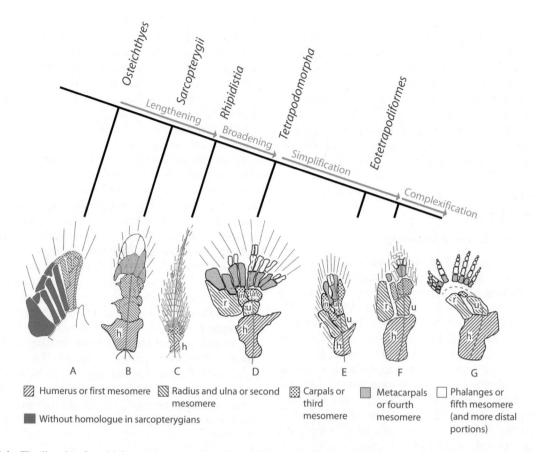

**FIGURE 4.4** The diversity of osteichthyan paired appendages (fins and limbs). Modified from Laurin (2010: figure 4.3). Pre-axial (anterior) is to the left. Homology is well-established only for the first two or three mesomeres; more distally, the speculative hypothesis of maximal homology (Laurin et al. 2000) is illustrated. The position of the metapterygial axis is indicated by a red line (in the e-book; grey in the printed version; dashed when position is uncertain). From left to right, the illustrated appendages are from: the actinopterygian *Acipenser sturio* (sturgeon), the actinistian *Latimeria chalumnae* (the coelacanth), the dipnoan *Neoceratodus forster*, an unidentified rhizodontid, the tristichopterid *Eusthenopteron foordi*, *Tiktaalik roseae* and the stegocephalian *Acanthostega gunnari*. The first three taxa are extant, whereas the last four are from the Devonian.

importantly, these fins share a central (metapterygial) axis with endoskeletal radials on both sides (pre- and postaxial). In dipnoans, in which the central axis is particularly well developed, this structure is called a biserial fin, or archipterygium, but even the Devonian tetrapodomorphs *Tiktaalik* (Shubin et al. 2006) and *Elpistostege* (Cloutier et al. 2020) have a similar structure. Other Devonian tetrapodomorphs, such as the tristichopterid *Eusthenopteron foordi*, only have radials preaxially (Laurin 2006). These endoskeletal radials are associated distally with various kinds of dermal fin rays (called ceratotrichia in chondrichthyans, lepidotrichia in actinopterygians and some other taxa, and camptotrichia in extant dipnoans). The endoskeletal radials may form simple rows, as in extant dipnoans, or some of them may be branched, as in rhizodontids and in *Elpistostege* (Cloutier et al. 2020). Conversely, the limb of extant tetrapods is characterized by unbranched digits that are inferred by most authors to be exclusively postaxial structures, following the digital arch model (Shubin and Alberch 1986), and dermal rays are absent.

Which of these differences would be implied by the apomorphy "digited limb present"? The loss of the dermal fin rays? The unbranched structure of the radials? The exclusively postaxial position of the radials branching off the metapterygial axis? Currently, these three changes appear to be simultaneous

(though lepidotrichia are known in the caudal fin of Devonian limbed vertebrates) and a substantial morphological gap continues to separate digited limb and fins, but this presumably simply reflects a gap in the fossil record of forms that document the fin-limb transition. Recent studies have filled this gap somewhat. Indeed, in early studies (for instance, Schultze 1977; Hinchliffe 1989, 2002; Vorobyeva 1991), the digited limb used to be called "pentadactyl limb" because the maximal number of digits observed in non-pathological tetrapods is five, and this number was inferred to be primitive for the crown-group, an inference that remains valid (Laurin 1998). However, it is now well-established that the Devonian stegocephalians were polydactylous, with six to eight digits (Lebedev 1986; Coates and Clack 1990; Lebedev and Coates 1995; Coates 1996), which is most parsimoniously interpreted as a primitive condition that preceded pentadactyly (Laurin et al. 2000; **Figure 4.5**). Given that most extant and extinct main sarcopterygian clades had more than five primary endoskeletal radials, polydactyly presumably represents a morphological intermediate condition between the sarcopterygian fin and the pentadactyl limb, a conclusion also supported by an optimization of the type of appendages on a phylogeny (**Figure 4.5**). It is plausible that we will ultimately discover other intermediate morphologies in Devonian tetrapodomorphs (in taxa such as *Elginerpeton* or

*Metaxygnathus*), either with unbranched, exclusively postaxial endoskeletal radials (which could then qualify as digits) associated with lepidotrichia, or branched, postaxial endoskeletal radials in a fin lacking dermal fin rays. Both structures would fill the remaining gap between fin and digited limb and might be considered by some (but not necessarily all) paleontologists as limbs.

Cloutier et al. (2020: figure 4.4) suggested that the pectoral fin of *Elpistostege* further fills this gap and that it includes some digits. This seems unlikely because, according to their own interpretation, at least some of these digits are both branched and preaxial structures. Both of these attributes preclude identifying these radials with digits, at least if the digital arch model is accepted (Shubin and Alberch 1986). But even if it were not, the branching nature of the prospective digits would still be problematic. Cloutier et al. (2020: 551) argued that "*Elpistostege* has two identifiable digits that are composed of **two** non-branching endoskeletal elements that articulate one-to-one proximodistally, and—potentially—three more digits that are each composed of a single preserved element." This interpretation seems highly questionable, for two reasons. First, the three Devonian stegocephalian taxa in which the distal portion of the limb is documented are all polydactylous, with six to eight digits, and Carboniferous stegocephalians have four or five digits per limbs (Laurin 2010). Given this early trend toward a decrease in digit number early in limb evolution, it seems unlikely that it started with two to five digits.

Second, digits are formed of several skeletal elements, even in Devonian stegocephalians (Lebedev 1986; Coates and Clack 1990; Lebedev and Coates 1995; Coates 1996); at best, if these elements were homologous with digits, they would most likely represent metacarpals. More importantly, a row composed of a single element can neither be branched nor unbranched; this character is simply inapplicable, and there is no way to tell if these radials are homologous to digits. Objectively, the pectoral fin of *Elpistostege* seems to contain up to seven rays of endoskeletal radials, but the four anterior rays are preaxial structures. Thus, there is no compelling evidence that the fin of *Elpistostege* contains digit precursors.

Despite this, Cloutier et al. (2020: 549) adopted an apomorphy-based definition of tetrapods as "all organisms from the first sarcopterygian to have possessed digits homologous with those in *Homo sapiens*," and, on that basis, suggested that "*Elpistostege* is **potentially** the sister taxon of all other tetrapods, and its appendages further **blur the line** between fish and land vertebrates." Delimiting a taxon based on a blurred character does not seem optimal. This other quote from Cloutier et al. (2020: 553) illustrates this caveat: "If one adopts an apomorphy-based interpretation of Tetrapoda[17,18] and considers the parallel, unbranched distal radials in the *Elpistostege* fin to be true digits, then *Elpistostege* would represent the earliest and most primitive known tetrapod." Obviously, the authors are unsure about the nomenclatural implications of their definition and their own data; will it be clearer for other authors? Even if we agreed that "digited limb" refers specifically to the position and/or structure of endoskeletal rays that allows them to be considered "digits" (rather than using the loss of lepidotrichia), and even if the position of the metapterygial axis were precisely known, uncertainties of interpretation of the pectoral fin of *Elpistostege* illustrate how apomorphy-based phylogenetic definitions are more difficult to apply than node- or branch-based definitions. The digited limb is an iconic, well-studied character of limbed vertebrates, but would it be a good idea to define a taxon name based on this character?

A similar case has been made with feathers and wings. As explained by Sereno (1998: 46), "Feathers certainly arose by way of a series of transformations, as recent discoveries are beginning to reveal (Chen et al. 1998), and so it may well be preferable to code these components separately (rather than to assume their correlation)." Low fossilization potential of feathers is also problematic. Sereno (1998: 46) stated that "We have little or no data regarding the presence or absence of feathers or their components in taxa immediately outside Aves, such as deinonychosaurian theropods." A few years later, Sereno (2005: 606) returned to this point when he criticized the apomorphy-based definition of *Avialae* based on feathered wings used for powered flight proposed by Gauthier and de Queiroz (2001: 25), with the hindsight offered by additional fossils of feathered dinosaurs that had been described in the meantime (reviewed in Norell et al. 2002):

> Their definition for *Avialae* also demonstrates how rapidly apomorphies can relocate with no change in phylogenetic relationships. At the time of their writing, it seemed safe to presume that nonvolant

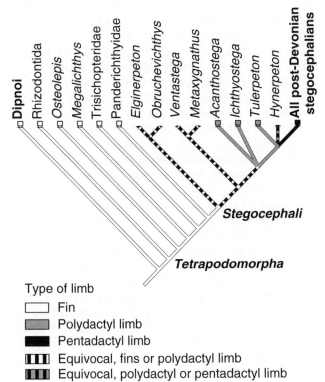

**FIGURE 4.5** Evolution of appendages among sarcopterygians (through parsimony optimization) with emphasis on the number of digits among Devonian taxa. Modified from Laurin et al. (2000: figure 1). Extant taxa are in bold type.

theropods like *Deinonychus* and kin would never be found with "feathered wings . . . used for powered flight," their preferred apomorphy encompassing *Archaeopteryx* and more derived birds.

While the ability for flight is sometimes difficult to infer in extinct taxa, it is now clear that some Mesozoic dinosaurs that have traditionally not been considered birds (such as *Velociraptor mongoliensis*) had secondary feathers anchored on the ulna. This is shown by traces of follicular ligaments whose presence is today correlated with flight capability (Turner et al. 2007). The history of feathered wings is even more complex than previously outlined because it now seems like bird ancestors at some point had four wings (two pairs) given that the hindlimb of the avialan *Pedopenna*, the dromaeosaur *Microraptor* and the troodontid *Anchiornis* all had long feathers that may well have provided lift and that in any case must have generated drag (Witmer 2009b).

Note that the problem here is not specific to the feathered wing; using feathers is equally problematic, as pointed out by Sereno (2005: 607):

> Despite these interpretive hurdles, *Avialae* is well behaved compared to another of their "apomorphy-based" taxa, *Avifilopluma*. This taxon was erected on the presence of "hollow-based, filamentous epidermal appendages produced by follicles" (Gauthier and de Queiroz 2001: 25), despite the fact that the filaments/feathers preserved on extinct nonavian dinosaurs are not demonstrably hollow-based or produced from follicles. Without any sense of concern, the authors remarked that their newly defined taxon might overlap with a half dozen others: "Avifilopluma might even contain all but the basalmost theropods . . . as well as taxa more distantly related to birds, such as herrerasaurs and *Eoraptor*" (Gauthier and de Queiroz 2001: 25). The "apomorphy-based" definitions proposed by Gauthier and de Queiroz (2001) underscore the need to restrict characters and their functional interpretations to diagnoses and interpretive discussion, respectively.

Subsequent discoveries showed that "hollow-based, filamentous epidermal appendages" indeed occurred among dinosaurs that are more distantly related to birds, such as the basal tyrannosaurid *Yutyrannus*, from the Early Cretaceous of northeastern China (Xu et al. 2012). Such epidermal appendages were later argued to diagnose *Dinosauria* because they may well be present in the basal ornithischian dinosaur *Tianyulong confuciusi* (Witmer 2009a). More recently, simple "monofilaments" and branching feathers with melanosomes of various shapes that may reflect color patterns were discovered in the Early Cretaceous pterosaur *Tupandactylus* (Benton 2022; Cincotta et al. 2022). This suggests a visual signaling function, in addition to thermal insulation of these simple feathers in pterosaurs. Thus, feathers may have appeared early in *Avemetatarsalia*, a taxon that includes pterosaurs and dinosaurs and that harks back to the Triassic.

The complex nature of feathers and their apparent stepwise evolution in avemetatarsalians, especially among non-avian theropods (Norell et al. 2002), raise problems for their use in apomorphy-based definitions. Recent works on this topic, prompted by the discovery of several exceptionally preserved dinosaur and a few pterosaur fossils, has documented several steps in the evolution of feathers. They were presumably initially simple, hollow quills (present in at least some pterosaurs and ornithiscian dinosaurs), before giving rise to down feathers (present in the pterosaur *Tupandactylus)*, pennaceous feathers, which were initially symmetrical, as in *Protarchaeopteryx robusta*, and presumably did not allow flight, and last but not least, asymmetrical pennaceous feathers allowing flight appeared (Chen et al. 1998: 151).

The evolutionary history of feathers is not a simple, linear, stepwise story of complexification of feathers; there may have been reversals. This is suggested, among others, by the discovery of scaly skin (and no trace of feathers) on at least part of the body (the skin is preserved only on the tail and hind leg) of the small coelurosaur *Juravenator* (Göhlich and Chiappe 2006). Yet, when this fossil was discovered, the prevailing evolutionary scenario postulated that simple, filamentous feathers had appeared no later than at the base of *Coelurosauria* (Xu 2006) and, as previously explained, a simple precursor of feathers is now thought to have appeared even earlier, at the base of *Dinosauria*, if not earlier (Witmer 2009a).

The increasingly complex evolutionary history of the feather that emerges illustrates some of the practical problems facing application of apomorphy-based definitions. The delimitation of *Avifilopluma*, as defined by Gauthier and de Queiroz (2001: 25), has arguably varied more than that of most rank-based taxa over the same period; this is hardly an inspiring example to illustrate the advantages of PN! Beyond their nomenclatural implications emphasized here, these discoveries of fossilized integumentary structures obviously help to discriminate between the hypotheses on the original function of feathers: aerodynamic, display (supported by color patterns found in the Early Cretaceous pterosaur *Tupandactylus)* or thermal insulation (also consistent with the feathers of *Tupandactylus)*.

Both of these examples illustrate why Sereno (2005: 597) stated that "Given perfectly complete specimens and a fully resolved phylogeny, taxonomists would still point to different nodes faced with definitions that are operationally dependent on 'complex apomorphies' such as 'feathers' or 'feathered wings.'" Similarly, Mishler (2010: 116) stated: "I personally think the use of apomorphy-based names is incoherent at any level, but that is another argument!" Indeed, apomorphy-based definitions may not deliver the full promise that "In phylogenetic nomenclature, taxon name definitions are based on ancestry and descent rather than the possession of subjective 'key' characters" (Brochu 2001: 1185). Bryant's (1994: 128) conclusion that "Apomorphy-based definitions should be avoided" still seems valid.

### 4.2.3 Crown-Group Names: Should They Be Used for Long-Established, Well-Known Names?

Since the early days of PN, there has been a push to define popular clade names as crown-groups (such as Meier and Richter

1992: 85). The rationale for this was explained by de Queiroz and Gauthier (1992: 468):

> Despite general agreement that Aves includes *Archaeopteryx*, biologists commonly use the name "Aves" ("birds") when making generalizations that apply to extant birds alone. Thus, supposed meaning and actual use are inconsistent. Furthermore, that inconsistency persists despite the existence of a less well-known name coined for the specific purpose of making the relevant distinction. Most biologists continue to use the widely known name "Aves" when referring to the taxon explicitly associated (e.g. 21) with the name "Neornithes." A comparable situation holds for the fossil taxon *Ichthyostega*, the widely used name "Tetrapoda," and Gaffney's (39) less well-known name "Neotetrapoda." . . . But even when the original names are explicitly defined so as to include at least some of the fossil outgroups, as they most often are, the majority of comparative biologists ignore the fossils. Consequently, the original names tend to be used as if they refer to crown clades, and new names coined specifically for the crown clades seldom gain wide use except among paleontologists.

This rationale was further explained as such by de Queiroz and Gauthier (1992: 469):

> Restricting widely used names to crown clades standardizes their meanings in a way that is most useful to the largest number of comparative biologists. Although it will entail changes in the taxonomy of various fossils—*Archaeoptery*, for example, will no longer be considered part of the taxon named "Aves"—this emphasis on extant organisms is not meant to imply that extant organisms are more important than fossils for establishing relationships. On the contrary, it is clear that phylogenetic relationships are best analyzed by considering both fossil and Recent organisms (26, 44). Nevertheless, most biologists study extant organisms, if for no other reason than that many aspects of the biology of extinct organisms are not only unknown but perhaps unknowable.

Crown clades are thus particularly significant for systematics, but not because of their stability in taxonomic content. Lee (1996a) showed that the taxonomic content (list of contained less inclusive taxa) of crown-clades is no more and no less stable than that of more or less inclusive clades. Thus, stability of content is not a reason to prefer to define popular taxon names as crown clades.

Arguments about the naming of crown clades have divided the systematic community, as noted by Bryant (1994: 124). While many neontologists have agreed with the idea of tying popular names to crown-clades, paleontologists have been less enthusiastic. For instance, Sereno (1998: 47) argued that Darwin conceptualized at least some well-known taxa as including at least part of a stem-group, based on this quote (Darwin 1859: 333):

> those groups, which have within known geological periods undergone much modification, should in the older formations make some slight approach to each other; so that the older members should differ less from each other in some of their characters than do the existing members of the same groups.

However, Darwin did not name any taxa in that paragraph, and it is unclear if he meant that the same name should be applied to what we now name crown and total clades. Also, this passage suggests that Darwin may have conceptualized taxa as total clades, rather than apomorphy-based ones. Sereno (1998: 47) argued that this quote from a textbook by Jollie also justified an apomorphy-based definition: "Many taxonomists are beginning to ask, what is a mammal? An **arbitrary line** must be drawn across a continuous and overlapping series of species separating mammal from reptile." However, the arbitrary nature of the delimitation is actually a good reason not to use apomorphies to define taxon names, or to look for phenotypic gaps in the fossil record (which simply reflect its incompleteness) to delimit taxa.

Thus, in addition to the arguments already evoked (de Queiroz and Gauthier 1992), a potential advantage of defining well-known taxon names as crown-clades is that this provides a simple, objective delimitation for the taxa most frequently mentioned in the scientific literature. Let's continue with the example of *Mammalia* (see also Section 4.1). Some stem-mammal taxa are very similar to crown mammals and display mammalian key characters. Thus, they have often been called "mammals" by systematists who did not associate *Mammalia* with a crown-group. For instance, *Diarthrognathus* has long been known to have a "double" jaw joint. A first joint was present between quadrate and articular (the primitive, sometimes erroneously named "reptilian") jaw joint, and a new joint had appeared between dentary and squamosal. This second joint is characteristic of mammals (Crompton 1963), which lack the old, "reptilian" joint (because the relevant elements are integrated into the mammalian middle ear). The cynodont *Morganucodon* (known from the late Triassic to the early Jurassic) has frequently been considered a mammal (i.e., Crompton 1963; Kermack et al. 1981; Ruta et al. 2013), even though it is outside the crown group (Luo et al. 2016: figure 6.1). In fact, the synapsid fossil record is sufficiently complete that delimiting *Mammalia* at a morphological gap is difficult, and the name *Mammalia* has been applied to multiple nested clades (Rowe and Gauthier 1992). Furthermore, several key mammalian characters do not fossilize readily (such as mammary glands, fur), so using them in an apomorphy-based definition would result in an uncertain placement of most extinct synapsids, even though their phylogenetic position is reasonably well-constrained (see Section 4.2.2). This combination of factors strengthens the case for a crown-clade definition of the name *Mammalia*, but a similar case could be made for many other taxa, such as *Aves* and *Tetrapoda*. More "traditional" concepts, such as Lucas' (1992: 371) definition of *Mammalia* as "the monophyletic taxon whose members share biologically significant evolutionary novelty(ies)" do not unambiguously identify any clade; indeed, this definition could identify *Nematoda*, *Primates*, *Arthropoda* and many other taxa, even though

Lucas (1992) stated that he meant the clade *Mammaliaformes*. Ironically, to make us understand what he means, Lucas (1992), who is critical of PN, had to refer to a name coined in the context of PN by one of its practitioners (Rowe 1988)!

Unsurprisingly, the best way to define various other taxa has been debated. One such taxon is *Tetrapoda*, which has had a long, confusing history harking back to Aristotle (Sues 2019). Anderson (2001) proposed an apomorphy-based definition based on the limb, whereas Laurin (2002) argued for a crown-clade definition (Laurin and Anderson 2004), which is now established under the *PhyloCode* (Laurin 2020). Recent usage of this term in the scientific literature exemplifies ambiguity in usage that can justify using this term for the crown-group. For instance, Pardo et al. (2020: 3) summarized thus the meaning of the name "*Tetrapoda*" most frequently encountered in the paleontological literature: "The term 'Tetrapoda' is generally applied to digited members of the total group, a usage that is equivalent to Stegocephalia (Laurin et al. 2000), whereas members of the crown-group are sometimes referred to Neotetrapoda (Sues 2019)." This is coherent with Pardo's usage of the name *Tetrapoda* in other publications (such as Pardo et al. 2017b). However, on the same page, in a discussion of molecular dating constraints, Pardo et al. (2020) wrote "the oldest calibration point (Tetrapoda) has a minimum age of 337 Ma," which clearly designates the age of the crown-group, given that the limb with digits is documented in the Famennian (360 Ma), that ichnological evidence of such limbs is even older and that the total group of tetrapods is older still. Thus, while Pardo et al. (2020) professed a preference for an implicitly apomorphy-based definition of *Tetrapoda* (unsurprisingly so, given that J. Anderson is a co-author of that paper and that he has expressed strong support for this idea), when using the name in a discussion of calibration constraints for molecular dating, they naturally shifted to a crown-based definition. This is logical to the extent that molecular systematists mostly date the divergences between extant taxa, rather than the appearance of morphological features.

Similar debates have involved the taxa *Reptilia* and *Sauropsida*. Modesto and Anderson (2004: 819) proposed a total-group definition of *Reptilia*: "the most inclusive clade containing *Lacerta agilis* Linnaeus 1758 and *Crocodylus niloticus* Laurenti 1768, but not *Homo sapiens* Linnaeus 1758." Note that this clade has often been called *Sauropsida*, a named that Huxley proposed in 1864 and this name has been used reasonably frequently (a search for "*Sauropsida*" done on March 18, 2021 recovered 3,540 articles), but using *Reptilia* for that clade would effectively prevent use of *Sauropsida*. Partly for this reason, Laurin and Reisz (2020a) proposed a crown-clade definition of *Reptilia*, which is now established under the *PhyloCode*: "The smallest crown clade containing *Testudo graeca* Linnaeus 1758 (*Testudines*), *Iguana iguana* Linnaeus 1758 (*Lepidosauria*), and *Crocodylus* (originally *Lacerta*) *niloticus* Laurenti 1768 (*Archosauria*)."

These debates need to be put into their proper perspective. Ultimately, the choice of name/definition combination matters little as long as only one is considered valid under the *PhyloCode* and that this choice does not depart strongly from previous usage (in all fields of biology, not only paleontology). The ultimate goal of nomenclature is efficiency in communication. Given that discussion of many (perhaps most) attributes of major clades are necessarily based on extant taxa (because many characters do not normally fossilize), using crown-clade definitions of these names ensures that most statements about these taxa are correct. Of course, many (probably most) attributes of these crown-clades appeared before these crown-clades, but, in most cases, we will never know where exactly in the Tree of Life.

If we agree to use well-known names for crown-clades, there may be ambiguity because several well-known names could conceivably be applied to a given crown-clade. Thus, Bryant (1994) gave the example of the crown-group of rhinoceroses, to which the names *Rhinocerotini*, *Rhinocerotidae* and *Rhinocerotoidea* could conceivably be applied. He argued that in such cases, the name that refers to the least inclusive clade (in this case, *Rhinocerotini*) should be used because this makes the other names available for more inclusive clades and, in this case, such names are needed because rhinoceroses have an abundant fossil record. This was implemented into the *PhyloCode* (Recommendation 10G).

Any nomenclatural principle regarding crown-clades must rest on a clear concept of what "extant" means. Indeed, Bryant (1994, 1996) pointed out that extinction could change the delimitation of taxa. For instance, Bryant (1996: 179) pointed out that the previously published definition of *Carnivora* (as "the most recent common ancestor of living carnivorans and all its descendants") referred to a clade only if one or more carnivoran taxa remain extant, and the delimitation could change depending on which taxa become extinct. This problem was solved in the *PhyloCode* by considering that "extant" refers to the time the taxon is established by default (Article 9.9 and Note 10.5.3). Thus, extinction of taxa in the future does not affect their delimitation.

### 4.2.4 Possible Compromise for Popular Names

Insufficient time has passed since the advent of PN to know to what extent the crown-clade definitions of major, popular taxa like *Aves* and *Mammalia* will gain acceptance. We have seen that apomorphy-based definitions are problematic for several reasons (see Section 4.2.2), but the literature suggests that many high-ranking taxa are conceptualized as apomorphy-based clades linked to conspicuous apomorphies, such as the feather or digited limb. This was explained by Sereno (1998: 47–48):

> For most biologists, past or present, higher taxa such as "Mammalia" and "Aves" are associated with "key" characters or sets of such characters—an apomorphy-based definition. A furry, mouse-sized creature with differentiated teeth that lived during the Jurassic will always be considered a "mammal" and if pressed allotted to Mammalia by the majority of extant biologists. Likewise, *Archaeopteryx*, with feathers, wings and the capacity for flight, will always be considered a "bird" within Aves. Altering this equation for well established monophyletic taxa will engender greater confusion than encouraging neontologists to use Neornithes or insert an appropriate modifier ("living" or "recent") before Aves, when doing so carries any significance.

If necessary, a compromise approach could be attempted in the future that avoids some of the pitfalls inherent with apomorphy-based definitions. This compromise could be to delimit major taxa using their historical delimitation, which often reflects the appearance of a prominent apomorphy, in the fossil record, but to use historical inclusion of the basalmost taxa to define clade boundaries. Thus, *Aves* could be defined as the smallest clade that includes *Archaeopteryx lithographica* and *Passer domesticus*, for instance, because *Archaeopteryx* has been known since 1861, it has always been included in the taxon *Aves* except in some recent paper that used PN and the taxon *Aves* has generally not included more distant Mesozoic relatives of the crown-group. Similarly, *Tetrapoda* could be redefined, if need be, using *Ichthyostega stensioei*, which has been known since 1932 and has usually been included in the taxon *Tetrapoda* by paleontologists, again with the exception of relatively recent papers using PN. A potential problem in this case is that *Acanthostega*, erected in 1952 and also normally included in *Tetrapoda* (except among some proponents of PN), may be more rootward and would be excluded from *Tetrapoda*, despite the fact that it has limbs. Thus, this proposed compromise would entail some subjective decisions about which specifier to use. To delimit *Mammalia* an emblematic taxon like *Morganucodon watsoni* (erected in 1949) could be selected. However, all three names are now established under the *PhyloCode* for crown clades, so these examples were used to illustrate general principles, rather than to argue that these names should be redefined. Also, this compromise would not solve the discrepancy between paleontological and neontological usage of these names, which is one of the main advantages of adopting crown-clade definitions for such names (see previous).

## 4.3 Unregulated Phylogenetic Nomenclature

Much work was done in PN before the *PhyloCode* took effect. This was a period (from the late 1980s to April 2020) of experimentation to determine how this new kind of nomenclature stood the test of time, and it provided opportunities to learn from mistakes. For instance, the taxa *Anthracosauria* and *Temnospondyli*, which were historically anchored to embolomeres (a taxon of Permo-Carboniferous stegocephalians), especially *Anthracosaurus russelli* Huxley 1863, and Permo-Triassic temnospondyls, such as *Mastodonsaurus* and *Trematosaurus*, respectively, were initially defined by practitioners of PN on the basis of assumed affinities between these Paleozoic or Triassic core taxa and extant taxa (*Amniota* and *Lissamphibia*, respectively). Thus, de Queiroz and Gauthier (1992: 475) defined *Temnospondyli* as "the clade including *Amphibia* plus all known and unknown tetrapods sharing a more recent common ancestor with *Amphibia* than with *Amniota*," which clearly designates the lissamphibian total group. This definition became problematic when Laurin and Reisz (1997) proposed a tetrapod phylogeny in which temnospondyls were stem-tetrapods. Under this topology, which is (still) accepted by a minority of paleontologists but is supported by recent studies based on large datasets (Marjanović and Laurin 2019; Laurin et al. 2022), *Temnospondyli* as defined

by de Queiroz and Gauthier (1992) excludes all the taxa traditionally considered temnospondyls, while lissamphibians and the Paleozoic lepospondyls, which have traditionally been excluded from *Temnospondyli*, are included. Likewise, the first phylogenetic definition of *Anthracosauria* referred to the amniote total group rather than the core taxa (embolomeres, especially *Anthracosaurus russelli*), which would be excluded from it under the same recent phylogenies (also see the discussion of Article 11.10 of the *PhyloCode* in Section 4.4.1). Thus, definitions of names must be based on the lower-ranking taxa that were originally included, rather than assumed affinities to their closest extant relatives. This principle is one of the foundations of RN, but it is also required in PN, as the examples of the names *Temnospondyli* and *Anthracosauria* have shown.

Several early phylogenetic definitions used higher taxa as specifiers. For instance, Gauthier et al. (1988) indicated that "*Tetrapoda* is restricted to the most recent common ancestor of extant *Lissamphibia* and *Amniota*, and all of its descendants." Here, the specifiers are thus *Lissamphibia* and *Amniota*. However, the use of higher taxa as specifiers proved problematic, especially when the composition or affinities of these specifiers shifted in new phylogenies, as shown by the example of *Temnospondyli* mentioned previously. Other problems with the use of higher taxa as specifiers is that their composition can vary between studies, some can become paraphyletic and some are not defined using PN. For these reasons, Bryant (1996) argued that nominal species should be used as specifiers, instead of higher taxa, a suggestion that was implemented in the *PhyloCode* (Article 11).

Thus, well before the *PhyloCode* was implemented, there was a trend to use lower-ranking taxa (nominal genera or species) as specifiers. However, given that not all taxa are equally well-known or their affinities not equally well-understood, care must be taken in selecting these specifiers. Furthermore, communication can be facilitated by selecting as specifiers taxa familiar to many systematists, and these can be reused in several definitions. Thus, Sereno (2005: 598) already concluded that "nested referencing using species (or specimens tied to a species nomen) has proven to constitute a major step toward clarity and simplicity in phylogenetic definitions." An example of nested referencing is the node-stem triplet (**Figure 4.6**), an important concept formalized by Sereno (1998: 51) thus:

A node-stem triplet involves three taxa—a node-based taxon composed of two stem-based subordinate taxa—with complementary definitions (Fig. I) [**Figure 4.6**]:

*Node:* Taxon A, taxon B, their most recent common ancestor and all descendants.
*Stem:* All [taxa] closer to taxon A than to taxon B.
*Stem:* All [taxa] closer to taxon B than to taxon A.

It should be noted that despite the terminology used by Sereno (1998), node-stem triplets need not involve a crown-clade and its two main total-clades. Any node-based clade and its two main branch-based clades can be named and defined as a node-stem triplet.

No matter how much care is put into formulating a phylogenetic definition, unforeseen changes in the composition

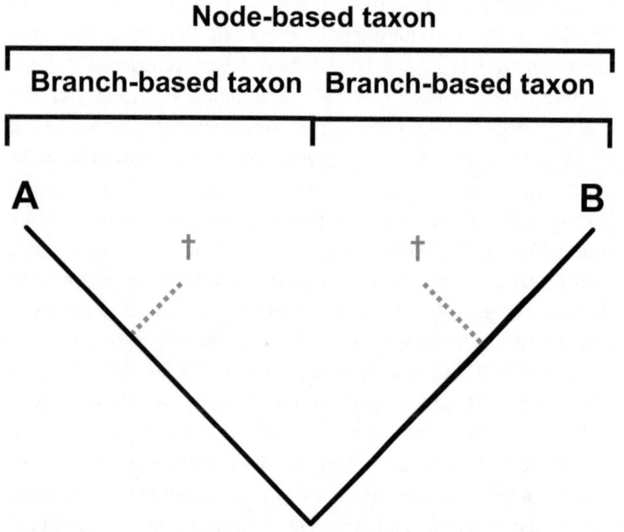

## Node-based taxon

## Branch-based taxon    Branch-based taxon

**A**                                          **B**

†            †

**FIGURE 4.6** Tree used to illustrate the node-stem triplet concept, first proposed by Sereno (1998). Such a triplet is composed of a node-based (minimum) clade and of the two branch-based (maximal) clades that can be recognized therein. In the illustrated example, the two specifiers (A and B) are extant taxa and the largest clade is thus a crown-clade, but the specifiers may also be extinct taxa and the largest clade needs not be extant.

of the defined taxon can occur if the phylogeny changes. Therefore, Bryant (1996: 187) suggested that the application of these name/definition combinations should be limited to "appropriate phylogenetic contexts." He provided the example of *Pinnipedia* (the taxon that encompasses the aquatic carnivorans), which had previously been defined (Wolsan 1993: 369) as "The most recent common ancestral species of otariids, odobenids, and phocids, plus all of its descendants." Bryant (1996) argued that under phylogenies that imply a diphyletic origin of aquatic carnivorans (that is, if phocids and otariids became adapted to an aquatic lifestyle independently of each other and did not form a clade exclusive of terrestrial carnivorans), the name *Pinnipedia* should not apply. This idea gained significant support among proponents of PN and has been integrated into the *PhyloCode*, notably through the inclusion of qualifying clauses (Article 11.12), use of additional specifiers (Article 11.13) or definitions restricting a taxon to be part of another taxon (Article 11.14).

Not all proponents of PN are in favor of establishing a code to regulate it. Thus, Sereno (2005: 614) emphasized the importance of considering previously proposed phylogenetic definitions. Sereno (2005: 615) even suggested:

> Despite all of the discussion and recommendations above, there is no single procedure to construct a phylogenetic definition for a particular taxon any more than there is one way to perform a phylogenetic analysis for a particular clade. One approach has been to downplay that fact and move to establish in perpetuity some definitions over others with the aim of achieving a unitary taxonomy. Another approach outlined below holds that those

phylogenetic definitions that best maintain historical continuity and current utility will gain currency among systematists, an outcome achieved by consensus rather than by the imposition of a formalized code and centralized authority.

To support this point, Sereno (2005: 615) pointed out that the eight phylogenetic definitions of the taxon *Aves* that he had found collectively identified only two clades, even though *Theropoda* (to which *Aves* belongs) is the very first taxon to which PN was applied (Gauthier 1986). However, this is a single case, and Sereno (2005: 617) also admitted that the presence of three competing phylogenetic definitions in a same book (Weishampel et al. 2004) of both *Ornithischia* and *Saurischia* was problematic. He suggested that such problems resulted largely from lack of awareness of previous phylogenetic definitions, which led various authors to unintentionally publish slightly different phylogenetic definitions that often referred to the same clades, and that such problems could be avoided or minimized by the development of "an on-line database of phylogenetic definitions that facilitates access to definitional history." Recently, he stated that his favorite approach is still to do PN without a code (Sereno, personal communication from February 11, 2022).

Indeed, centralized databases are essential for biodiversity research. As summarized by Sereno et al. (2005: 2):

> Darwin (1859) famously predicted that taxonomy would come to reflect knowledge of genealogy, and that knowledge is currently undergoing rapid advance. Modern taxonomists, nonetheless, are in the unenviable position of riding a tidal wave of interest in biodiversity and its evolutionary history while lacking the tools to efficiently log, locate, and share basic information regarding suprageneric taxa. Such information includes original authorship, bibliographic citation, phylogenetic definition, taxonomic composition, taxonomic history, and estimated temporal duration.

To partly remedy this problem, Sereno et al. (2005) developed *TaxonSearch*, an online paleobiodiversity database. Contrary to most online biodiversity databases that focus on species, *TaxonSearch* focuses on suprageneric taxa. It currently features a single compilation entitled "Stem Archosauria," which actually includes "all suprageneric taxa [within *Archosauria*] excluding those within the archosaurian crown clades, *Crocodylia* and *Neornithes*, and those within *Pterosauromorpha*" (Sereno et al. 2005: 8). Thus, it does not include stem-archosaurs, but rather, the stem of *Crocodylia* and most of the stem of *Aves*. *TaxonSearch* was designed to accommodate other compilations, but as of this writing (Feb. 22, 2022), no new compilations have been added. This database seems to enjoy moderate success, as shown by 169 Google Scholar citations for "*TaxonSearch*," mostly of papers on early archosaurs, but this is to be compared with the 1,920 Google Scholar citations of the *PhyloCode* (Cantino and de Queiroz 2020) and the 473 citation for *Phylonyms* (de Queiroz et al. 2020), despite that *TaxonSearch* became available much earlier (all searches carried out on Oct. 23, 2022). Admittedly, the *Phylonyms* citations are so far found overwhelmingly in its various chapters (which cite other chapters), but this is not unexpected given that it was published recently (April 30, 2020). Thus, the brief history of PN suggests that most of its practitioners favor use of a code; in any case, Bryant (1994: 129) suggested that "Stability regarding the use of names is probably best achieved through codes of nomenclature." This is hardly surprising given that the history of RN (see Section 2.1) suggests that nomenclature without a code generates chaos.

There seems to be a broad consensus that biological nomenclature needs to be regulated, but history shows that developing codes to that effect takes several years. Thus, Taylor (2007) proposed a series of recommendations to guide the development of phylogenetic nomenclature before the *PhyloCode* was introduced. Contrary to Sereno et al.'s (2005) *TaxonSearch* initiative, it was not meant to obviate the need for a code; it was just a temporary fix. Taylor (2007) explained that the pre-*PhyloCode* years were an "experimental period," discussed the advantages and shortcomings of various definitions that had been proposed and suggested that "definitions proposed now, in the last days of the pre-PhyloCode era, should be treated in the same way as those of ten and fifteen years ago—with respect, but not with deference." Taylor's (2007) recommendations were sensible enough: not to recognize "accidental [unintended] definitions"; to "Be generous in recognizing deliberate but malformed definitions"; to recognize all kinds of phylogenetic definitions (including those based on an apomorphy) and those using all kinds of specifiers, provided that intent was clear; and, finally, to "Use priority of synonyms and homonyms as a guideline, not as a rule." This last recommendation means that before the *PhyloCode* took effect, there was no strict rule of priority, but, even then, if a name had been given several equally compelling definitions, the oldest one should be preferred. However, if the first name-definition combination happened to be objectionable, it should not be preferred over junior synonyms or homonyms.

## 4.4 The *PhyloCode*

### 4.4.1 Introduction to the *PhyloCode*

The *PhyloCode* aims at facilitating communication about taxa. As the preface of that code states (p. vii):

> The development of the *International Code of Phylogenetic Nomenclature* (referred to here as the *PhyloCode*) grew out of the recognition that the current rank-based systems of nomenclature, as embodied in the current botanical, zoological, and bacteriological codes, are not well suited to govern the names of clades. Clades (along with species) are the entities that make up the tree of life, and for this reason they are among the most theoretically significant biological entities above the organism level. Consequently, clear communication and efficient storage and retrieval of biological information require names that explicitly and unambiguously refer to clades and do not change over time. The current rank-based codes fail to provide such names for clades.

In other words, the way names regulated by the rank-based codes are defined does not ensure unambiguous, stable delimitation, even if the phylogeny does not change and if no additional taxa are discovered. The *PhyloCode* represents a sharp departure from the RN codes in many ways, but it also shares with them many similarities (Cantino and de Queiroz 2020: ix): the goal of providing methods to unambiguously identify the correct name of each taxon, the aim of not infringing upon judgment of taxonomists, the use of precedence (chronological priority) to resolve problems of synonymy and homonymy, and a conservation mechanism to allow later-established synonyms or homonyms to be given precedence when this promotes nomenclatural stability. Some parts of the *PhyloCode* are inspired from the *BioCode* (Cantino and de Queiroz 2020: xv).

The *PhyloCode* (Cantino and de Queiroz 2020) is organized into a preface, a preamble, principles, 22 articles organized into 11 chapters that form the bulk of that code, a glossary, three appendices and an index. In this respect, it is reminiscent of the rank-based codes. However, the names of the chapters and articles reveal substantial differences from rank-based codes. For instance, the titles of chapters IV ("Establishment of Clade Names") and V ("Selection of Accepted Clade Names") clearly show that this code regulates clade names.

Given the basic information on PN already provided, only brief highlights based on selected passages of the *PhyloCode* are required here. The "Principles" section includes seven items. Principle 4 is about stability; it stipulates that "The names of taxa should not change over time. As a corollary, it must be possible to name newly discovered taxa without changing the names of previously discovered taxa." This alludes to the problem of taxon name changes caused by rank changes under RN that is explained in Section 3.4.1, "Nomenclatural instability," in this book.

The first chapter, composed of three articles, is about taxa. Article 1 explains that only clade names are regulated by this code, and Article 2 mentions (in note 2.1.2) that "It is not necessary that all clades be named." Indeed, given that phylogeny is never known with certainty, it is a good idea to name clades that are well-corroborated by evidence (such as supporting synapomorphies). The sole figure present in the *PhyloCode* shows how hybridization and species fusion can lead to partially overlapping clades (which reflects reticulate evolution).

Chapter II, composed of two articles, is entitled "Publication." Article 4.2 explains that electronic publications are acceptable, but subject to the following conditions:

> electronic text with or without images or sound in Portable Document Format (PDF) in an online publication (however, not just in supplementary material; see Note 7.2.2); in both cases with an International Standard Serial Number (ISSN) or an International Standard Book Number (ISBN).

Future progress is anticipated in Article 4.4, which states that "Should Portable Document Format (PDF) be succeeded, a successor international standard format approved and communicated by the Committee on Phylogenetic Nomenclature would be acceptable." Article 4.6 explains that dissemination of electronic documents through storage media (such as CDs), theses and abstracts of papers presented in scientific meetings, even if published in peer-reviewed journals, do not qualify as publications under this code.

Chapter III, composed of Articles 6 to 8, is about names. Article 6.1 states that "Established clade names are those that are published in accordance with Article 7 of this code." Recommendation 6.1A states that all scientific names should be italicized, which is coherent with most rank-based codes, except for the *Zoological Code*. Recommendation 6.1B indicates that "it may be desirable to distinguish these names from supraspecific names governed by the rank-based codes, particularly when both are used in the same publication," and Example 1 suggests using the letters "P" (bracketed or in superscript) and "R" for "*PhyloCode*" and "Rank-based codes," respectively. Another possibility is to mention "clade" and a given rank for the same purpose. Article 6.2 defines "preexisting names" (those that are regulated by rank-based codes and can be redefined under the *PhyloCode*); these include, for zoological names, potentially valid names as well as names in use but not really governed by the code, such as those above the superfamily rank. Once such a name has been defined in conformity with the *PhyloCode*, it is a "converted name" (Article 6.3). Article 7 describes requirements for establishment, and Article 8 describes registration into the *RegNum* database.

Chapter IV deals with the establishment of clade names. Article 9.1 stipulates that clade names "may be established through conversion of preexisting names or introduction of new names." Article 9.2 explains that to be established:

> converted clade names must be clearly identified as such in the protologue by the designation "converted clade name" or "nomen cladi conversum." New clade names must be identified as such by the designation "new clade name" or "nomen cladi novum."

The associated definition must be in English or Latin (Article 9.3). Note 9.4 provides conventions for abbreviated definitions (each definition must be provided both in full and in abbreviated language). For instance, the sign "~" signifies "but not" (for external specifiers). In cases of disagreements between the plain text definition and the abbreviated definition, the latter should be preferred to interpret the author's intent (Recommendation 9.4A). Article 9.5 explains how to formulate "minimum clade" definitions (formerly known as node-based definitions). Branch-based definitions are now called "maximum-clade" definitions (Article 9.6). Recommendation 9.7B stresses that if an apomorphy-based definition is used, and if the apomorphy is a complex character that may have evolved in a stepwise manner, the author should specify which aspect(s) of the apomorphy are considered defining. However, the authors of the *PhyloCode* obviously recognize that possibility for ambiguity despite Recommendation 9.7B because Article 9.8 stipulates:

> If the author of an apomorphy-based definition based on a complex apomorphy did not identify in the protologue which aspect(s) of that apomorphy must be present in order for an organism to be considered

to belong to the clade whose name is defined by that apomorphy (Rec. 9.7B), or if an aspect that the author did identify is later found to be a complex apomorphy itself, then subsequent authors are to interpret the definition as applying to the clade characterized by the presence of all of the components of the complex apomorphy described by the author of the definition (see Rec. 9.7A) or present in the taxa or specimens that the author of the definition considered to possess that apomorphy. Similarly, if multiple apomorphies are used in an apomorphy-based definition, subsequent authors are to interpret the definition as applying to the clade characterized by the presence of all of those apomorphies.

In light of the problems unique to apomorphy-based definitions outlined in Section 4.2.2, Recommendation 9.7B and Article 9.8 are indeed necessary to prevent nomenclatural instability linked to apomorphy-based definitions.

In addition to the three basic definition types, the *PhyloCode* allows more complex definitions that may be useful in some situations. For instance, Note 9.9.2 mentions that:

> A **maximum-crown-clade** definition (formerly known as a **branch-modified or a stem-modified node-based** definition) may thus take the form "the crown clade originating in the most recent common ancestor of A and all extant organisms or species that share a more recent common ancestor with A than with Z" or "the largest crown clade containing A but not Z", where A is an extant internal specifier and Z is an external specifier.

Such definitions could be useful, among others, when only part of the extant biodiversity of a clade is known, or when the phylogeny of that clade is poorly resolved, but when the affinities of that clade to other, more distant clades are better-resolved.

Another complex kind of definition is the apomorphy-modified crown-clade definition, formerly known as the "apomorphy-modified node-based definition." It may take the form "the crown clade originating in the most recent common ancestor of A and all extant organisms or species that inherited M synapomorphic with that in A" or "the crown clade for which M, as inherited by A, is an apomorphy relative to other crown clades" or "the crown clade characterized by apomorphy M (relative to other crown clades) as inherited by A," in which A is an extant specifier species or specimen and M is a specifier apomorphy (Note 9.9.2). In such definitions, the apomorphy is used to specify a crown-clade, which may mitigate, to an extent, the ambiguity inherent in apomorphy-based definitions.

Article 9.10 explains how total-clade definitions can be formulated. One of the simplest possibilities is simply in the form "the largest total clade containing A but not Z," where A and Z are internal and external specifiers, respectively. Article 9.10 also mentions that additional internal and/or external specifiers may be added. It indicates that such additional internal specifiers may be useful "if the author intends for the name not to apply to any clade in the context of particular phylogenetic hypotheses."

Some elementary phylogenetic quality control is provided by Recommendation 9.13A, which stipulates that "A reference phylogeny should be derived via an explicit, reproducible analysis."

Article 9.14 ensures that phylogenetic definitions can be understood reasonably easily by requiring that:

> Any specimen citation [in the protologue] must include the name of a species or clade (less inclusive than the one whose composition is being described) to which the specimen can be referred, unless the clade whose composition is being described does not contain any named species or clades.

This makes the definitions intelligible without requiring consultation of the catalogue of the collection that includes the named specimens.

Recommendation 9B explains that "Conversion of pre-existing names to clade names should only be done with a thorough knowledge of the group concerned, including its taxonomic and nomenclatural history and previously used diagnostic features." This is justified by the fact that conversion of preexisting names has deep nomenclatural repercussions that are arguably more numerous than erecting new taxa under RN.

Article 10 is entitled "Selection of clade names for establishment." It stresses that "Clade names are generally to be selected in such a way as to minimize disruption of current and/or historical usage (with regard to composition, diagnostic characters, or both) and to maximize continuity with existing literature" (Article 10.1). Furthermore, Article 10.1 allows conversion of names that refer (or referred to in the past) to paraphyletic groups under RN to name a clade stemming from the same last common ancestor. This is necessary to promote continuity in the literature because many well-known names that formerly applied to paraphyletic groups can thus be converted, rather than rejected; a few examples include *Osteichthyes*, *Sarcopterygii*, *Reptilia*, *Synapsida* and *Dinosauria*.

Recommendation 10.1 explains that the most widely used name of a clade should be chosen when several alternatives exist, except if a **panclade name** (see Section 4.4.3.1) is to be proposed. As explained in Article 10.3, panclade names, which designate total groups, are formed by adding the prefix "*Pan-*" in front of the crown-group name, which retains its capital first letter (such as *Pan-Mammalia*). This once again highlights a tension between promoting nomenclatural stability or an "integrated" nomenclatural system. Some of this could have been relieved by a greater reliance on informal (rather than formal) panclade names, which are also recommended by the *PhyloCode*; to distinguish these from formal panclade names, their informal versions should neither be capitalized nor italicized (Recommendation 10.3A).

Also, to promote an integrated nomenclatural system (though in a slightly less visible way than with formal panclade names), Recommendation 10.1B indicates that:

> The name that is more commonly used than any other name to refer to (e.g., discuss or describe) a particular crown clade should generally be defined

as applying to that crown clade, even if the name is commonly considered to apply to a clade that includes extinct taxa outside of the crown.

This has been justified on various grounds in numerous debates about how to name crown-groups and how to apply some well-known taxon names, such as *Tetrapoda, Aves* and *Mammalia* (see Section 4.2.3).

Still to better integrate nomenclature, Article 10.8 indicates that:

> If the name of a crown clade refers etymologically to an apomorphy, and a new name (as opposed to a converted name) is to be established for the clade originating with that apomorphy by adding an affix to the name of the crown clade, the prefix *Apo-* must be used.

Thus, the meaning of taxon names starting by "*Pan-*" or "*Apo-*" under the *PhyloCode* is reasonably intuitive.

Article 10.10 forbids conversion of names of species groups and specific or infraspecific epithets. This is presumably partly motivated by the fact that such epithets are often not unique, and some are indeed very common. Some well-known examples include "*californiensis,*" "*grandis*" and "*japonicus.*"

Recommendation 10D deals with conversion of cross-code homonyms. Only one such homonym can be converted without modification. For the others, prefixes must be used to differentiate them from the first-converted homonym. These prefixes include "*Phyto-*" for plants; "*Phyco-*" for algae other than cyanobacteria; "*Myco-*" for fungi (these three sets of names are governed by the *Botanical Code*); "*Zoo-*" for animals; "*Protisto-*" for unicellular, non-photosynthetic eukaryotes (both governed by the *Zoological Code*); and "*Prokaryo-*" or "*Bacterio-*" for prokaryotic organisms.

The *PhyloCode* provides the following example to clarify this recommendation: Under RN, the name *Prunella* refers both to a bird genus and to an angiosperm genus. If the name *Prunella* were first converted for a bird clade (approximating the genus *Prunella* in composition), and if botanists wanted to name a clade approximating the angiosperm genus *Prunella* in composition, if no other widely used name were available, the appropriate name for that clade under the *PhyloCode* would be *Phyto-Prunella*.

Recommendation 10G states that:

> When establishing a name for a crown clade that, under rank-based nomenclature, corresponds to a **monogeneric "higher" taxon**, the **genus name** should be converted for that clade rather than any of the suprageneric names that have been applied to it. Doing so will permit the use of the "higher" taxon names for larger clades that extend beyond the crown.

The rationale for this recommendation is straightforward, but an example is provided (see the example on rhinoceroses in Section 4.2.3).

Article 11 is entitled "Specifiers and qualifying clauses," and thus, deals with the core of phylogenetic definitions.

Specifiers "are species, specimens, or apomorphies cited in a phylogenetic definition" (Article 11.1). Under the *PhyloCode*, when species names are used as specifiers, these names refer to the type-specimen. However, complications can arise when the name of the species that includes that specimen changes, or when a new type is selected, and these processes remain regulated by RN codes. Article 11.6 clarifies that:

> When a type specimen is used as a specifier, it retains its status as a specifier even if a different type for the species name that it typified is subsequently designated under the relevant rank-based code, or if the species name that it typifies is no longer accepted because the species has been re-circumscribed and the name relegated to synonymy.

Given the tight link between type-specimens and species names under RN codes, the specimens selected as specifiers under the *PhyloCode* are normally type-specimens. However, in some situations, other specimens may be used as specifiers. This is explained in Article 11.7:

> Specimens that are not types may not be used as specifiers unless: (1) the specimen that one would like to use as a specifier cannot be referred to a named species, so that there is no type specimen that could be used instead; or (2) the clade to be named is nested entirely within a species; or (3) the clade to be named includes part of a non-monophyletic species and its descendants, but the type of the non-monophyletic species is either excluded from that clade or it is not possible to determine whether it is included.

The *PhyloCode* also provides rules (Article 11.10) to ensure that the contents of taxa named or converted under that code do not differ too drastically from those of homonyms under RN. This point is also addressed briefly in the preface (Cantino and de Queiroz 2020: viii):

> Mechanisms are also provided to reduce certain types of nomenclatural divergence relative to the rank-based systems. For example, if a clade name is based on a genus name, the type of the genus under the appropriate rank-based code must be used as an internal specifier under the PhyloCode (Article 11.10, Examples 1 and 2).

Article 11.10 is thus aimed at preventing the excessive and confusing divergence in taxon delimitation that would arise between names regulated by the *PhyoCode* and those regulated by the RN codes, if type-species of the latter became excluded from names regulated by the *PhyloCode*. It reads:

> In the interest of avoiding confusion, a clade name should not be based on the name of another taxon that is not part of the named clade. Therefore, when a clade name is converted from a preexisting name that is typified under a rank-based code or is a new or converted name derived from the stem of a typified name, the definition of the clade name must use the type species of that preexisting typified name or of the genus name from which it is derived (or the type specimen of that species) as an internal specifier.

The importance of this article can be illustrated by another example (four are provided in the *PhyloCode*, but a different one will be used here). Gauthier et al. (1988: 139) defined *Anthracosauria* as "the sister-group of Amphibia within Tetrapoda." In the context of their phylogeny, *Anthracosauria* included *Amniota, Diadectomorpha, Solenodonsaurus, Seymouriamorpha* and *Anthracosauridae*. The latter is based on the genus *Anthracosaurus*. Note that this definition did not explicitly include *Anthracosaurus* within *Anthracosauria*. This would not have been problematic if the reference phylogeny had proven uncontroversial, but some recent analyses (for instance, Marjanović and Laurin 2019; Laurin et al. 2022) suggest that the sister-group of *Amphibia*, now called *Pan-Amniota* (Laurin and Smithson 2020), includes *Amniota* and *Diadectomorpha*, but probably not *Seymouriamorpha* and *Anthracosauridae*. Thus, under the original definition of *Anthracosauria* (which fortunately has no standing under the *PhyloCode*), *Anthracosaurus* and its close relatives are no longer part of *Anthracosauria*. Article 11.10 of the *PhylocCode* prevents such undesirable nomenclatural outcomes.

In addition, Recommendation 11A states that:

> Definitions of converted clade names should be stated in a way that attempts to capture the spirit of traditional use to the degree that it is consistent with the contemporary concept of monophyly. Consequently, such a definition should not necessitate (though it may allow) the inclusion of subtaxa that have traditionally been excluded from the taxon designated by the preexisting name, as well as the exclusion of subtaxa that have traditionally been included in the taxon. To accomplish this goal, internal specifiers of converted clade names should be chosen from among the taxa that have been considered to form part of a taxon under traditional ideas about the composition of that taxon, and they should not include members of subtaxa that have traditionally been considered not to be part of the taxon.

The *PhyloCode* provides as an example the name *Dinosauria*, which originally included only Mesozoic taxa, such as *Megalosaurus* and *Iguanodon*. Even though it is now well-established that birds are dinosaurs, it would be a bad idea to define the name *Dinosauria* using some bird taxa as specifiers. Instead, some of the taxa that were included early (ideally, when the taxon was erected) should be used as internal specifiers. Thus, even under the very unlikely scenario in which the dinosaurian affinities of birds become refuted, the historical contents will remain tied to the name *Dinosauria*.

Thus, the claims by some opponents of the *PhyloCode* who argued that its use would result in drastic change in taxon delimitation (such as Foer 2005) are baseless.

Recommendation 11C stipulates that ichnotaxa (trace fossils, such as trackways and burrows) are "taxonomically ambiguous types," and thus should not be used as specifiers. For most taxa, this seems like sound advice, and this presumably reflects the parataxonomy that is used for the nomenclature of ichnofossils.

Recommendation 11F effectively recommends naming node-stem triplets (see Section 4.3) when we want to maintain a hierarchy of the kind "taxon A is composed of taxa B and C."

Chapter V opens with Article 12, about precedence, where it is clarified that the latter is based on date of publication rather than date of registration. Article 13 deals with homonymy. Note 13.1.1. explains that "Homonyms result when an author establishes a name that is spelled identically to, but **defined differently** than, an earlier established name." This is reminiscent of the RN codes, but in PN, taxa may differ conceptually (by being different kinds of clades), even if their known content is the same. To cover this, Article 13.2 stipulates that phylogenetic definitions are considered to be different if they cite different specifiers, if they have different qualifying clauses or if they are of different kinds (for instance, minimum-clade or maximum-clade); these clarifications are also useful to assess synonymy.

Synonymy (covered in Article 14) under the *PhyloCode* occurs when two or more names refer to the same clade. Note that this requirement is more stringent than identical taxonomic content; it requires the names to share the same definition type (such as node- or branch-based). Thus, two redundant taxa are not necessarily synonymous under the *PhylocCode*. This is because two redundant clades of different types (such as node- and branch-based) may no longer be redundant when additional taxa are discovered (**Figure 4.3**), and such clades differ by other attributes, such as geological age of origin.

Article 14.1 distinguishes two kinds of synonyms: homodefinitional (based on the same definition) and heterodefinitional (based on different definitions). Homodefinitional synonyms are synonyms no matter which reference phylogeny is used (Note 14.1.1), whereas names that have different definitions are (heterodefinitional) synonyms only under some phylogenetic contexts. For instance, if an author defined *Amniota* as the smallest crown-clade that includes *Homo sapiens* Linnaeus 1758 (*Synapsida*) and *Testudo graeca* Linnaeus 1758 (*Testudines*), and another author defined *Reptilia* as the smallest clade that includes *Testudo graeca* Linnaeus 1758 (*Testudines*) and *Crocodylus* (originally *Lacerta*) *niloticus* Laurenti 1768 (*Diapsida*), under all recent phylogenies, these definitions would refer to distinct clades and no synonymy would be involved. However, under the phylogeny proposed by Gaffney (1980: figure 4.1A), in which turtles are the sister-group of a clade that includes mammals and diapsids, both definitions would refer to the same clade and would thus be heterodefinitional synonyms.

Priority works a little differently in the *PhyloCode* than in other codes. Article 14.3 indicates that "if two or more synonyms have the same publication date," the first registered (that is, with the lowest registration number) has precedence. Article 14.4 indicates that in most circumstances, panclade names have priority over other names that were not explicitly established as applying to a total clade even if the panclade names were established later, which is another way of promoting an "integrated" nomenclature.

Like the RN codes, the *PhyloCode* allows **conservation** of junior homonyms and synonyms, in justified cases (Article 15.1). This is to prevent an obscure name (used rarely in the literature) from replacing a much more frequently used name— typically one that was converted later. This requires approval of the CPN (Article 15.2).

No matter how careful systematists are in their nomenclatural work, errors creep in, among other reasons because new phylogenies may produce topologies that were not considered when formulating definitions. Thus, some mechanisms must allow revisions when required, and these must strike the right balance between facilitating such necessary revisions while discouraging arbitrary and poorly justified revisions that simply reflect personal preferences. The *PhyloCode* allows two kinds of emendations of phylogenetic definitions, unrestricted and restricted (see Article 15 of the *PhyloCode*), which are a bit reminiscent of Sereno et al.'s (2005) first- and second-order revisions. An unrestricted emendation may involve changes in specifiers or qualifying clauses, but must retain the same clade category (such as crown clade, total clade or neither) and definition type (such as node-based or branch-based) as in the original definition. Note 15.11.1 of the *PhyloCode* explains that unrestricted emendation "is a mechanism to prevent undesirable changes in the application of a particular name (in terms of clade conceptualization) when the original definition is applied in the context of a revised phylogeny." By contrast, restricted emendation is "intended to change the application of a particular name through a change in the conceptualization of the clade to which it refers. Restricted emendations may involve changes in definition type, clade category, specifiers, and/or qualifying clauses" (Article 15.10). Note 15.10.1 explains that "A restricted emendation is a mechanism to correct a definition that fails to associate a name with the clade to which it has traditionally referred, even in the context of the reference phylogeny adopted by the original definitional author." Obviously, restricted emendations imply more substantial changes than unrestricted emendations. This is why only restricted emendations require approval by the CPN (Article 15.8). However, some constraints also apply to unrestricted emendations. Thus, Recommendation 11I stipulates that:

> It is preferable, though not required, that the emendation be published by the original definitional author(s) (Art. 15.14). If such an emendation is published by anyone other than the original definitional author(s), the intent of the original author(s) should be considered carefully and addressed in the prologue of the emendation (see Note 15.11.4 and Arts. 15.12 and 15.13).

These articles and recommendations are aimed at minimizing arbitrary and frequent emendations that would lack strong justification.

Chapter VI, composed solely of Article 16, deals with hybrids. Article 16.1 explains that "Hybrid origin of a clade may be indicated by placing the multiplication sign (x) in front of the name." This is one of the sections of the *PhyloCode* that deals with non-divergent evolution. Article 16.2 explains further how such names may be formed: the names of both source clades are used, with the multiplication sign in between.

Chapter VII deals with orthography. Article 17.1 stipulates that to be established, names "must be a single word and begin with a capital letter." Furthermore, these names must contain only letters used in English. Thus, name formation is simpler than under the rank-based codes, which have requirements concerning Latin grammar for some names (especially species names) if Latin roots are used (see Sections 2.4.2 and 2.4.3).

Chapter VIII is entitled "Authorship." It differs most from the RN codes by distinguishing between the nominal author (who first established the name, including under RN codes) and the definitional author, who published the phylogenetic definition.

Chapter IX deals with citation. Article 20.2 explains that the nominal author must be cited without enclosing symbols, whereas authors of the original definition are cited within brackets, and those of an emended definition, between braces. Article 20.3 further stipulates that:

> If more than one set of authors is cited, they are to be cited in the following order: nominal author(s) of the preexisting or new name (including a replacement name); author(s) of the original definition; author(s) of an emended definition.

Given that the *PhyloCode* ignores absolute ranks, attribution of authorship to a name differs a bit from that of the RN codes. Thus, as explained under Recommendation 20.4A:

> If a preexisting name was used in association with more than one rank or composition, and authorship is cited, the nominal author(s) cited should be the original author(s) of the name, as spelled for the purpose of conversion.

By contrast, under some RN codes, the principle of coordination attributes authorship to the same author(s) for various names that differ by their suffix within some nominal series. Thus, under the *Zoological Code*, the author of a family name is considered to also be the author of the superfamily, subfamily, tribe and subtribe names, even if these were explicitly erected (that is, named with the corresponding endings for the first time) later by other authors. But, under the *PhyloCode*, in this situation, the author of the name ranked as subfamily or superfamily (to mention two of the relevant cases) under the *Zoological Code* would be attributed authorship of the converted name (if it ended in "-oidea" or "-inae," respectively), rather than the author of the family name.

Chapter X is composed solely of Article 21, entitled "Provisions for species names." The fact that the *PhyloCode* does not govern species names, which was previously explained (Section 4.1), is covered in Article 21.1. This chapter also deals with the first part of the binomial names (genus names, but not interpreted as such under the *PhyloCode*, given that it disregards ranks). Thus, Recommendation 21.3B stipulates that when erecting new species, "a genus name that is also an established clade name (or is simultaneously being established as a clade name) under this code should be selected if

possible." This should prevent incorporation of names of paraphyletic genera into the new species names. The *PhyloCode* allows a more flexible use of species names than the RN codes. Thus, Article 21.4 reads:

> Subsequent to a species binomen becoming available (*ICZN*) or validly published (*ICNAFP, ICNP*) under the appropriate rank-based code, the second part of the species binomen may be treated as the de facto name of the species under this code, termed a species uninomen. In this context, the species uninomen may be combined with the names of one or more clades the species is part of, in place of or in addition to the genus name.

**(see Rec. 21A)**

To avoid ambiguity in the species name and facilitate communication with users of other (RN) codes, Note 21.4B.1 stipulates that "Under this code, if the genus name is not used in combination with the specific name or epithet, both the author and year of the specific name or epithet should be cited." This is especially useful when referring to species names based on common epithets, such as "*californiensis*." Furthermore, Recommendation 21A indicates that:

> When a species uninomen is combined with more than one genus name and/or clade name, hierarchical relationships among the taxa designated by those names can be indicated in a variety of ways, but the taxa should be listed in order of decreasing inclusiveness from left to right.

Examples on how to do this are provided in Example 1, which provides a non-exhaustive list of possible formats for a squamate species that has been attributed to the nominal genera *Anolis* and *Norops*. In this example, if *Norops* were established for a clade included in *Anolis*, possible formats for the species name under the *PhyloCode* would include: *Anolis/auratus* Daudin 1802, *Norops: auratus* Daudin 1802; *Anolis/Norops/auratus* Daudin 1802; or *Anolis Norops auratus* Daudin 1802.

The reliance on the rank-based codes to regulate species names implies that the *PhyloCode* can coexist with these other codes. This was noted by other systematists, who already established various taxon names under both kinds of nomenclature. For instance, Maxwell and Rymer (2021) and Maxwell (2021) converted several mollusk taxon names and erected two more, and these names, to which they also attributed absolute ranks, are established under the *PhyloCode* as well as under the *Zoological Code*.

The last chapter (XI) details governance. It discusses the ISPN and its committees, especially the CPN (see Section 4.1), and, to a lesser extent, the Registration Committee (which manages the registration database). The governance is democratic because, as indicated in Article 22.4, "The members of the CPN will be elected by the membership of the ISPN."

The *PhyloCode* also includes a sizeable glossary (pp. 111–121), three appendices and a detailed index (pp. 131–149). The first appendix (A) details the registration procedures in

*RegNum* and data requirements. Appendix B is entitled "Code of Ethics." Notably, it states that:

> Authors should not publish a new name or convert a preexisting one if they have reason to believe that another person intends to establish a name for the same clade (or that the clade is to be named in a posthumous work).

Appendix C presents a table showing the "Equivalence of nomenclatural terms among codes." It includes terms from the *BioCode*, the *Prokaryotic Code*, the *Botanical Code* and the *Zoological Code*, in addition to the *PhyloCode*. These appendices are short; they occupy only seven pages altogether.

The *PhyloCode* is freely available online through the ISPN website (at http://phylonames.org/code/). A pdf version can be purchased. It facilitates search by keyword in the text, but the online index also allows the most important occurrences of most relevant keywords to be located quickly. In the pdf version, the large font size facilitates reading the code.

All names established under the *PhyloCode* (Cantino and de Queiroz 2020) have to be registered in *RegNum* (www.phyloregnum.org/?). This has several advantages. One of the main ones is that this automatically yields an exhaustive database of taxon names and definitions. Thus, the old problem of favoring strict priority over use, which still awaits a satisfactory resolution among proponents of the *Zoological Code* and *Botanical Code*, should be less acute. It applies mostly when a systematist converts a name, but the articles and recommendations of the *PhyloCode* are fairly clear about this: stability is the main goal when selecting a name/definition combination to be established, so the choice is clearly for use rather than strict priority when a name is selected for conversion, although once a name has been established under the *PhyloCode*, strict priority should be enforced. The availability of *RegNum* should prevent any name listed therein from being forgotten, which eliminates the situation that most frequently requires systematists to make the difficult choice of use vs priority to resolve cases of synonymy or homonymy among established names.

### 4.4.2 Similarities and Differences with the Rank-Based Codes

As stated in the preface of the *PhyloCode* (Cantino and de Queiroz 2020: xii–xiii):

> An advantage of the *PhyloCode* over the rank-based codes is that **it applies at all levels** of the taxonomic hierarchy. In contrast, the zoological code does not extend its rank-based method of definition above the level of superfamily, and the botanical code extends that method of definition only to some names above the rank of family (automatically typified names) and **the principle of priority is not mandatory for those names**. Consequently, at higher levels in the hierarchy, the rank-based codes permit multiple names for the same taxon as well as alternative applications of the same name. Thus, as phylogenetic studies continue to reveal many deep clades, there is an increasing potential for **nomenclatural chaos** due to synonymy and homonymy. By imposing rules of precedence on clade names at all levels of the hierarchy, the *PhyloCode* will improve nomenclatural clarity at higher hierarchical levels.

Other differences concern fewer cases but are still fairly important. Like the *Code for Cultivated Plants*, the *PhyloCode* accommodates hybrids; in this case, this is done by allowing hybrids to belong to two clades that otherwise do not overlap. Indeed, Note 2.1.3 states that "Clades are often either nested or mutually exclusive; however, phenomena such as speciation via hybridization, species fusion, and symbiogenesis can result in clades that are partially overlapping." The other rank-based codes are more problematic for hybrids because a given species cannot belong simultaneously (in the context of a given taxonomy and nomenclature) to two genera, and a given genus cannot belong simultaneously to two families. This is not only a theoretical problem; inter-generic hybrids are known in various taxa. For instance, among baboons, hybrids have been reported between *Theropithecus gelada* and *Papio hamadryas*, which belong to genera that have been distinct for several million years (Jolly et al. 1997). Example are also found among birds (Marini and Hackett 2002) and angiosperms (Li et al. 1995), among others.

The *PhylocCode* uses **specifiers**, which are vaguely analogous with types of the rank-based codes, to define taxon names. In the *PhyloCode*, specifiers can be specimens, species or even apomorphies, and, contrary to the rank-based codes, these specifiers can be specifically excluded from the taxon that they define to delimit it.

Contrary to what some opponents (Foer 2005) argued, the aim of the *PhyloCode* is not to rename everything. On the contrary, the aim is to provide a precise, stable delimitation to each named taxon (Cantino and de Queiroz 2020: viii):

> The objective of the *PhyloCode* is not to replace existing names but to provide an alternative system for governing the application of both existing and newly proposed names. In developing the *PhyloCode*, much thought has been given to minimizing disruption of the existing nomenclature. Thus, rules and recommendations have been included to ensure that most names will be applied in ways that approximate their current and/or historical use. However, names that apply to clades will be redefined in terms of phylogenetic relationships rather than taxonomic rank and therefore will not be subject to the subsequent changes that occur under the rank-based systems due to changes in rank.

Furthermore, due credit is given to the authors who erected taxa under RN by distinguishing between the nominal authors (those who erected the taxa, typically under RN, except for very recently erected taxa) and definitional authors (those who proposed the phylogenetic definition in the conformity with the *PhyloCode*). This distinction had already been suggested by Sereno (2005: 611).

Contrary to the rank-based codes, the *PhyloCode* allows systematists to define clade names in a way that will ensure that the taxon to which they refer includes or excludes some less inclusive taxa. This can be done either by using additional internal and/or external specifiers (beyond the minimum required for such definitions), or by using qualifying clauses. How this can be done and why this is useful to respect the historical usage of a taxon can be illustrated through the example of the taxon *Lissamphibia*, which Laurin et al. (2020a) defined as:

> The smallest crown clade containing *Caecilia tentaculata* Linnaeus 1758, *Andrias japonicus* (Temminck 1836), *Siren lacertina* Österdam 1766, and *Rana temporaria* Linnaeus 1758 but not *Homo sapiens* Linnaeus 1758 or *Eryops megacephalus* Cope 1877 or *Diplocaulus salamandroides* Cope 1877.

Note that for a definition of a crown-clade, no external specifier is required. However, the concept of *Lissamphibia* has always excluded Paleozoic stegocephalians, as the history of this taxon shows. Haeckel (1866) initially included only anurans and urodeles in *Lissamphibia* (gymnophionans were included in *Phractamphibia*, along with Paleozoic and Triassic temnospondyls and baphetids), and since Gadow (1898), *Lissamphibia* has always included the three large clades of extant amphibians (*Anura, Urodela* and *Gymnophiona*), often the Jurassic to Pliocene albanerpetontids, which were at some point confused with urodeles (Gardner and Böhme 2008), but not Paleozoic taxa. Notably, the presumed Paleozoic amphibians (temnospondyls and lepospondyls) were always excluded from *Lissamphibia*. The phylogeny of extant taxa nearly always showed that *Lissamphibia* is the sister-group of *Amniota* (Irisarri et al. 2017; **Figure 4.7A**), but paleontological trees are less congruent; most recent ones (based on parsimony analyses) only agree in placing the lepospondyls closer to amniotes than to temnospondyls (**Figure 4.7B**). Many recent stegocephalian phylogenies would make a monophyletic *Lissamphibia* that excludes temnospondyls, lepospondyls and amniotes, as this group is traditionally conceptualized, even in the absence of external specifiers. This includes phylogenies supporting the "temnospondyl hypothesis" (Mann et al. 2019; **Figure 4.7C**) and the "lepospondyl hypothesis" (for instance, Marjanović and Laurin 2019; Laurin et al. 2022; **Figure 4.7D**). However, if external specifiers had not been included in the definition, other phylogenies would have led to a strange delimitation of *Lissamphibia*, either because temnospondyls (Pardo et al. 2017a; **Figure 4.7H**), temnospondyls and lepospondyls (Milner 1993; Carroll 2007; **Figure 4.7F**), or temnospondyls, lepospondyls and amniotes (Fröbisch et al. 2007; Anderson et al. 2008; **Figure 4.7E, G**) would also be included. Thus, to ensure that amniotes or Paleozoic stegocephalians (such as temnospondyls and lepospondyls) do not become part of *Lissamphibia* as new phylogenies are published, three external specifiers were added by Laurin et al. (2020a): the amniote *Homo sapiens*, the Permian temnospondyl *Eryops megacephalus* and the Permian lepospondyl *Diplocaulus salamandroides*. A similar result could have been obtained by including these three species in a qualifying clause. The definition including such a clause could have been:

> The smallest crown clade containing *Caecilia tentaculata* Linnaeus 1758, *Andrias japonicus* (Temminck 1836), *Siren lacertina* Österdam 1766,

and *Rana temporaria* Linnaeus 1758, provided that it does not include *Homo sapiens* Linnaeus 1758 or *Eryops megacephalus* Cope 1877 or *Diplocaulus salamandroides* Cope 1877.

### 4.4.3 Controversies within the PhyloCode Community

#### 4.4.3.1 Panclade Names

The *PhyloCode* includes a naming convention designed to:

> promote an integrated system of names for crown and total clades. The resulting pairs of names (e.g., *Testudines* and *Pan-Testudines* for the turtle crown and total clades, respectively) enhance the cognitive efficiency of the system and provide hierarchical information.

**(Cantino and de Queiroz 2020: xi)**

While the principle of adding the prefix "Pan" before a crown-clade to designate a total clade is straightforward, this convention has proven controversial because panclade names can compete with established names, some of which are ancient and well-known. This is acknowledged in the *PhyloCode*, which states (Cantino and de Queiroz 2020: xix) that this convention:

> was introduced and vigorously discussed. Some participants [of the Paris *ISPN* meeting] were reluctant to make these conventions mandatory because doing so would result in replacing some names that had already been explicitly defined as the names of total clades (e.g., replacing *Synapsida* by *Pan-Mammalia*). A compromise that made exceptions for such names was acceptable to the majority of the participants, and it served as the basis for the set of rules and recommendations that was eventually adopted by the CPN (Recommendation 10.1B and Articles 10.3–10.8 in version 3 of the *PhyloCode*, and after some subsequent modifications, Recommendation 10.1B and Articles 10.3–10.7 in the current version).

One of the compromises evoked in the quote is found in Article 10.6, which states that "If there is a preexisting name that has been applied to a particular total clade, that name may be converted or a panclade name may be established instead." So far so good, but Article 14.4 is more problematic; it states that:

> If a panclade name (Art. 10.3) and a name that was not explicitly established as applying to a total clade are judged to be heterodefinitional synonyms (Art. 14.1), the panclade name has precedence even if it was established later (except in cases covered by Art. 10.7).

This represents a departure from the initial goal of this naming convention (Lauterbach 1989). As explained by Meier and Richter (1992), the goal of this proposal was to remove an ambiguity that affected many taxonomic papers by distinguish

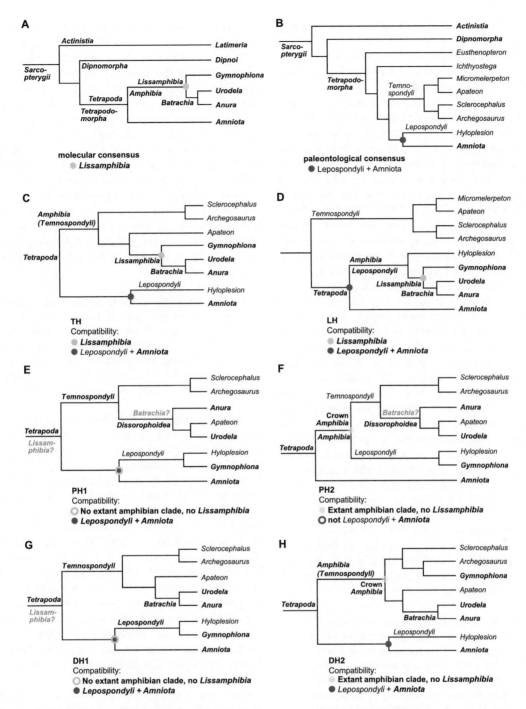

**FIGURE 4.7** Stegocephalian phylogeny and definition of Lissamphibia. The consensus among neontologists (A) is that *Lissamphibia* forms the sister-group of *Amniota* among extant taxa (e.g., Irisarri et al. 2017). Among paleontologists, there is no such consensus about *Lissamphibia*, but there is a near-consensus that lepospondyls are more closely related to amniotes than to temnospondyls (B). Six hypotheses about stegocephalian phylogeny and the origin of extant amphibians that have been upheld by paleontologists are shown next (C–H), some of which show a monophyletic *Lissamphibia* (C–D), whereas others do not (E–H). These are the "temnospondyl hypothesis" (C), the "lepospondyl hypothesis" (D), two variants of the "polyphyly hypothesis" that imply a triphyletic origin of extant amphibians (E–F) and two other variants of the "polyphyly hypothesis" that imply diphyly of extant amphibians (G–H). These phylogenies have recently been supported by various studies, such as Mann et al. (2019; C), Marjanović and Laurin (2019; D), and Fröbisch et al. (2007; E). Milner (1993) and Carroll (2007: fig. 75) collectively present an old "polyphyly hypothesis" (F). Anderson et al. (2008; G) and Pardo et al. (2017a; H) presented recent variants of the "polyphyly hypothesis." In the phylogenies in which *Lissamphibia* as defined by Laurin et al. (2020a) cannot be recognized, the smallest clade that includes all extant amphibians but not amniotes is labeled as "Crown *Amphibia*," when it exists; if external specifiers had not been included in the definition to make the name inapplicable under some phylogenies, the name *Lissamphibia* would have applied to these nodes, but this would have been undesirable because *Lissamphibia* has never been conceptualized as including amniotes, lepospondyls or temnospondyls. Extant taxa are in bold type. The taxon *Batrachia* is shown in grey and followed by a question mark in trees in which it includes some Paleozoic taxa that have normally been excluded from it.

*Source:* Modified from Laurin et al. (2022: figure 1).

between the crown-clade and the total clade when both bore the same name. However, Article 14.4 of the *PhyloCode* can result in well-established names being displaced by new (or recent and little-used) taxon names. An example is provided by *Synapsida* and *Pan-Mammalia*.

*Synapsida* was erected by Osborn (1903a), along with *Diapsida*; these two taxa, which Osborn (1903b) considered subclasses, represented the basal subdivisions of the class *Reptilia* (at the time conceptualized as a paraphyletic group). *Synapsida* was diagnosed by a combination of characters, the most important one being a single, lower temporal fenestra, whereas *Diapsida* encompassed taxa with two (upper and lower) temporal fenestrae. Osborn (1903a, 1903b) included in *Synapsida* the cotylosaurs (diadectomorphs, which are currently considered the sister-group of amniotes, along with procolophonids and pareiasaurs, now considered parareptiles), "Anomodontia," in which he placed taxa now considered to be therapsids (cynodonts and dicynodonts), turtles and sauropterygians (placodonts and plesiosaurs). He indicated that *Anomodontia* gave rise to mammals. Osborn (1903a, 1903b) placed nearly all other amniotes in *Diapsida*, namely the mesosaurs (now considered basal sauropsids or parareptiles), *Pelycosauria* (a paraphyletic group now considered basal synapsids), dinosaurs, ichthyosaurs, pterosaurs, squamates and *Crocodylia*. He indicated that some diapsids gave rise to birds. Thus, the initial composition of *Synapsida* differed rather strongly from the current one by including turtles, placodonts and plesiosaurs, and by excluding "pelycosaurs." This surprising exclusion resulted from a misinterpretation of the temporal region of some "pelycosaurs," which prevailed at the time (Baur and Case 1897: 118) and was sorted out later (Williston 1917: 65). Similarly, the inclusion of *Cotylosauria* in *Synapsida* seems to reflect Osborn's (1903b) acceptance of paraphyletic taxa because he wrote that "The remote common stock uniting the two subclasses [*Synapsida* and *Diapsida*] is not the Proganosauria but the Cotylosauria with a solid skull roof" (Osborn 1903b: 455). This is confirmed by another, congruent quote (Osborn 1903b: 456–457):

> Certain of the Cotylosauria (*Diadectes*) show rudimentary supratemporal openings; according to the recent observations of Case these are variable in the Permian Cotylosaurs of Texas, in certain cases being present on one side and not on the other. These rudimentary openings support the theory of fenestration as well as the theory that the **Cotylosauria are the source of both the Diapsida and Synapsida**.

The composition of *Synapsida* shifted gradually. Williston (1917) moved the "pelycosaurs" among the *Theromorpha*, within *Synapsida*, and removed turtles and cotylosaurs, which he placed among the *Anapsida*. From then on, *Synapsida* included only the presumed extinct relatives of mammals, even though Williston (1917) included among these the sauropterygians and placodonts, which are now considered diapsids. *Synapsida* acquired a delimitation closer to the current one a few years later when Broom (1924) removed sauropterygians and placodonts form this taxon. From then on, this taxon would have a very stable delimitation, as the subclass of reptiles that included mammalian relatives, and with the latter correctly identified as

the therapsids and more basal synapsids, which many authors (for instance, Berman et al. 1992; Angielczyk and Kammerer 2018; Mann and Paterson 2020) still call "pelycosaurs" (despite the paraphyly of this group having been demonstrated many decades ago). Broom's (1924: figure 13) evolutionary tree also shows that mammals and other synapsids were the sister-group of all other extant (and most extinct) amniotes. Even in taxonomies that diverged strongly from the established consensus, such as Olson's (1947: figure 4.8) view that entails amniote diphyly (his *Parareptilia* included seymouriamorphs, which had aquatic, gilled larvae and turtles, among others), *Synapsida* retained a delimitation similar to Broom's (1924).

It could be argued that since Williston (1917), *Synapsida* was delimited as if it were conceptualized as the stem-group of *Mammalia*, even though this is not stated explicitly in these studies. However, this quote from Williston (1917: 65) comes very close:

> This group or subclass, which, with due modifications of the original concept, may properly bear the name Synapsida given to it by Osborn, includes scores of well-known genera of the orders Theromorpha, Therapsida, and doubtless also the Sauropterygia. It is the group that **gave origin to the mammals**, and **has long since been extinct**.

This text suggests that Williston (1917) conceptualized **Synapsida** as a paraphyletic group that included stem-mammals. This quote does not state that all extinct taxa more closely related to mammals than to other extant taxa were included in *Synapsida*, but the rest of the paper is compatible with this inference because no taxa that Williston (1917) considered closely related to mammals were excluded from *Synapsida*. Broom (1924) is less explicit, but nothing in that paper indicates departure from Williston's (1917) conceptualization of *Synapsida*. Note that this name should not have been defined explicitly as a total group under the *PhyloCode* (as would be required by its Article 14.4 to have priority over the panclade name *Pan-Mammalia*) because mammals were originally excluded from *Synapsida*. This reflects the fact that at that time, taxa were delimited based as much on phenotype as on phylogenetic affinities, and paraphyletic taxa (including *Pisces*, *Amphibia* and *Reptilia*, as delimited at the time) were extremely common.

Many taxa were arguably conceptualized as total taxa (and more generally, branch-based taxa), stem-groups of extant taxa or part of the stem-group of extant taxa well before the advent of PN. Several examples are provided by Sereno (2005), and, often, taxa had been re-delimited as if they were triplet consisting of one crown-group and two branch-based (sometimes total) groups. These include *Dinosauria* (node-based) subdivided into *Ornithischia* and *Saurischia* (both branch-based), and *Testudines* or *Casichelydia* (crown-group) subdivided into *Cryptodira* and *Pleurodira* (both total groups). To these, we could add *Amniota* (crown-group), subdivided into *Synapsida* and *Sauropsida* (both total groups; Laurin and Reisz 2020b, 2020c) and *Aves*, subdivided into *Palaeognathae* and *Neognathae* (Wolff 1991: 91). *Therapsida* has long been conceptualized as the crownward part of the mammal stem (such as Romer 1956: 682).

Does etymology suggest that *Synapsida* was conceptualized as an apomorphy-based taxon? The name *Synapsida* makes explicit reference to the temporal fenestration, and Osborn (1903b: 455, footnote 8) was explicit about this by giving the etymology of the term: "σύν [syn], together, αψις [apsis], an arch." However, the change in contents of *Synapsida* over time (previously reviewed) does not suggest that the taxon was conceptualized as an apomorphy-based taxon (that is, linked to a particular character) because the discovery that various parareptiles and mesosaurs had a lower temporal fenestra did not result in proposals to include them in *Synapsida* (for instance, Watson 1954: 437; Daly 1969). More importantly, in Osborn (1903a, 1903b), *Synapsida* included *Cotylosauria*, and the latter included mostly taxa that lack the lower temporal fenestra, such as pareiasaurs and diadectomorphs. Furthermore, this quote from Osborn (1903a: 276) refutes an apomorphy-based conceptualization: "In the ancestral *Synapsida*: (1) The roof of the skull is solid (*Cotylosauria*), or there is a single large supratemporal opening, the infratemporal opening being rudimentary or wanting." This quote is followed by a list of four more diagnostic characters of *Synapsida*, which suggests that temporal fenestration was not of overwhelming importance in his taxonomy. Thus, it could be argued that the name *Synapsida* made reference to a useful diagnostic, but not a defining, character.

*Synapsida* is a well-known taxon; a Google Scholar search conducted on January 14, 2022, yielded 4,470 references that mention this taxon and a search for the informal "synapsids" yielded 4,240 results. By contrast, a Google Scholar search performed the same day for "*Pan-Mammalia*" yielded 22 references and the informal "pan-mammals" yielded 19 hits. Thus, replacing *Synapsida* with *Pan-Mammalia* is clearly against nomenclatural stability (though it promotes an integrated, intuitive nomenclatural principle).

With these considerations in mind, Laurin and Reisz (2020c) defined *Synapsida* as "The largest clade that includes *Cynognathus crateronotus* Seeley 1895 but not *Testudo graeca* Linnaeus 1758, *Iguana iguana* (Linnaeus 1758), and *Crocodylus niloticus* Laurenti 1768." This definition was deliberately not formulated in terms of a total group because mammals were initially (Osborn 1903a, 1903b; Williston 1917) excluded from *Synapsida*, as mentioned previously. Thus, the internal specifier, the mid-Triassic cynodont therapsid *Cynognathus crateronotus* Seeley 1895, is one of the core taxa that was always thought to be part of *Synapsida*. This definition does in fact delimit the total clade of mammals under all recently proposed phylogenies (and all or nearly all those published in over a century). However, Rowe (2020a) defined *Pan-Mammalia* as "The total clade of the crown clade *Mammalia* Linnaeus 1758." Under any plausible phylogeny, both definitions refer to the same clade and must be considered heterodefinitional synonyms, and given Article 14.4 of the *PhyloCode*, *Pan-Mammalia* must have precedence, with the consequence that the well-known name *Synapsida* should no longer be used by proponents of the *PhyloCode*.

The name *Synapsida* could be redefined, and an appeal could be made to the CPN (Committee on Phylogenetic Nomenclature) to conserve it, but this is only moderately compelling. Using an apomorphy-based definition (such as

"the clade diagnosed by the appearance of a lower temporal fenestra synapomorphic with that of *Cynognathus crateronotus* Seeley 1895") might lead to the undesirable effect that the clade would come to encompass all amniotes because the fenestra may have appeared at the base of *Amniota* and have disappeared various times subsequently (Piñeiro et al. 2012: figure 4.7). This adverse effect could be overcome by adding a qualifying clause or at least one external specifier, but this would fail to capture the way the name was used before and, under some phylogenies, its delimitation would no longer match the previous conceptualization. This would happen, for instance, under the topology proposed by Berman et al. (1992), in which diadectomorphs are closer to "*Pelycosauria*" than to eureptiles (**Figure 4.8**). Under an apomorphy-based definition, *Synapsida* would exclude diadectomorphs. However, considering that diadectomorphs were included in *Synapsida* by Osborn (1903a, 1903b), this neither matches the total group conceptualization suggested by the way the taxon was delimited over time, nor (less importantly) the original delimitation.

### 4.4.3.2 Stable vs Integrated Nomenclatural Systems

The example of *Synapsida* and *Pan-Mammalia* shows that the current edition of the *PhyloCode* arguably goes too far in favoring panclade names for total taxa at the expense of nomenclatural stability and continuity with the older literature. This issue has long been controversial among proponents of PN. It may be useful to quote a passage from the report of the second ISPN meeting (Laurin and Cantino 2007: 112):

> Some discussants would prefer that **all** total clade names begin with the prefix Pan- as part of an integrated system of crown and total clade names, thereby communicating hierarchical information while reducing the number of totally different clade names that one would have to learn. Others prefer that the *PhyloCode* not require or even recommend that total clade names take a particular form, in order to maximize both nomenclatural **freedom** and the **simplicity** of the code. The set of rules and recommendations in the current draft was intended as a compromise between these conflicting viewpoints, but it is **still controversial**.

Yet, Mike Keesey's suggestion to have the word "pan" represent a function to form informal combinations, such as "pan(mammalia)" that would not compete with formal names, seems to have been forgotten, even though it had been well-received at the second ISPN meeting (Laurin and Cantino 2007: 112). It would be good to discuss this again in a future ISPN meeting. After all, as Minelli (2001: 168) reminded us, "Rules can still evolve but a Code, historically, follows and consolidates practice. It does not establish it from scratch." Given that the *PhyloCode* was finally implemented after about three decades of systematic practice under PN, Minelli's statement can reasonably be considered to apply to most of the *PhyloCode*, but the panclade naming convention is arguably an exception to this rule, as exemplified by the case of *Synapsida*

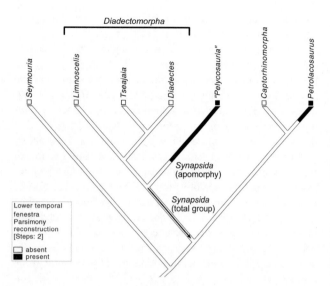

**FIGURE 4.8** Amniote phylogeny proposed by Berman et al. (1992: figure 13). They recognized a formal taxon *"Pelycosauria"* that has long been known to be paraphyletic and thus, is not recognized formally in this book. Note that under an apomorphy-based definition of *Synapsida*, this taxon would include only *"Pelycosauria,"* but not diadectomorphs.

(but many other examples could have been selected, such as *Sauropsida*, *Lepodosauromorpha* and *Archosauromorpha*).

The panclade naming convention and the controversies about how to name large extant clades (as crown-clades or not; see Section 4.2.3) are part of a debate about how to reach the best consensus between two contradictory aims of the *PhyloCode*: stabilizing nomenclature and promoting an integrated, arguably more intuitive nomenclature. Sereno (2005: 614) argued that prior use of a name and how systematists re-delimited the taxon when the phylogeny changed or when new taxa were discovered should influence how a name is defined, and this seems like a good idea compatible with the *PhyloCode*'s goal of respecting historical usage of taxon names. It could be argued that from this perspective, Article 14.4 (about precedence of panclade names) goes too far into promoting an integrated nomenclatural system at the expense of nomenclatural continuity and freedom, and, perhaps, at the expense of acceptance of the *PhyloCode* in the systematic community.

Fortunately, some legacy names that were arguably conceptualized as total clades will continue being used in a meaning close to their historical usage under the *PhyloCode*. For instance, *Amphibia* has arguably had three meanings in the pre-cladistic literature: a paraphyletic group that included all limbed vertebrates except for amniotes, the total group that includes extant amphibians and the crown-group of extant amphibians. The first meaning, which was arguably implied most frequently in the pre-cladistic literature, cannot be established under the *PhyloCode* because it is not a clade. It would have been difficult to choose between both other meanings had there not been another name (*Lissamphibia*) that was always at least implicitly linked to the amphibian crown (see previous and **Figure 4.7**). Thus, under the *PhyloCode*, *Lissamphibia* refers to the amphibian crown (Laurin et al. 2020a), whereas *Amphibia* refers to a total group (Laurin et al. 2020b).

## 4.5 Established Names under the *PhyloCode*

### 4.5.1 First Established Names (Phylonyms)

*Phylonyms* (de Queiroz et al. 2020), the monograph that includes the first batch of names established under the *PhyloCode*, includes a table of contents, an introduction, a main section that includes the entries on numerous taxa (arranged in an order that vaguely reflects complexity, with prokaryotic organisms listed first and amniotes last) and an index. The introduction explains the long history of the genesis of this monograph, which spans about two decades. Fortunately, participants of the second workshop on phylogenetic nomenclature, which convened at the Yale Peabody Museum in July 2002, decided to scale down its scope (compared to the very ambitious initial plans); otherwise, we would still be awaiting its publication. *Phylonyms* is designed to provide "well-vetted definitions of many widely known names," which include, for instance, *Euglenozoa*, *Ascomycota*, *Viridiplantae*, *Tracheophyta* and *Porifera*, among many others.

Each entry includes the name, its authorship (both nominal and definitional), registration number (in the *RegNum* database), the definition, etymology, reference phylogeny, composition (in the taxonomic sense), diagnostic apomorphies, synonyms, comments and the literature cited. Of these, the comments section is by far the longest. It includes details about the history of our ideas on the taxon, evidence and support for the clade, reasons for choosing the name if it is converted rather than new, conceptualization of the clade (especially, how it is delimited under alternative phylogenies), preferred precedence relative to other names established in *Phylonyms* if they became synonyms under alternative phylogenies, and information about the fossil record and relevant divergence time estimates. The section on reference phylogeny obviates the need for figures in this monograph (none are provided).

The 285 taxonomic entries (one per defined name) are arranged into eight sections. These entries range in size from two pages (for instance, *Ciliophora*, *Intramacronucleata* and *Rhodophyta*) to seventeen (such as *Foraminifera*), but most occupy three to five pages. The first section includes the basalmost taxa, ranging from *Pan-Biota* (which refers to the total group of all Earth-based life forms) through unicellular eukaryotic taxa and fungi. Unfortunately, no contribution defines a single name among *Archaea* or *Bacteria*, which is undoubtedly the largest gap in the taxonomic coverage of this work. Section two includes algae (such as *Rhodophyta*, *Chlorophyta*); section three includes most *Embryophyta*, except for the *Angiospermae*, which make up section four. Section five includes *Metazoa*, except for *Vertebrata*, which are included in sections six (which contains mostly anamniotic taxa), seven (synapsids) and eight (*Reptilia*). The book closes with a sizeable index, which covers pages 1,297 through 1,324. As expected for such a work, the index includes mostly taxon names, but other words and expressions are also included, such as "ribosomal RNA," "ribosomes" and "secondary endosymbiosis," among many others.

### 4.5.2 Other Established Names

On August 19, 2022, the official journal of the ISPN, entitled the *Bulletin of Phylogenetic Nomenclature*, published its first paper (Johnson et al. 2022), which converted six names (and defined a new one) of thalattosuchians, a taxon composed of Mesozoic aquatic crocodylomorphs. However, names have also become established under the *PhyloCode* in papers published in other journals. For instance, Lemierre and Laurin (2021) converted the names of three anuran taxa in the journal *Bionomina*, which focuses on biological nomenclature, but without a focus on phylogenetic nomenclature. Similar papers were also published in more classical systematics journals that do not emphasize nomenclature. For instance, Maxwell (2021) converted more than 30 names of mollusk taxa (*Neostromboidae* and less inclusive taxa) in the malacology journal *The Festivus*. Joyce et al. (2021) established 113 clade

names for turtles and their close relatives in a monograph published in the *Swiss Journal of Paleontology*. Of these names, 79 had been defined before publication of *Phylonyms* and 34 had never been defined, 15 of which were new names. Madzia et al. (2021) established 76 preexisting ornithischian dinosaur clade names and erected five new clade names in the journal *PeerJ* (**Figure 4.9**). Avian nomenclature of this nature is represented by at least two papers, both in the journal *Vertebrate Zoology*: Chen and Field (2020) converted 20 names and coined two new ones, all of avian taxa (*Caprimulgimorphae* and less inclusive clades therein), while Sangster and Mayr (2021) established a new avian taxon (*Feraequornithes*).

This non-exhaustive sample shows that in the short period that followed publication of *Phylonyms*, systematists have started using routinely the *PhyloCode* in their work. It is neither possible nor necessary to cover all studies that have established names under the *PhyloCode* because a Google Search

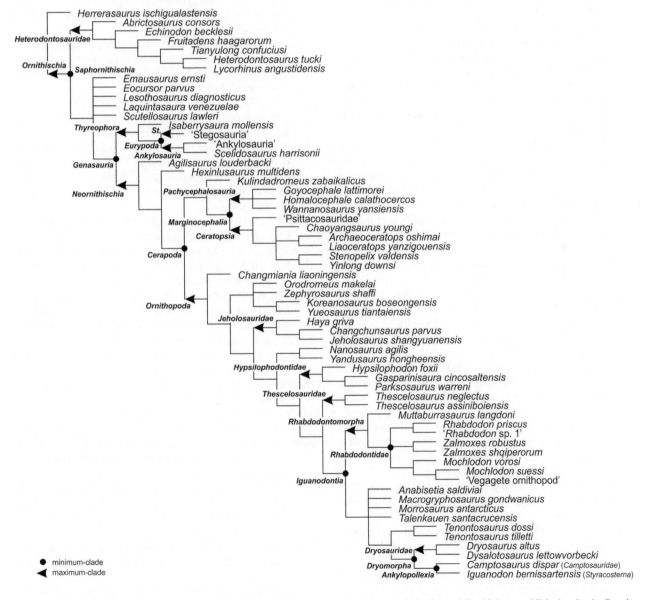

**FIGURE 4.9** Phylogeny and nomenclature of ornithischian. Reproduced from Madzia et al. (2021: figure 4.1), which was published under the Creative Commons CC-BY 4.0 license (www.creativecommons.org/licenses/by/4.0/).

carried out on October 25, 2022, for "PhyloCode, established" and limited to papers published in 2020 or later yielded 170 results. The examples provided here all come from the first (most relevant) of 17 pages of results yielded by the search. Admittedly, fewer than 170 papers established names under the *PhyloCode* in the 30 months or so that followed publication of *Phylonyms* because no paper from the last (least relevant) page (of 10 results each) appears to establish names under the *PhyloCode*, but at least one from page 8 does (Zverkov 2022), plus at least two from page 6 (out of four papers downloaded to verify). The number and variety of taxa established in such a short time by many authors in various journals bodes well for the future of the *PhyloCode*, but the inherent qualities of this code are its greatest assets for the future.

# REFERENCES

Alverson, W. S., B. A. Whitlock, R. Nyffeler, C. Bayer, and D. A. Baum. 1999. Phylogeny of the core Malvales: Evidence from ndhF sequence data. American Journal of Botany 86:1474–1486.

Anderson, J. S. 2001. The phylogenetic trunk: Maximal inclusion of taxa with missing data in an analysis of the Lepospondyli (Vertebrata, Tetrapoda). Systematic Biology 50:170–193.

Anderson, J. S., R. R. Reisz, D. Scott, N. B. Fröbisch, and S. S. Sumida. 2008. A stem batrachian from the Early Permian of Texas and the origin of frogs and salamanders. Nature 453:515–518.

Angielczyk, K. D. and C. F. Kammerer. 2018. 5. Non-mammalian synapsids: The deep roots of the mammalian family tree; pp. 117–198 in F. Zachos and R. Asher (eds.), Mammalian evolution, diversity and systematics. De Gruyter, Berlin.

Baum, D. A., W. S. Alverson, and R. Nyffeler. 1998. A durian by any other name: Taxonomy and nomenclature of the core Malvales. Harvard Papers in Botany 3:315–330.

Baur, G. and E. C. Case. 1897. On the morphology of the skull of the pelycosauria and the origin of the mammals. Anatomischer Anzeiger 13:109–120.

Benton, M. J. 2000. Stems, nodes, crown clades, and rank-free lists: Is Linnaeus dead? Biological Reviews of the Cambridge Philosophical Society 75:633–648.

Benton, M. J. 2022. A colourful view of the origin of dinosaur feathers. Nature 604:630–631.

Berman, D. S., S. S. Sumida, and R. E. Lombard. 1992. Reinterpretation of the temporal and occipital regions in *Diadectes* and the relationships of diadectomorphs. Journal of Paleontology 66:481–499.

Bremer, K. 2000. Phylogenetic nomenclature and the new ordinal system of the angiosperms; pp. 125–133 in B. Nordenstam, G. El-Ghazaly, and M. Kassas (eds.), Plant systematics for the 21st century. Portland Press, London.

Brochu, C. A. 1997. Synonymy, redundancy, and the name of the crocodile stem-group. Journal of Vertebrate Paleontology 17:448–449.

Brochu, C. A. 1999. Phylogenetics, taxonomy, and historical biogeography of Alligatoroidea. Journal of Vertebrate Paleontology 19:9–100.

Brochu, C. A. 2001. Progress and future directions in archosaur phylogenetics. Journal of Paleontology 75:1185–1201.

Brochu, C. A. and C. D. Sumrall. 2001. Phylogenetic nomenclature and paleontology. Journal of Paleontology 75:754–757.

Broom, R. 1924. On the classification of the reptiles. Bulletin of the American Museum of Natural History 51:39–65.

Bryant, H. N. 1994. Comments on the phylogenetic definition of taxon names and conventions regarding the naming of crown clades. Systematic Biology 43:124–130.

Bryant, H. N. 1996. Explicitness, stability, and universality in the phylogenetic definition and usage of taxon names: A case study of the phylogenetic taxonomy of the Carnivora (Mammalia). Systematic Biology 45:174–189.

Cantino, P. D. 1998. Binomials, hyphenated uninomials, and phylogenetic nomenclature. Taxon 47:425–429.

Cantino, P. D. and K. de Queiroz. 2020. International code of phylogenetic nomenclature (PhyloCode): A phylogenetic code of biological nomenclature. CRC Press, Boca Raton, Florida, xl + 149 pp.

Cantino, P. D., R. G. Olmstead, and S. J. Wagstaff. 1997. A comparison of phylogenetic nomenclature with the current system: A botanical case study. Systematic Biology 46:313–331.

Carroll, R. L. 2007. The Palaeozoic ancestry of salamanders, frogs and caecilians. Zoological Journal of the Linnean Society 150:1–140.

Chen, A. and D. Field. 2020. Phylogenetic definitions for *Caprimulgimorphae* (*Aves*) and major constituent clades under the International code of phylogenetic nomenclature. Vertebrate Zoology 70:571–585.

Chen, P.-J., Z.-M. Dong, and S.-N. Zhen. 1998. An exceptionally well-preserved theropod dinosaur from the Yixian Formation of China. Nature 391:147–152.

Cincotta, A., M. Nicolaï, H. B. N. Campos, M. McNamara, L. D'Alba, M. D. Shawkey, E.-E. Kischlat, J. Yans, R. Carleer, and F. Escuillié. 2022. Pterosaur melanosomes support signalling functions for early feathers. Nature 604:684–688.

Clack, J. A. 2002. Gaining ground: The origin and evolution of tetrapods. Indiana University Press, Bloomington, 369 pp.

Cloutier, R., A. M. Clement, M. S. Lee, R. Noël, I. Béchard, V. Roy, and J. A. Long. 2020. *Elpistostege* and the origin of the vertebrate hand. Nature 579:549–554.

Coates, M. I. 1996. The Devonian tetrapod *Acanthostega gunnari* Jarvik: Postcranial anatomy, basal tetrapod interrelationships and patterns of skeletal evolution. Transactions of the Royal Society of Edinburgh 87:363–421.

Coates, M. I. and J. A. Clack. 1990. Polydactyly in the earliest known tetrapod limbs. Nature 347:66–69.

Crompton, A. 1963. The evolution of the mammalian jaw. Evolution 17:431–439.

Daly, E. 1969. A new procolophonoid reptile from the Lower Permian of Oklahoma. Journal of Paleontology 43:676–687.

Darwin, C. 1859. On the origin of species by means of natural selection or the preservation of favoured races in the struggle for life. John Murray, London, 502 pp.

Dayrat, B., P. D. Cantion, J. A. Clarke, and K. de Queiroz. 2008. Species names in the PhyloCode: The approach adopted by the international society for phylogenetic nomenclature. Systematic Biology 57:507–514.

de Queiroz, K., P. D. Cantino, and J. A. Gauthier, eds. 2020. Phylonyms: A companion to the PhyloCode. CRC Press, Boca Raton.

de Queiroz, K. and J. Gauthier. 1990. Phylogeny as a central principle in taxonomy: Phylogenetic definitions of taxon names. Systematic Zoology 39:307–322.

de Queiroz, K. and J. Gauthier. 1992. Phylogenetic taxonomy. Annual Review of Ecology and Systematics 23:449–480.

de Queiroz, K. and J. Gauthier. 1994. Toward a phylogenetic system of biological nomenclature. Trends in Ecology and Evolution 9:27–31.

Ereshefsky, M. 2001. The poverty of the Linnaean hierarchy: A philosophical study of biological taxonomy. Cambridge University Press, Cambridge, x + 317 pp.

Eriksson, T., M. J. Donoghue, and M. S. Hibbs. 1998. Phylogenetic analysis of *Potentilla* using DNA sequences of nuclear ribosomal internal transcribed spacers (ITS), and implications for the classification of Rosoideae (Rosaceae). Plant Systematics and Evolution 211:155–179.

Felsenstein, J. 2004. Inferring phylogenies. Sinauer Associates, Sunderland, 664 pp.

Foer, J. 2005. Pushing *Phylocode*: What if we decide to rename every living thing on Earth? Discover, April, 46–51.

Foote, M. 1996. On the probability of ancestors in the fossil record. Paleobiology 22:141–151.

Fröbisch, N. B., R. L. Carroll, and R. R. Schoch. 2007. Limb ossification in the Paleozoic branchiosaurid *Apateon* (Temnospondyli) and the early evolution of preaxial dominance in tetrapod limb development. Evolution & Development 9:69–75.

Gadow, H. 1898. A classification of *Vertebrata*, recent and extinct. Adam and Charles Black, London.

Gaffney, E. S. 1980. Phylogenetic relationships of the major groups of amniotes; pp. 593–610 in A. L. Panchen Academic Press, London.

Gardner, J. D. and M. Böhme. 2008. Review of the *Albanerpetontidae* (*Lissamphibia*), with comments on the paleoecological preferences of European Tertiary albanerpetontids; pp. 178–218 in J. T. Sanskey and S. Baszio (ed.), Vertebrate microfossil assemblages: Their role in paleoecology and paleobiogeography. Indiana University Press, Indianapolis.

Gauthier, J. 1986. Saurischian monophyly and the origin of birds; pp. 1–55 in K. Padian (ed.), The origin of birds and the evolution of flight. California Academy of Sciences, San Francisco.

Gauthier, J. and K. de Queiroz. 2001. Feathered dinosaurs, flying dinosaurs, crown dinosaurs, and the name "Aves"; pp. 7–41 in J. Gauthier and L. F. Gall (eds.), Peabody museum of natural history. Yale University, New Haven.

Gauthier, J., A. G. Kluge, and T. Rowe. 1988. The early evolution of the *Amniota*; pp. 103–155 in M. J. Benton (ed.), The phylogeny and classification of the tetrapods, vol. 1: Amphibians, reptiles, birds. Clarendon Press, Oxford.

Ghiselin, M. T. 1984. "Definition," "character," and other equivocal terms. Systematic Zoology 33:104–110.

Göhlich, U. B. and L. M. Chiappe. 2006. A new carnivorous dinosaur from the Late Jurassic Solnhofen archipelago. Nature 440:329–332.

Griffiths, G. C. D. 1974. On the foundations of biological systematics. Acta Biotheoretica 23:85–131.

Griffiths, G. C. D. 1976. The future of Linnaean nomenclature. Systematic Biology 25:168–173.

Haeckel, E. 1866. Generelle Morphologie der Organismen. Reimer, Berlin.

Härlin, M. 1998. Taxonomic names and phylogenetic trees. Zoologica Scripta 27:381–390.

Hennig, W. 1966. Phylogenetic systematics. University of Illinois Press, Urbana, Chicago, London, 263 pp.

Hennig, W. 1969. Die Stammesgeschichte der Insekten. Kramer, Frankfurt am Main, 436 pp.

Hennig, W. 1981. Insect phylogeny. John Wiley & Sons, Chichester, xi + 514 pp.

Hibbett, D. S. and M. J. Donoghue. 1998. Integrating phylogenetic analysis and classification in fungi. Mycologia 90:347–356.

Hillis, D. M., C. Moritz, B. K. Mable, and R. G. Olmstead, eds. 1996. Molecular systematics, vol. 23. Sinauer Associates, Sunderland, MA.

Hinchliffe, J. R. 1989. Reconstructing the archetype: Innovation and conservatism in the evolution and development of the Pentadactyl limb; pp. 171–179 in D. B. Wake and G. Roth (eds.), Complex organismal functions: Integration and evolution in vertebrates. John Wiley & Sons Ltd, Chichester, England.

Hinchliffe, J. R. 2002. Developmental basis of limb evolution. International Journal of Development Biology 46:835–845.

Holmgren, N. 1933. On the origin of the tetrapod limb. Acta Zoologica 14:185–295.

Holtz, T. R. 1996. Phylogenetic taxonomy of the Coelurosauria (Dinosauria: Theropoda). Journal of Paleontology 70:536–538.

Irisarri, I., D. Baurain, H. Brinkmann, F. Delsuc, J.-Y. Sire, A. Kupfer, J. Petersen, M. Jarek, A. Meyer, and M. Vences. 2017. Phylotranscriptomic consolidation of the jawed vertebrate timetree. Nature Ecology & Evolution 1:1370.

Jarvik, E. 1963. The composition of the intermandibular division of the head in fish and tetrapods and the diphyletic origin of the tetrapod tongue. Kunglia Svenska Vetenskapakademiens Handlingar (4) 9:1–74.

Jarvik, E. 1986. The origin of the Amphibia; pp. 1–24 in Z. Roăek (ed.), Studies in herpetology. Charles University, Prague.

Johnson, M. M., M. T. Young, A. Brignon, and S. L. Brusatte. 2022. Addition to "the phylogenetics of *Teleosauroidea* (*Crocodylomorpha; Thalattosuchia*) and implications for their ecology and evolution". Bulletin of Phylogenetic Nomenclature 1:1–7.

Jolly, C. J., T. Woolley-Barker, S. Beyene, T. R. Disotell, and J. E. Phillips-Conroy. 1997. Intergeneric hybrid baboons. International Journal of Primatology 18:597–627.

Joyce, W. G., J. Anquetin, E.-A. Cadena, J. Claude, I. G. Danilov, S. W. Evers, G. S. Ferreira, A. D. Gentry, G. L. Georgalis, and T. R. Lyson. 2021. A nomenclature for fossil and living turtles using phylogenetically defined clade names. Swiss Journal of Palaeontology 140:1–45.

Judd, W. S., R. W. Sanders, and M. J. Donoghue. 1994. Angiosperm family pairs: Preliminary phylogenetic analyses. Harvard Papers in Botany 1:1–51.

Judd, W. S., W. L. Stern, and V. I. Cheadle. 1993. Phylogenetic position of *Apostasia* and *Neuwiedia* (Orchidaceae). Botanical Journal of the Linnean Society 113:87–94.

Kermack, K. A., F. Mussett, and H. W. Rigney. 1981. The skull of *Morganucodon*. Zoological Journal of the Linnean Society 71:1–158.

Kron, K. A. 1997. Exploring alternative systems of classification. Aliso 15:105–112.

Lanham, U. 1965. Uninomial nomenclature. Systematic Zoology 14:144.

Laurin, M. 1991. The osteology of a Lower Permian eosuchian from Texas and a review of diapsid phylogeny. Zoological Journal of the Linnean Society 101:59–95.

Laurin, M. 1998. A reevaluation of the origin of pentadactyly. Evolution 52:1476–1482.

Laurin, M. 2002. Tetrapod phylogeny, amphibian origins, and the definition of the name Tetrapoda. Systematic Biology 51:364–369.

Laurin, M. 2006. Scanty evidence and changing opinions about evolving appendages. Zoologica Scripta 35:667–668.

Laurin, M. 2010. How vertebrates left the water. University of California Press, Berkeley, xv + 199 pp.

Laurin, M. 2020. *Tetrapoda*; pp. 759–764 in K. de Queiroz, P. D. Cantino, and J. A. Gauthier (eds.), Phylonyms: A companion to the PhyloCode. CRC Press, Boca Raton, Florida.

Laurin, M. and J. S. Anderson. 2004. Meaning of the name Tetrapoda in the scientific literature: An exchange. Systematic Biology 53:68–80.

Laurin, M., J. W. Arntzen, A. M. Báez, A. M. Bauer, R. Damiani, S. E. Evans, A. Kupfer, A. Larson, D. Marjanović, H. Müller, L. Olsson, J.-C. Rage, and D. Walsh. 2020a. *Lissamphibia*; pp. 773–778 in K. de Queiroz, P. D. Cantino, and J. A. Gauthier (eds.), Phylonyms: A companion to the PhyloCode. Boca Raton, Florida: CRC Press.

Laurin, M., J. W. Arntzen, A. M. Báez, A. M. Bauer, R. Damiani, S. E. Evans, A. Kupfer, A. Larson, D. Marjanović, H. Müller, L. Olsson, J.-C. Rage, and D. Walsh. 2020b. *Amphibia*; pp. 765–771 in K. de Queiroz, P. D. Cantino, and J. A. Gauthier (eds.), Phylonyms: A companion to the PhyloCode. CRC Press, Boca Raton, Florida.

Laurin, M. and H. N. Bryant. 2009. Third meeting of the international society for phylogenetic nomenclature: A report. Zoologica Scripta 38:333–337.

Laurin, M. and P. D. Cantino. 2004. First international phylogenetic nomenclature meeting: A report. Zoologica Scripta 33:475–479.

Laurin, M. and P. D. Cantino. 2007. Second meeting of the international society for phylogenetic nomenclature: A report. Zoologica Scripta 36:109–117.

Laurin, M., M. Girondot, and A. de Ricqlès. 2000. Early tetrapod evolution. Trends in Ecology and Evolution 15:118–123.

Laurin, M., O. Lapauze, and D. Marjanović. 2022. What do ossification sequences tell us about the origin of extant amphibians? Peer Community Journal 2: e12.

Laurin, M. and R. R. Reisz. 1995. A reevaluation of early amniote phylogeny. Zoological Journal of the Linnean Society 113:165–223.

Laurin, M. and R. R. Reisz. 1997. A new perspective on tetrapod phylogeny; pp. 9–59 in S. Sumida and K. Martin (eds.), Amniote origins: Completing the transition to land. Academic Press, San Diego.

Laurin, M. and R. R. Reisz. 2020a. *Reptilia*; pp. 1027–1031 in K. de Queiroz, P. D. Cantino, and J. A. Gauthier (eds.), Phylonyms: A companion to the PhyloCode vol. CRC Press, Boca Raton, Florida.

Laurin, M. and R. R. Reisz. 2020b. *Amniota*; pp. 793–797 in K. de Queiroz, P. D. Cantino, and J. A. Gauthier (eds.), Phylonyms: A companion to the PhyloCode. CRC Press, Boca Raton, Florida.

Laurin, M. and R. R. Reisz. 2020c. *Synapsida*; pp. 811–814 in K. de Queiroz, P. D. Cantino, and J. A. Gauthier (eds.),

Phylonyms: A companion to the PhyloCode. CRC Press, Boca Raton, Florida.

Laurin, M. and T. R. Smithson. 2020. *Pan-Amniota*; pp. 789–792 in K. de Queiroz, P. D. Cantino, and J. A. Gauthier (eds.), Phylonyms: A companion to the PhyloCode. CRC Press, Boca Raton, Florida.

Lauterbach, K.-E. 1989. Das Pan-Monophylum: Ein Hilfsmittel für die Praxis der Phylogenetischen Systematik. Zoologischer Anzeiger 223:139–156.

Lebedev, O. A. 1986. The first record of a Devonian tetrapod in the USSR. Doklady-Earth Science Sections 278:220–222.

Lebedev, O. A. and M. I. Coates. 1995. The postcranial skeleton of the Devonian tetrapod *Tulerpeton curtum* Lebedev. Zoological Journal of the Linnean Society 114:307–348.

Lee, M. S. Y. 1996a. Stability in meaning and content of taxon names: An evaluation of crown-clade definitions. Proceedings of the Royal Society of London, Series B 263:1103–1109.

Lee, M. S. Y. 1996b. The phylogenetic approach to biological taxonomy: Practical aspects. Zoologica Scripta 25:187–190.

Lee, M. S. Y. 1998. Ancestors and taxonomy. Trends in Ecology and Evolution 13:26.

Lemierre, A. and M. Laurin. 2021. Conversion of the names *Pyxicephaloidea*, *Pyxicephalidae* and *Pyxicephalinae* (*Anura*, *Ranoidea*) into phylogenetic nomenclature. Bionomina 25:81–92.

Li, Z., H. Liu, and P. Luo. 1995. Production and cytogenetics of intergeneric hybrids between *Brassica napus* and *Orychophragmus violaceus*. Theoretical and Applied Genetics 91:131–136.

Lucas, S. G. 1992. Extinction and the definition of the class Mammalia. Systematic Biology 41:370–371.

Luo, Z.-X., J. A. Schultz, and E. G. Ekdale. 2016. Evolution of the middle and inner ears of mammaliaforms: The approach to mammals; pp. 139–174 in J. A. Clack, R. R. Fay, and A. N. Popper (eds.), Evolution of the vertebrate ear: Evidence from the Fossil record. Springer, Cham, Switzerland.

Madzia, D., V. M. Arbour, C. A. Boyd, A. A. Farke, P. Cruzado-Caballero, and D. C. Evans. 2021. The phylogenetic nomenclature of ornithischian dinosaurs. PeerJ 9:e12362.

Mann, A., J. D. Pardo, and H. C. Maddin. 2019. *Infernovenator steenae*, a new serpentine recumbirostran from the "Mazon Creek" Lagerstätte further clarifies lysorophian origins. Zoological Journal of the Linnean Society 187:506–517.

Mann, A. and R. S. Paterson. 2020. Cranial osteology and systematics of the enigmatic early "sail-backed" synapsid *Echinerpeton intermedium* Reisz, 1972, and a review of the earliest "pelycosaurs". Journal of Systematic Palaeontology 18:529–539.

Marini, M. Â. and S. J. Hackett. 2002. A multifaceted approach to the characterization of an intergeneric hybrid manakin (*Pipridae*) from Brazil. The Auk 119:1114–1120.

Marjanović, D. and M. Laurin. 2019. Phylogeny of Paleozoic limbed vertebrates reassessed through revision and expansion of the largest published relevant data matrix. PeerJ 6:e5565.

Martin, J. E., D. C. Blackburn, and E. O. Wiley. 2010. Are node-based and stem-based clades equivalent? Insights from graph theory. Plos Currents 2:RRN1196.

Maxwell, S. J. 2021. Registration of Neostromboidae clades in the RegNum of the PhyloCode, and Errata. The Festivus 53:192–209.

Maxwell, S. J. and T. Rymer. 2021. Are the ICZN and PhyloCode that incompatible? A summary of the shifts in Stromboidean taxonomy and the definition of two new subfamilies in Stromboidae (Mollusca, Neostromboidae). The Festivus 53:44–51.

Meier, R. and S. Richter. 1992. Suggestions for a more precise usage of proper names of taxa: Ambiguities related to the stem lineage concept. Journal of Zoological Systematics and Evolutionary Research 30:81–88.

Michener, C. D. 1963. Some future developments in taxonomy. Systematic Zoology 12:151–172.

Milner, A. R. 1993. The Paleozoic relatives of lissamphibians. Herpetological Monographs 7:8–27.

Minelli, A. 2001. Zoological nomenclature: Reflections on the recent past and ideas for our future agenda. Bulletin of Zoological Nomenclature 58:164–169.

Mishler, B. D. 1999. Getting rid of species?; pp. 307–315 in R. Wilson (ed.), Species: New interdisciplinary essays. MIT Press, Cambridge, MA.

Mishler, B. D. 2010. Species are not uniquely real biological entities; pp. 110–122 in F. J. Ayala and R. Arp (eds.), Contemporary debates in philosophy of biology vol. Wiley-Blackwell, New York.

Modesto, S. P. and J. S. Anderson. 2004. The phylogenetic definition of Reptilia. Systematic Biology 53:815–821.

Moore, G. 1998. A comparison of traditional and phylogenetic nomenclature. Taxon 47:561–579.

Nelson, G. and N. I. Platnick. 1981. Systematics and biogeography: Cladistics and vicariance. Columbia University Press, New York, xi + 567 pp.

Nixon, K. C. and J. M. Carpenter. 2000. On the other "phylogenetic systematics". Cladistics 16:298–318.

Norell, M., J. Qiang, G. Kegin, Y. Chongxi, Y. Zhao, and W. Lixia. 2002. 'Modern' feathers on a non-avian dinosaur. Nature 416:36–37.

O'Hara, R. J. 1997. Population thinking and tree thinking in systematics. Zoologica Scripta 26:323–329.

Olson, E. C. 1947. The family Diadectidae and its bearing on the classification of reptiles. Fieldiana Geology 11:1–53.

Osborn, H. F. 1903a. On the primary division of the Reptilia into two sub-classes, *Synapsida* and *Diapsida*. Science 17:275–276.

Osborn, H. F. 1903b. The reptilian subclasses *Diapsida* and *Synapsida* and the early history of the *Diaptosauria*. Memoirs from the American Museum of Natural History 1:449–507.

Papavero, N., J. Llorente-Bousquets, and J. M. Abe. 2001. Proposal of a new system of nomenclature for phylogenetic systematics. Arquivos de Zoologia Museu de Zoologia da Univeristade de São Paulo 36:1–145.

Pardo, J. D., K. Lennie, and J. S. Anderson. 2020. Can we reliably calibrate deep nodes in the tetrapod tree? Case studies in deep tetrapod divergences. Frontiers in Genetics 11:506749.

Pardo, J. D., B. J. Small, and A. K. Huttenlocker. 2017a. Stem caecilian from the Triassic of Colorado sheds light on the origins of Lissamphibia. Proceedings of the National Academy of Sciences 114:E5389–E5395.

Pardo, J. D., M. Szostakiwskyj, P. E. Ahlberg, and J. S. Anderson. 2017b. Hidden morphological diversity among early tetrapods. Nature 546:642–645.

Piñeiro, G., J. Ferigolo, A. Ramos, and M. Laurin. 2012. Cranial morphology of the Early Permian mesosaurid *Mesosaurus tenuidens* and the evolution of the lower temporal fenestration reassessed. Comptes rendus Palevol 11:379–391.

Pleijel, F. 1999. Phylogenetic taxonomy, a farewell to species, and a revision of Heteropodarke (Hesionidae, Polychaeta, Annelida). Systematic Biology 48:755–789.

Romer, A. S. 1956. Osteology of the reptiles. University of Chicago Press, Chicago, 772 pp.

Roth, B. 1996. Homoplastic loss of dart apparatus, phylogeny of the genera, and a phylogenetic taxonomy of the Helminthoglyptidae (Gastropoda: Pulmonata). Veliger 39:18–42.

Rowe, T. 1988. Definition, diagnosis, and origin of Mammalia. Journal of Vertebrate Paleontology 8:241–264.

Rowe, T. B. 2020a. *Pan-Mammalia*; pp. 801–810 in K. de Queiroz, P. D. Cantino, and J. A. Gauthier (eds.), Phylonyms: A companion to the PhyloCode. CRC Press, Boca Raton, Florida.

Rowe, T. B. 2020b. *Mammalia*; pp. 859–865 in K. de Queiroz, P. D. Cantino, and J. A. Gauthier (ed.), Phylonyms: A companion to the PhyloCode. CRC Press, Boca Raton, Florida.

Rowe, T. B. and J. Gauthier. 1992. Ancestry, paleontology, and definition of the name Mammalia. Systematic Biology 41:372–378.

Ruta, M., J. Botha-Brink, S. A. Mitchell, and M. J. Benton. 2013. The radiation of cynodonts and the ground plan of mammalian morphological diversity. Proceedings of the Royal Society B: Biological Sciences 280:20131865.

Sangster, G. and G. Mayr. 2021. *Feraequornithes*: A name for the clade formed by *Procellariiformes, Sphenisciformes, Ciconiiformes, Suliformes* and *Pelecaniformes* (*Aves*). Vertebrate Zoology 71:49–53.

Schander, C. 1998a. Types, emendations and names: A reply to Lidén & al. Taxon 401–406.

Schander, C. 1998b. Mandatory categories and impossible hierarchies: A reply to Sosef. Taxon 47:407–410.

Schander, C. and M. Thollesson. 1995. Phylogenetic taxonomy-some comments. Zoologica Scripta 24:263–268.

Schultze, H.-P. 1977. The origin of the tetrapod limb within the rhipidistian fishes; pp. 541–544 in M. Hecht, P. C. Goody, and B. M. Hecht (eds.), Major patterns in vertebrate evolution. Plenum Publishing Corporation and NATO Scientific Affairs Division, New York and London.

Sereno, P. C. 1998. A rationale for phylogenetic definitions, with application to the higher-level taxonomy of Dinosauria: Neues Jahrbuch für Geologie und Paläontologie. Abhandlungen 210:41–83.

Sereno, P. C. 1999. Definitions in phylogenetic taxonomy: Critique and rationale. Systematic Biology 48:329–351.

Sereno, P. C. 2005. The logical basis of phylogenetic taxonomy. Systematic Biology 54:595–619.

Sereno, P. C., S. McAllister, and S. L. Brusatte. 2005. TaxonSearch: A relational database for suprageneric taxa and phylogenetic definitions. PhyloInformatics 8:1–25.

Shubin, N. H. and P. Alberch. 1986. A morphogenetic approach to the origin and basic organisation of the tetrapod limb. Evolutionary Biology 20:318–390.

Shubin, N. H., E. B. Daeschler, and F. A. Jenkins, Jr. 2006. The pectoral fin of *Tiktaalik roseae* and the origin of the tetrapod limb. Nature 440:764–771.

Strickland, H. E., J. S. Henslow, J. Phillips, W. E. Shuckard, J. B. Richardson, G. R. Waterhouse, R. Owen, W. Yarrell, L. Jenyns, C. Darwin, W. J. Broderip, and J. O. Westwood. 1842. Report of a committee appointed "to consider of the rules by which the Nomenclature of Zoology may be established on a uniform and permanent basis. Annals and Magazine of Natural History 11:1–17.

Stuessy, T. F. 2000. Taxon names are not defined. Taxon 49:231–232.

Sues, H.-D. 2019. Authorship and date of publication of the name Tetrapoda. Journal of Vertebrate Paleontology 39:e1564758.

Sundberg, P. and F. Pleijel. 1994. Phylogenetic classification and the definition of taxon names. Zoologica Scripta 23:19–25.

Swann, E. C., E. M. Frieders, and D. J. McLaughlin. 1999. *Microbotryum, Kriegeria* and the changing paradigm in basidiomycete classification. Mycologia 91:51–66.

Swofford, D., G. J. Olsen, P. J. Waddell, and D. M. Hillis. 1996. Phylogenetic inference; pp. 407–514 in D. M. Hillis, C. Moritz, and B. K. Mable (eds.), Molecular systematics. Sinauer Associates, Sunderland, MA.

Taylor, M. P. 2007. Phylogenetic definitions in the pre-PhyloCode era; implications for naming clades under the PhyloCode. PaleoBios 27:1–6.

Turner, A. H., D. Pol, J. A. Clarke, G. M. Erickson, and M. A. Norell. 2007. A basal dromaeosaurid and size evolution preceding avian flight. Science 317:1378–1381.

Vorobyeva, E. I. 1991. The fin-limb transformation: Palaeontological and embryological evidence; pp. 339–345 in J. R. Hincliffe, J. M. Hurle, and D. Summerbell (eds.), Developmental patterning of the vertebrate limb. Plenum

Press in Cooperation with NASO Scientific Affairs Division, New York and London.

Watson, D. M. S. 1954. On *Bolosaurus* and the origin and classification of reptiles. Bulletin of the Museum of Comparative Zoology 111:299–449.

Weishampel, D. B., P. Dodson, and H. Osmólska, eds. 2004. The *Dinosauria*. University of California Press, Berkeley, 2nd ed.

Williston, S. W. 1917. The phylogeny and classification of reptiles. Contributions from the Walker Museum 2:61–71.

Witmer, L. M. 2009a. Fuzzy origins for feathers. Nature 458:293–295.

Witmer, L. M. 2009b. Feathered dinosaurs in a tangle. Nature 461:601–602.

Wolff, R. G. 1991. Functional chordate anatomy. D. C. Heath and Company, Lexington, MA, 752 pp.

Wolsan, M. 1993. Phylogeny and classification of early European Mustelida (Mammalia: Carnivora). Acta Theriologica 38:345–384.

Wyss, A. R. and J. Meng. 1996. Application of phylogenetic taxonomy to poorly resolved crown clades: A stem-modified node-based definition of Rodentia. Systematic Biology 45:559–567.

Xu, X. 2006. Scales, feathers and dinosaurs. Nature 440:287–288.

Xu, X., K. Wang, K. Zhang, Q. Ma, L. Xing, C. Sullivan, D. Hu, S. Cheng, and S. Wang. 2012. A gigantic feathered dinosaur from the Lower Cretaceous of China. Nature 484:92–95.

Zverkov, N. 2022. A problem of naming of the families of Late Jurassic and Cretaceous ichthyosaurs. Paleontological Journal 56:463–470.

# 5

# Comparisons with the Nomenclature of Other Sciences and Beyond

## 5.1 Introduction

We saw in Chapter 2 that absolute ranks, the so-called Linnaean categories, are inherent to rank-based nomenclature (RN). This may simply reflect historical contingency because the human mind handles many kinds of hierarchies, but many of them lack absolute levels. Mishler (1999) gave the example of a grocer who might classify salt as a spice and spices as food items; this requires no absolute ranks to work. He further argued that RN is unusual in assigning absolute ranks to the taxonomic hierarchy and that these ranks are superfluous. Comparisons with other fields performed in the following aim at determining how ranks are used, how types are selected, how the named entities are delimited and how precise such delimitations are. Such comparisons may show to what extent the nomenclatural rules of other fields and recent trends therein bring them closer to RN or phylogenetic nomenclature (PN). This, in turn, may suggest some form of spontaneous consensus about the properties of a good nomenclatural system. I already published such comparisons several years ago (Laurin 2008), but only with chemical elements, stratigraphy and geopolitics. The present analysis considers other fields (biogeography and folk taxonomy), performs more detailed comparisons and provides more references.

It could be objected that this exercise is of limited value because taxa are hierarchical historical entities that arguably evolved mostly through divergent evolution (see Section 6.4), whereas the entities discussed in the following (except in the section on folk taxonomies) are arranged differently, such as linearly (stratigraphic units, or even taxa when they were organized under the *scala naturae*), spatially (biogeographic areas, geopolitical units) or according to some properties (atoms). I see two problems with this reasoning.

First and most importantly, if the scientific communities who study these entities have optimized their nomenclature to achieve similar goals (most importantly, should technical terms be precisely delimited or should they have fuzzy boundaries?) or used similar devices (like absolute ranks), this may suggest general ideas about what nomenclature should be (or perhaps common problems and solutions). This question is interesting in its own right, but it is also relevant to the debate between proponents of PN and RN.

Second, our ideas about how entities are organized evolve and the current consensus is sometimes challenged. For instance, time, which is an important criterion to classify stratigraphic units, and, arguably, paleobiogeographical ones, is usually conceptualized linearly, but a hierarchical concept

of time was proposed by Zaragüeta Bagils et al. (2004), which they argue is useful to assess the stratigraphic fit of phylogenetic trees. Similarly, the concept of paralogy of biogeographical units, popularized by Nelson and Ladiges (1996), implies a hierarchy. Taxa are now thought to be mostly hierarchically organized (see Section 6.4), but back in the 18th century, several systematists (Linnaeus and Antoine-Laurent de Jussieu, among others) conceptualized affinities of taxa using a map metaphor (see Section 1.3). More importantly, with the discovery of the phenomenon of horizontal gene transfer, which is fairly widespread among prokaryotic organisms, many authors model biodiversity as a web (or network) of life for the simplest forms of life (see Section 6.4), and this is reminiscent of a map. Thus, the fact that most of the entities discussed in this chapter do not share the hierarchical organization typical of eurkaryotic taxa does not negate the interest of these comparisons.

## 5.2 Stratigraphy

### 5.2.1 Similarities between Stratigraphic and Rank-Based Nomenclatures

Stratigraphic nomenclature shows interesting parallels with RN because the entities that it classifies are typically ranked. The *North American stratigraphic code* (North American Commission on Stratigraphic Nomenclature 2021: table 2) recognizes the following categories of lithostratigraphic entities, from highest to lowest: Supergroup, Group, Formation, Member, Submember and Bed. The recognized ranks of geochronologic entities include the following (also listed from highest to lowest rank): Eon, Era, Period, Epoch, Age and Chron. Other types of units (such as biostratigraphic, pedostratigraphic, allostratigraphic and magnetostratigraphic units) are regulated by these codes, but these need not be detailed here.

Although the discussion of stratigraphic nomenclature will emphasize the *North American stratigraphic code* (abbreviated as *NAsc*) because this is one of the most recent and influential stratigraphic codes, such formalizations have been implemented in older regional or national codes, such as the *Code of stratigraphic nomenclature for Norway* (Bjørlykke 1961: 230).

Like RN, early stratigraphic works recognized type units that do not precisely delimit chronostratigraphic units, but considerable effort has been made in recent decades to define precise boundaries, which makes stratigraphic nomenclature more similar to PN; this is detailed in Section 5.2.4.

DOI: 10.1201/9781003092827-6

Another parallel between stratigraphic nomenclature and RN is the unnecessary debate between splitters and lumpers. This was evoked for stratigraphic nomenclature by Benison et al. (2015: 628), who reported confusion in the nomenclature of the Nippewalla Group of Kansas (Early and Middle Permian): "some confusion in stratigraphic names has resulted from 'lumping' or 'splitting' since these rocks were first studied (Cragin 1896; Norton 1939; Swineford 1955; Fay 1964)." See Section 5.2.2 for more information about the Nippewalla Group.

As under RN, inconsistent use of names, coupled with vague definitions and a lack of explicit delimitation, hampered communication (in geochronology and biostratigraphy). An early example is provided by Pendery (1963: 1828) in his revision of mid-Permian strata from the Southwest USA. He declared that "Basic stratigraphy of the Blaine Formation, extending from Kansas to north-central Texas, is relatively simple; yet a confusing nomenclature has resulted from various 'type sections' at geographically separated localities." Pendery (1963: 1836) further explained that:

> At present, three different concepts of the term Blaine exist: the 'Blaine of Oklahoma and Kansas'; Lloyd and Thompson's 'Blaine of Texas' (Flowerpot-Blaine-Dog Creek Formations); and Roth's 'Blaine of the Pease River Group' (Van Vacter Gypsum Member and Mangum Dolomite Member of the Blaine Formation).

More recently, Nelson et al. (2013) documented the great nomenclatural confusion that still prevails in the geological nomenclature of Permian sedimentary rocks in Texas. Nelson et al. (2013: 312) stated that:

> These rocks host some of the most productive and, therefore, data-rich oil fields in North America, yet their lithostratigraphy is a nomenclatural quagmire. Names based on fully terrestrial outcrops more than 300 km away from the basin have been applied to totally dissimilar marine rocks that may not even be the same age as their terrestrial namesakes. This practice falsely implies that accurate correlations have been achieved from the outcrop to the subsurface. It also obscures key facies relationships worked out by pioneering geologists such as P.B. King nearly a century ago.

They also argued that "The barrier [to solving these nomenclatural problems] in Texas is 70 years of misuse and malaise-based indifference." These geological nomenclatural problems in Texas are not atypical of similar problems elsewhere. Thus, Benison et al. (2015: 628) reported confusion in the nomenclature of the Nippewalla Group of Kansas (Early and Middle Permian):

> The Nippewalla Group's six formations are fairly similar to one another in lithologies (Figs. 3, 4; Benison 1997a, 1997b). This has led to challenges in reconciling the stratigraphy of surface exposures. For example, two buttes in different townships in Barber

County, Kansas, have been identified as "Flowerpot Mound," and each is considered the type locality of the Flowerpot (aka Flower-pot) Shale (Norton 1939; Fay 1964; Buchanan and McCauley 1987). Some possibly correlative units with similar lithologies have different stratigraphic names in Kansas and Oklahoma (e.g., Norton 1939; Jordan and Vosburg 1963; Fay 1964; Holdoway 1978; Johnson 1991).

Political factors complicate stratigraphic nomenclature unnecessarily and contribute to the kind of nomenclatural confusion that also plagues RN. This was documented for Triassic strata of the Dockum Group in Texas and New Mexico. Thus, Lehman (1994: 37) stated:

> The Upper Triassic Dockum Group consists of continental red beds exposed around the southern High Plains of western Texas and eastern New Mexico. Although these strata are contiguous between the two states, different stratigraphic nomenclature is used in Texas and New Mexico.

These nomenclatural differences (and to a lesser extent, those between the Pease River, El Reno and Nippewalla groups evoked previously) do not result primarily from lithological differences across state lines. Lehman (1994: 37) indicated that:

> This paper portrays a case study in the problem of a perceived state boundary "fault" between Texas and New Mexico. This case study illustrates what happens when stratigraphers across state boundaries do not communicate or lack regard for one another, and when paleontologists apply formal names to strata without actually mapping them.

More specifically, (Lehman 1994: 39) explained that "Traditionally, separate nomenclature has been adopted for many units on either side of the state because these were separate centers of investigation" and he drew an explicit parallel with biological nomenclature on this basis:

> So, in spite of the obvious physical continuity of these strata, by the beginning of the 1980s separate stratigraphies evolved in Texas and New Mexico without regard for the priority of established names on either side of the state line, and without any serious attempt to trace named strata from one state to -the [sic] other. This practice would be analogous in paleontology to naming a new species without comparing it to similar or identical taxa, without documenting how it differs from them, and without determining if it had already been named.

The legalistic aspects, as well as the war of influence between groups of stratigraphers, is reminiscent of similar problems in RN, although the RN codes seem to have been more successful than the stratigraphic codes in preventing nomenclatural chaos (Lehman 1994: 46):

> The stratigraphic code allows for endless legalistic harangue, but hopes to enforce respect where it

does not already exist. Where egos collide, the code provides a means of independent review to evaluate specific contentions. Such an evaluation may have little effect, however, if stratigraphers map their own states, without crossing the fence to see what is on the other side.

Lehman (1994: 47) concluded that: "The practice of naming strata should be every bit as exacting and rigorous as that of naming new species."

These nomenclatural problems developed despite the long-established practice of defining chronostratigraphic units by a type section that allowed comparisons, and the regulation of these activities by stratigraphic codes (American Commission on Stratigraphic Nomenclature 1961; North American Commission on Stratigraphic Nomenclature 2021). Perhaps part of the problem was that type sections in many cases did not delimit chronostratigraphic units precisely, which is reminiscent of how types are used in RN, but this changed in the last decades in stratigraphic nomenclature (see Section 5.2.4).

## 5.2.2 Differences between Stratigraphic and Biological Nomenclatures

Like all analogies, the comparison between biological and stratigraphic nomenclatures has its limits, and there are important differences between both kinds of nomenclatures. Thus, the linear succession of stratigraphic units, rather than hierarchical relationships between taxa, suggests that the nomenclatural systems must differ on some key points. Such a key difference is that a lithostratigraphic unit of a given level may belong to more than one unit of the next higher rank. For instance, the Middle Permian Blaine Formation forms part of the Pease River Group in the Midland Basin of Texas, as well as part of the El Reno Group and the Nippewalla Group in the Anadarko Basin of Oklahoma and Kansas, respectively (Nelson and Hook 2005; Foster et al. 2014). This is neither oversight nor anomaly, and it occurs at several levels in stratigraphic nomenclature. Thus, Article 25 of the *NAsc* states that "A member may extend laterally from one formation to another," and Article 28, note (b), stipulates that "The formations making up a group need not necessarily be everywhere the same." By contrast, a given genus cannot belong to two families, even though there may be uncertainty about its affinities and hence, about the family to which it should be assigned under RN.

Perhaps the greatest difference between biological and stratigraphic nomenclature is that the latter regulates several kinds of units (lithostratigraphic, geochronologic, biostratigraphic, pedostratigraphic and so on), and there is no simple relationship between these. Thus, the limits of lithostratigraphic units need not (and most often do not) coincide with the limits of geochronologic units, nor is their age necessarily entirely comprised within such a unit. For instance, the Chickasha Formation of Oklahoma (a Permian lithostratigraphic unit) occupies part of the Roadian (a chronostratigraphic unit), whereas the San Angelo Formation of Texas may straddle the Kungurian-Roadian boundary (Laurin and Hook 2022). This is a bit as if codes of biological nomenclature regulated not

only taxon names, but also their ecological classification (position in the trophic chain or habitat use), although this is also an imperfect parallel.

The various kinds of units previously evoked show that the stratigraphic codes changed more than the RN codes over time, obviously because they were revised to take advantage of scientific and technical progress, contrary to some of the RN codes. For instance, the *NAsc* recognizes magnetostratigraphic units, which developed from pioneering research carried out in the 1960s (such as Cox et al. 1963), and these units were incorporated into stratigraphic codes in the 1970s (IUGS 1979). By contrast, the *Zoological Code* and *Botanical Code* completely ignored advances in phylogenetics linked to the advent of cladistics (Hennig 1966) or molecular systematics, which hark back to Zuckerkandl and Pauling (1965), and which both developed rapidly over the last several decades (e.g., Swofford 2002; Goloboff et al. 2008; Ronquist et al. 2012a, 2012b; Sauquet 2013). These developments deeply transformed systematics, except for zoological and botanical RN (bacteriologists and virologists revised more substantially their nomenclatural codes to take advantage of scientific progress, as discussed in Chapter 2).

Other differences between biological and stratigraphic nomenclatures are that the latter may reflect local conditions (for lithostratigraphic units, not for chronostratigraphic units), whereas the former always applies globally. Thus, Remark (c) of Article 28 of the *NAsc* indicates that:

> The wedge-out of a component formation or formations may justify the reduction of a group to formation rank, retaining the same name. When a group is extended laterally beyond where it is divided into formations, it becomes in effect a formation, even if it is still called a group.

This change in rank of lithostratigraphic units is not unique to the *NAsc;* similar provisions were already included in the *International Stratigraphic Guide* (Murphy and Salvador 1999). These changes of rank of lithostratigraphic units according to their thickness are without parallels in RN. An imaginary parallel would be, for instance, to consider that a taxon is a family in an area (or time) where it includes several species belonging to a few genera, but to consider that the same taxon is only a genus in another area (or time) in which fewer of its species occur. Such a notion would seem odd to any systematist.

Another aspect in which the nomenclature of biostratigraphic units differs from biological nomenclature is evoked in Remark (e) of Article 54 of the *NAsc*, which stipulates that "Biostratigraphic units do not have stratotypes in the sense of Article 3, item (iv), and Article 8. Nevertheless, it is desirable to designate a reference section in which the biostratigraphic unit is characteristically developed." Thus, biostratigraphic units are not really typified under stratigraphic nomenclature. However, this is specific to biostratigraphic units. Lithostratigraphic units do have formal stratotypes (*NAsc* Article 22b), as do lithodemic units (formed of intrusive and/or metamorphic rocks; *NAsc* Article 31b), magnetostratigraphic units (*NAsc* Article 43b), chronostratigraphic units (*NAsc* Article 78) and polarity-chronostratigraphic units (*NAsc* Article 86).

Some nomenclatural problems facing modern geology are challenging and specific to geology, such as the sharp differences in aspect between exposures and drill cores of the same lithological units caused by evaporite dissolution near the surface. Benison et al. (2015: 642) thus raised the problem of types for such formations:

> How to use outcrops as reference sections when they are poorly preserved compared to subsurface correlatives? Three possible solutions include: (1) the stratotype (aka "type section") could be based on core only; (2) the outcrops and cores, which have different lithologies, yet represent the same environment, could be assigned different stratigraphic names and descriptions; or (3) the stratigraphic names and formal descriptions could include both outcrop and core descriptions. We acknowledge that the first two options above are extreme and impractical, and would certainly lead to further complications. As stated in the *North American Stratigraphic Code* (2005, p. 1566), "If the originally specified stratotype is incomplete, poorly exposed, structurally complicated, or unrepresentative of the unit, a principal reference section or several reference sections may be designated to supplement, but not to supplant, the type section." Therefore, the third option, using cores to constitute a reference section for previously defined type sections in outcrop, is the only viable solution for stratigraphic units containing evaporites.

The relationship between names and attributes is handled differently in stratigraphic and biological nomenclatures. Thus, Benison et al. (2015: 630) reported that "Despite the name, XRD and petrographic observations shows [sic] that **there is no dolomite** in the Cedar Springs Dolomite Member" (emphasis mine throughout the book unless stated otherwise). However, such errors in the names of lithostratigraphic units do not invalidate these names, which can and should be changed. This can be achieved by replacing the inappropriate lithic designation by the word "Formation" or "Member" (if these words did not originally occur in the name), or by a more appropriate lithic designation, while keeping the geographic designation (Article 18 of the *NAsc*). For instance, the Cedar Springs Dolomite Member could be renamed the Cedar Springs Calcite Member (because calcite is the most abundant mineral in this unit). On the contrary, in biological nomenclature, an inappropriate specific epithet cannot be changed if it is subsequently found to be inappropriate, or if it was originally inappropriate. For example, the angiosperm *Rumex britannica* actually occurs in North America rather than Great Britain, but the *Botanical Code* forbids renaming it *Rumex americana*, which would be a more appropriate name (but such a change would create nomenclatural problems and complicate literature searches, which is why specific epithets cannot be changed beyond what is required to maintain the agreement in grammatical gender with their genus name; see Section 2.4.3).

A last difference is that in stratigraphy, priority is given less emphasis than in RN or PN. Practical considerations can lead to redefinition of boundaries in well-justified cases (Remane et al. 1996: 78).

## 5.2.3 Relationships between Stratigraphic and Rank-Based Nomenclatures

Despite these many differences between stratigraphic and biological nomenclatures, the stratigraphic codes do not ignore biological nomenclature. Thus, Remark "a" of Article 54 of the *NAsc* indicates the following:

> The name of a biozone consists of the name of one or more distinctive taxa or parataxa (for trace fossils) found in the biozone, followed by the word "Biozone." (e.g., *Turborotalia cerrozaulensis* Biozone or *Cyrtograptus lundgreni-Testograptus testis* Biozone). The name of the species whose lowest occurrence defines the base of the zone is the most common choice for the biozone name. Names of the nominate taxa, and hence the names of the biozones, **conform to the rules of the international codes of zoological or botanical nomenclature** or, in the case of trace fossils, internationally accepted standard practice.

## 5.2.4 Similarities between Stratigraphic and Phylogenetic Nomenclatures

Geologists tried long ago to introduce a nomenclature that would delimit geological strata and time. Thus, the *Code of stratigraphic nomenclature for Norway* (Bjørlykke 1961: 230) stated:

> The boundaries of systems, series, and stages are based primarily on lithological or palaeontological boundaries in a type section. The boundaries of zones and subzones primarily are based on palaeontological boundaries, whereas those of substages primarily are based on lithological boundaries.

This quote attests that already by the 1960s, Norwegian geologists were trying to define the boundaries of stratigraphic units, and this preoccupation was shared by other contemporary geologists; Remane et al. (1996) indicated that "The Silurian-Devonian Boundary Committee was the first to put into practice (in 1960) the principle to define chronostratigraphic units by their lower boundary only, which thus becomes automatically the upper boundary of the underlying unit."

In the last decades, the delimitation of chronostratigraphic units was made more precise and permanent by the introduction of **GSSPs** (*Global Stratotype Section and Points*), in which a golden spike precisely marks the bottom of a chronostratigraphic unit (see also Article 10, entitled "Boundaries," of the *NAsc*). The first version of the official guidelines issued by the ICS (International Commission on Stratigraphy) was published by Cowie (1986) and an update was published by Remane et al. (1996). The upper end of such units is defined by the lower bond of the next (younger) unit. For example, the base of the Roadian is defined by its *GSSP* and dates from about 272.3 Ma, according to recent geochronological studies (Lucas and Shen 2018). The top of the Roadian is defined by the *GSSP* at the base of the Wordian (the next stage in the Guadalupian, Middle Permian), which dates from about 268.8 Ma. All *GSSPs* must be approved by a vote of the ICS (Remane et al. 1996: 78).

GSSPs work well for the Phanerozoic, but they would be difficult to implement for earlier periods of Earth's history, notably because of the scarce fossil record, which hampers stratigraphic correlations. Thus, the ICS recognizes an alternative system for this more remote past: the *Global Standard Stratigraphic Age (GSSA)*. This is a theoretical postulate, independent of the dating method used (Remane et al. 1996: 78).

This system of precise stratigraphic delimitation using *GSSPs* and *GSSAs* is reminiscent of PN in allowing great precision and stability in the definition and delimitation. In both cases, this precision is limited only by our knowledge: by the precision of stratigraphic correlations in the case of chronostratigraphic units and by our understanding of phylogenies in the case of taxa (Laurin 2008).

### 5.2.5 Stratigraphic Nomenclature: Conclusions

The previous comparisons between stratigraphic and biological nomenclatures highlighted a number of parallels with RN, especially the pervasive use of ranks in various kinds of stratigraphic units. However, there are also fascinating differences with RN, especially the effort over the several last decades to precisely delimit chronostratigraphic units using *GSSPs*, which started early in the 1960s. A similar preoccupation for precise delimitation of taxa did not occur before the rise of PN in the late 1980s, more than two decades later. Perhaps more importantly, in stratigraphy, these developments (of *GSSPs*) were integrated early on into the main codes, and these developments did not generate excessive controversy; Remane et al. (1996: 77) stated that "the Revised Guidelines were approved by the full Commission with an **overwhelming majority**, with **only one opposing vote**." Yet, the revisions in these guidelines appear bold, especially when viewed from a systematist's perspective. This can be illustrated by this quote from Remane et al. (1996: 78):

> Our main task for a number of years will be to develop **precise boundary definitions** for traditional chronostratigraphic units. Most of them were defined in the last century by their characteristic fossil contents, and their boundaries coincided with spectacular biostratigraphic and lithologic changes. These were 'natural' boundaries, in perfect agreement with the catastrophist philosophy of that time. In reality, rapid faunal turnovers are to a certain extent artefacts due to **stratigraphic gaps** or condensation. **Most of the classic type-localities are thus unsuitable for a precise boundary definition**: we have to look for new sections where sedimentation is continuous across the boundary interval; but then boundaries will rarely correspond to a lithologic change.

The described shift in approach thus amounts to abandoning much of the typification that was done in the 19th century to look for new type-localities suitable to define precise boundaries. By contrast, the RN codes used by most systematists (zoologists and botanists) have not changed significantly in still claiming that nomenclature does not delimit taxa (see Section 2.4.1), and do not even require monophyly of taxa. Given the reluctance of the regulating bodies of the RN codes to implement substantial changes, development of PN had to occur through the development of a separate code (the *PhyloCode*; see Chapter 4). It could be argued that stratigraphers managed this transition much better than systematists.

---

## 5.3 Geography

### 5.3.1 Biogeographical Units

#### 5.3.1.1 Biogeography, Introduction

Geography can be considered from various perspectives: human (geopolitical), biological (biogeographical) and physical (geological). In this book, we will consider only the first two, starting with biogeography, given that this is a scientific field that overlaps with biology and geography.

Biogeography assesses the geographic distribution of taxa and their communities in space and time (Servais et al. 2013; Cecca and Zaragüeta i Bagils 2015), as well as the relationships between areas of endemism, a concept introduced by de Candolle, who also distinguished between climatic and soil factors on the one hand, and other factors, such as historical ones, to explain floristic distributions (de Candolle 1820). It is often divided along taxonomic lines into fields such as zoogeography (for animals), phytogeography (for plants) and mycogeography (for fungi). Paleobiogeography is another important subdivision of biogeography. It incorporates data from the fossil record and from plate tectonics to study the distribution of early taxa and communities, as well as application of these data to study the distribution of continental plates and other geographical features. Phylogeography uses phylogenetic data to infer ancestral biogeographic areas of various taxa. Paleobiogeography and phylogeography can also be seen as two forms of historical biogeography, which traces the history of biogeographic areas.

#### 5.3.1.2 ICAN, a Code for Biogeography

An *International code of area nomenclature (ICAN)* was developed under the auspices of the Systematic and Evolutionary Biogeographical Association (SEBA) after some biogeographers expressed a need for such a code a few years earlier (Viloria 2004). Despite its name, this code was designed for biogeographic areas (areas of endemism), rather than other branches of geography (Ebach et al. 2008).

Ebach et al. (2008) explained that biogeographic areas need to be defined carefully, but focus exclusively on spatial delimitation. This is unfortunate because the distribution of taxa can shift over time and phenomena like continental drift and erosion change basic geography over time, but the *ICAN* includes no rules or recommendations about geological time and how it should be used in definitions or delimitation of bio-geographic areas, as noted by Zaragüeta-Bagils et al. (2009). Indeed, among the examples used by Ebach et al. (2008), Gondwana was formed and destroyed by continental drift. This ancient supercontinent incorporated India for much, but not all, of its history. As a result, most authors include India in Gondwana (e.g., Philippe et al. 2003), but others do not (such as Barker et al. 2007). As a result, there is ambiguity in what Gondwana refers to in the literature. But ambiguity in biogeo-graphical areas in simpler situations (even when considering only the present time) also results from vague nomenclature. The *ICAN* gives the example of "Mediterranean," which Ebach et al. (2008: 1154) argue can refer to the Mediterranean sea, sometimes including also the Black Sea and the Sea of Azov, or to the largest islands of the Mediterranean basin (for instance, Corsica, Crete, Mallorca, Lesvos and Sardinia), or to a larger area that incorporates "North Africa, the western Mediterranean, Balkans–Anatolia, Middle East, Caucasus, the Iranian Plateau and Central Asia," among others.

To improve the nomenclature of biogeographic areas, the *ICAN* requires using ranks (Section C, Article 1.3) and, more specifically, recommends the following ranks, listed in increasing hierarchical level (Section C, Article 1.2): district, province, dominion, region and realm. The same article also indicates that the prefix "sub-" may be used to increase the number of ranks, but no other prefix is mentioned. This sug-gests that no more than ten ranks could be used under the *ICAN*, which is low compared to the rank-based codes for taxa (see Chapter 2). Zaragüeta-Bagils et al. (2009) commented that "In ICAN, the use of ranks is quite arbitrary, because there may be many more areas (nodes) than ranks, at least in some methods such as area cladistics and Brooks parsimony analysis," which also suggests that the ten ranks allowed by the *ICAN* may not be sufficient.

The *ICAN* indicates (Section C, Article 2.1) that "An avail-able area name has a type-locality and either a published diagnosis or a description in a refereed publication." This is reminiscent of the RN codes for taxa, with the difference that the ICAN requires either a diagnosis or a description, whereas the RN codes require a diagnosis (and a description is almost always provided, but not strictly required by the codes).

Synonymy is assessed differently in the *ICAN* than under the RN codes. Ebach et al. (2008: 1154) indirectly suggest that the diagnosis (or description) of biogeographic areas should play a more important role than under RN: "The Northern Adriatic would be synonymized with the Central Adriatic, for example, if it were concluded that the presumed endemics live in both the Central and Northern Adriatic." This example does not mention type areas, which differs from RN because under the latter, taxa need not be synonymized simply because their diagnoses are similar; synonymy is required if taxa share the same type (objective synonymy), or if taxa are judged to be the same (subjective synonymy) and if they are of the same rank. The code itself (Section C, Article 2.8) states that "The prin-ciples of homonymy, synonymy and priority apply to the names **within any rank** that have valid diagnoses," which is ambiguous; it could mean that priority applies to names of a given rank or that it applies regardless of rank (hence, areas of different ranks could be considered synonyms), as under the

proposed rules for paleobiogeography (see Section 5.3.1.3). If the first interpretation is correct, ranks play a key role (as in RN) under this proposal, but under the second, more plausible interpretation (see below), ranks play a more minor role. Thus, under *ICAN*, diagnosis (or description) seems to play a greater role than type-locality to assess synonymy. Section C, Article 4.2 reinforces this impression. It states that "A name can be rejected if it has the same diagnosis, description, geographi-cal coordinate or distribution as an existing name, regardless if the type-locality differs." Thus, the *ICAN* does not appear to distinguish between objective and subjective synonymy (to use the nomenclature of the *Zoological Code* and *Prokaryotic Code*; these are called nomenclatural and taxonomic synon-ymy in the *Botanical Code*).

Initial reactions to the *ICAN* proposal were mixed. Zaragüeta-Bagils et al. (2009) welcomed this initiative, but voiced several concerns, most of which seem to be well-founded. For instance, Zaragüeta-Bagils et al. (2009) pointed out that:

> It is not stated in ICAN how higher-rank areas inherit the type-locality of names given to included areas. In a ranked taxonomy, the onomatophore of a species is a holotype, the onomatophore of a genus is its type species and that of a family is its type genus. This system guarantees the link between the name of any supraspecific taxon and the holotype without multiplying onomatophores. In ICAN, the type-locality of a higher-rank area can be different from those of included areas. This renders decisions about synonymy/homonymy virtually impossible.

While this criticism may go a little too far (decisions about synonymy are not necessarily impossible under the *ICAN*), it does point out at a fundamental difference in how typification is handled by the *ICAN* and the RN codes. In their response to this criticism, Parenti et al. (2009) indicated that "We agree that the use of classificatory ranks is arbitrary and gave sev-eral examples in our explanatory discussion of how in prac-tice these may be applied or even ignored." This is a rather puzzling statement because the *ICAN* is basically rank-based, so it is difficult to imagine how ranks could be ignored while using the *ICAN*. Indeed, Article 1.3 of Section C of the *ICAN* stipulates that "An area needs to have a rank to be named," although Parenti et al. (2009) may have meant that under the *ICAN*, synonymy applies regardless of rank (see previous). In addition, the most relevant example that Ebach et al. (2008) provided about the use of ranks pertained to New York, which may refer to a city or to a state (though both are geopolitical units, rather than biogeographic ones), and at least for such examples, it is difficult to imagine how a rank-based code of (bio)geography could work without ranks, though a rankless code of biogeography based on historical events and vaguely analogous to the *PhyloCode* for taxa might work, as I previ-ously suggested (Laurin 2016: 613).

Zaragüeta-Bagils et al. (2009) also regretted that the *ICAN* does not make "any reference to the temporal significance of named areas." Given that geography and taxon distributions change over time, this is problematic. Zaragüeta-Bagils et al.

(2009) mentioned the risk of "synonymization of area names based on studies concerning organisms of different geological ages," but more generally, temporal delimitation of biogeographic entities is required to avoid seeing similar areas bearing multiple names associated with different times, even in the absence of other substantial differences. In a subsequent study, Ebach seems to have realized the importance of time in biogeography, but this did not result in proposals to amend the *ICAN* (Dowding et al. 2019).

Unsurprisingly, Servais et al. (2013: 32) noted that the *ICAN* had "not yet gained a significant following," and a Google Scholar search showed only a modest rise in the citation rate of this code over the subsequent years. This rate rose from 3–7 citations/year from 2009 to 2014, to about 8–10 papers/year subsequently, for a total of 101 citations as of December 7, 2022. This is apparently a very small proportion of papers in this field because a Google Scholar search performed the same day for "biogeography" yielded about 18,000 references published in 2022 alone. Thus, a few recent studies, notably by co-authors of the *ICAN*, follow that code (such as Morrone et al. 2017), but most typically use other types of geographical units (such as geopolitical ones), rather than areas of endemism (e.g., Bardet 2012: fig. 2), or use physical paleogeography to define biogeographic areas of endemism without following the *ICAN* (such as Licht et al. 2022).

These approaches are not ideal because using biogeographic areas without a code implies nomenclatural confusion, whereas using geopolitical units (whose borders are typically well-defined, although they may change over time) creates biases in biodiversity assessments (Murphy 2021). Worse, even using paleogeographical or tectonic units (which are more natural than geopolitical ones) is not ideal and can be only a first step in the analysis because such units need not coincide with paleobiogeographical units. Servais and Sintubin (2009: 106) provided some present-day examples. For instance, Baja California (Mexico) does not form a distinct biogeographical province, but it is located on a different tectonic plate (the Pacific plate) than the North American continent. Furthermore, the distribution of paleobiogeographical units is also constrained by habitats (and hence, biofacies), and this complicates further linking such units with tectonic units (Servais and Sintubin 2009: 106).

### 5.3.1.3 Recommendations for Paleobiogeographic Nomenclature

There was also an attempt at regulating paleobiogeographic nomenclature (Cecca and Westermann 2003). This initiative, supported by the "Friends of Paleobiogeography," is not a code proper, but it includes four recommendations that appeared to have good support among the Friends following a survey conducted within this international group.

The first recommendation states that "Biogeographic Units/ Biochoremas are defined as dynamic units that **change in range and rank through time**. They are based on the overall endemism of biotas." The change in rank through time of biochoremas (the name of biogeographic units recommended

under that proposal) is without equivalent in RN because taxa do not change in rank over time, even if their biodiversity changes drastically through time. For instance, Lissamphibia was represented (by definition) by only two lineages at its origin, sometime between the Late Carboniferous and the Late Permian (Pyron 2011), but it now contains about 8,605 recognized extant species (data from Amphibiaweb, searched March 28, 2023). Despite this spectacular diversification, the rank of Lissamphibia does not vary over geological time; it is typically considered a subclass (e.g., Dubois et al. 2021: 5), from its origin to the present. This is reminiscent of the change in rank of stratigraphic units under the *NAsc*, except that the latter allows changes in rank through space, rather than time (see Section 5.2).

The second recommendation lists the ranks: "(Superrealm), Realm, Subrealm or Region, Province and Subprovince." It also states that "Ranking is according to **endemism, duration** and **range**." This is problematic to the extent that in the absence of a precise weighting scheme and quantification of these three criteria, ranking is presumably a highly subjective operation, as it is under RN.

The third recommendation indicates that typification is through type area and type age. Consideration of the time dimension is a notable difference from the *ICAN*, but is unsurprising because time is crucial to describe any paleontological entity.

The fourth recommendation is more complex and deserves to be quoted in full:

> The nomenclature of Biogeographic Units/ Biochoremas should follow the following guidelines: names should be geographic or geologic, not taxonomic. Synonymy and homonymy apply to all ranks, with priority from 1911, the publication year of Uhlig's seminal work on Jurassic and Cretaceous marine palaeobiogeography. The long misuse of names may serve to invalidate them. To provide nomenclature stability, names may only change at extreme change in distributional range or composition of biotas, or significant plate movements.

This recommendation thus makes four important points.

First, the names of biochoremas are inspired from geographic or geologic names rather than from taxonomic terms (in contrast to biozones, for instance).

Second, the statement that "Synonymy and homonymy apply to all ranks," combined with the decision that biochoremas change in rank through time implies that priority applies regardless of rank, an interpretation supported by several statements in Westermann (2000a, 2000b). If this interpretation is correct, ranks play a minor role in paleogeographical nomenclature, as (probably) under the *ICAN*, but in contrast to RN. This appears to correspond established usage because Servais et al. (2013: 25) stated that "some palaeobiogeographers use a range of biogeographical units without a precise order or rank. Terms such as 'province', 'realm' or 'subprovince' are often used interchangeably without any consistency." Contrary to RN, in which sister-group taxa should be of the same rank (Dubois et al. 2019: 29), there is no such relationships within biochoremas, as illustrated by this quote from Westermann (2000b: 53): "There was no global symmetry in the Mesozoic: no southern complement of equal rank to the 'Euroboreal' or 'Panboreal Superrealm' existed." Thus, while boreal biochoremas were included in a Superrealm according to Westermann (2000b), the meridional biochoremas ranked no higher than Realm, which is consistent with the fact that the Superrealm is not a compulsory rank in this proposal. In fact, one could even wonder if this proposal featured ranks simply because their use in paleobiogeography had been established by Uhlig in his influential paper published in 1911 (Westermann 2000b: 49). Such a weak role for ranks (linked to the change in rank of biochoremas through time) may yield the worst of both worlds and complicate nomenclatural decisions. Indeed, various passages in Westermann (2000b), including a section entitled "Ranking problems," show how the author struggled with nomenclatural problems linked with subjective ranks, and this can be illustrated by this representative quote (Westermann 2000b: 56): "If the East Pacific Realm is reduced to Subrealm rank (unless the 'Tethys—Panthalassa Superrealm' is recognized), the status of 'North-Cordilleran' at Realm-group rank becomes uncertain." Surely, a better nomenclature for biochoremas could be devised!

Third, the statement "The long misuse of names may serve to invalidate them" suggests that usage plays a greater role (compared to priority) than under RN, in which the initial definition is of paramount importance unless the name was subsequently forgotten for a long time and an alternative name gained general acceptance (see Chapter 2).

Fourth, the statement that "names may only change at extreme change in distributional range or composition of biotas, or significant plate movements" indicates an intention to stabilize delimitation using biological attributes or geological events. This mirrors an earlier statement by Westermann (2000a: 2), which indicated that some biochoremas "may disappear at mass extinction and major geotectonic or eustatic events." A concern for objective biochorema delimitation is also obvious in the statement (Westermann 2000a: 3) that "boundaries may be defined statistically by clusters of range endpoints of individual taxa." This again differs rather sharply from RN, which is not designed to stabilize taxon delimitation, as stated in Principle 2 of the *Zoological Code* (see Section 2.4.1).

The rules proposed by Cecca and Westermann (2003) were a small subset of the earlier proposal by Westermann (2000a), which provides many clarifications and enlightening background information. Westermann (2000a: 1) justified this initiative by the fact that "A host of biochore [= "biochorema"] names are being added annually, often without concern for historic validity, usefulness or the broader context," which is reminiscent of the situation of RN before the first codes were enforced. Some regulation is obviously needed to facilitate communication because, according to Westermann (2000a: 2), "Hundreds of Provinces have been named for the Mesozoic alone." Westermann (2000a: 2) also clarified that "To solve problems of classification in paleobiogeography, the *Friends* have to extend their work beyond the marine biomes to the freshwater and terrestrial biomes, and include the Present."

This indeed seems necessary to facilitate studying the emergence of extant biochoremas; also, all RN codes regulate simultaneously extant and extinct taxa, so it would make sense for biogeography to do the same, even though both initiatives have been supported by separate communities working on extant (*ICAN*) and ancient (Cecca and Westermann 2003) biochoremas, obviously without much communication with the other community.

Typification of biochoremas was dealt very briefly by Cecca and Westermann (2003), which left considerable ambiguity. Westermann (2000a: 4) was more explicit in stating that:

> The preferred choice of chorotype is a biochore of lower rank, i.e. equivalent to the designation of type species for genera in zoological (and botanical) nomenclature. For the lowest rank or if an enclosed lower-ranking biochore is unavailable, a geographic or geologic region is to be designated—just as there are holotypes (type specimens) in zoological nomenclature. The preferred choice of chronotype is the stage (age), analogous to the stratotype in stratigraphy.

However, it is unclear if this aspect of typification was covered in the survey performed among the Friends, so there is no guarantee that this represents more than the preferred choice of Westermann.

### 5.3.1.4 Persistent Absence of Regulation of Biogeographic Nomenclature

Westermann (2000a: 2) explained that the task of regulating biogeographical units "is made much more difficult by the curious fact that neither paleo- nor neobiogeographers are organized in any formal groupings or societies, nationally (so far as I know) or internationally—an exception among active disciplines." This indeed may explain why biogeographic nomenclature remains effectively unregulated because Servais and Sintubin (2009: 104) offered a similar explanation for the lack of standardization in the nomenclature of geodynamic and paleogeographical entities. Westermann had founded the "Friends of Paleobiogeography," but this informal association appears to have ceased to exist long ago. Similarly, the Systematic and Evolutionary Biogeographical Association (SEBA) was established in 2006 to promote the *ICAN*, but this association appears not to have been active in a long time given that its website was last updated on February 28, 2006, according to its website (http://biogeographyportal.free.fr/; consulted on December 22, 2022).

Servais et al. (2013: 32) commented positively on the proposal by Cecca and Westermann (2003); they stated that "Although not yet accepted as a standard, several palaeobiogeographers have expressed a consensus by using the terminology proposed by Westermann (2000a) and Cecca and Westermann (2003)." This may have been overly enthusiastic given that a Google Scholar search (performed on December 7, 2022) recovered only 42 citations of Cecca and Westermann (2003), whereas a search for "palaeobiogeography" for papers published after 2003 yielded about 16,900 results (the variant "paleobiogeography" yielded 15,100 citations). Thus, this proposal appears to have been accepted by a fairly small proportion of paleobiogeographers, or at least, to have been enforced in a small minority of the works of paleobiogeography (some of which are published by paleontologists whose core expertise lies elsewhere, especially in systematic paleontology). This is congruent with an earlier statement by Servais and Sintubin (2009: 106) that "most Palaeozoic palaeontologists continue to use terms that are not necessarily accepted generally" and that do not follow the recommendations of Cecca and Westermann (2003), notably by naming paleobiogeographic entities after taxa rather than geographical names.

The low acceptance rate of the simple recommendations on biogeographical nomenclature of Ebach et al. (2008) and those on paleobiogeographical nomenclature of Cecca and Westermann (2003) leaves this field in a situation reminiscent of RN in the 19th century before the advent of the codes (see Section 2.1). This situation was aptly summarized by Servais et al. (2013: 28):

> While many scientific disciplines have nomenclatural rules, palaeogeographers and palaeobiogeographers commonly use a terminology based mostly on traditional use rather than being based on a clearly established nomenclatural code. This is partly due to conceptual confusion between biotic entities (endemic areas, biogeographical units) and physical entities (plates, terranes etc.), which exists in most palaeobiogeographical studies. However, every scientific discipline requires a precise and simple terminology or nomenclature that should be used consistently by all scientists. **Only such a code allows a standardized terminology** that avoids misunderstandings and erroneous interpretations.

Unfortunately, a decade after this quote was published, no significant progress appears to have been made on this front.

## 5.3.2 Geopolitical Units

### 5.3.2.1 Ranks in Geopolitical Units

Geopolitical entities show an interesting parallel with RN in having long been ranked, even though the names of these ranks and their exact number are not universal. From the largest to the lowest ranks, these typically include empires (not present in all regions), countries (many of which, like France, were formerly kingdoms, and several of which still have royal families, like Spain, Sweden and the UK), and depending on countries, states, provinces or departments (which are roughly equivalent), cities and smaller units denoting neighborhoods within a given city.

Contrary to RN, objective criteria are available to attribute geopolitical units to at least some of these ranks. For instance, countries are typically the largest geopolitical units with enough independence from other such units to make their own laws, with the possible exception of countries that belong to empires, in which case this independence may be more limited (those belonging to the European Union are in a similar situation, even though the EU is not an empire). Lower-ranking

units (such as provinces) may also promulgate laws, but these must be compatible with those of their respective countries. Cities typically encompass a densely populated area separated from other such units by more sparsely populated areas (forests, cultivated fields and so on dotted with villages), although with the spectacular demographic growth of human populations in the last few centuries, some neighboring cities are now contiguous, or have fused, as happened to Paris, which has assimilated several previously neighboring cities, such as Montmartre. Cities normally have a city hall, and their authorities organize local infrastructure for waste disposal, health care and transportation (bus, trams and subways, for instance). By contrast, in RN, only the specific level can conceivably be argued to be attributed through objective criteria, such as gene flow, but this remains controversial (see Sections 3.3.2 and 6.2.2.1).

### 5.3.2.2 Typification of Geopolitical Units

Geopolitical units further resemble taxa under RN in having vague analogues of types, namely capitals, at least for countries and states (or provinces or departments). Capital cities, like types under RN, seem to play a defining role in geopolitical entities because despite widely fluctuating borders through history, such capitals are usually retained in a given state and, if lost, this is only temporarily, and much effort it made to regain them. For instance, since the 4th century, Paris has been the capital of France (though occupied by other countries for a few decades in the meantime), despite the large variations in the location of the French borders, which occurred notably because of the 100 Years War (1337–1453) with England, and over longer periods of time because of conflicts with the Germanic Empire and Germany. Similarly, the Byzantine Empire, also called the Eastern Roman Empire, has had Byzantium (renamed Constantinople in 330 AD) as the capital city, and despite the widely variable size of this empire over its millennium-long history (330–1453 AD), Constantinople was occupied by foreign powers for only a few years (1203–1261). Its definitive fall to the Ottomans (1453) marked the end of that empire.

The fact that capital cities can be even temporarily excluded from countries is a notable difference from RN because in the latter, a type cannot be excluded from its taxon, even as a brief, temporary solution. This difference can be explained partly because geopolitical units are also partly defined by their inhabitants and rulers. Thus, in the Middle Ages, France was a kingdom, and the French ruling family resided in the part of its kingdom over which it retained control through the 100 Years War. The fact that the French ruling family subsequently regained control of Paris and made it the capital of France could be argued to have enhanced historical continuity of the French state. Perhaps presence of the rulers is closer to types in geopolitics, but the fact that rulers normally live in the capital city hampers determining how typification (or rather, its equivalent because, as far as I know, geographers do not conceptualize territorial units through typification) is determined in geopolitics.

Another difference between geopolitical entities (especially countries) and taxa under RN is that countries occasionally change capitals, without this resulting necessarily from external influences (from other countries), and without giving cause to reconsider the identity of the country. For instance, through its history, Russia has had several capitals: Kiev (now capital of Ukraine), from to about the mid-10th century to the mid-12th century (when the Mongol invasion gave the final blow to the Kievan kingdom that had already declined sharply), Vladimir (about 200 km east of Moscow) from about the mid-14th century (after the eviction of the Mongols) to 1533, then Moscow, from 1533 to 1713, then Saint-Petersburg, from 1713 to 1918, and, finally, Moscow again from 1918 till now. Of course, Russia was not really a country throughout this entire period; rather, initially, the territory where European Russia and Ukraine are now located was then occupied by a constellation of small Slavic states, and both countries (Russia and perhaps more properly Ukraine) can be seen as somehow derived from Kievan Russia. Thus, Kievan Russia was not a proper country as now conceptualized. Nevertheless, the changes in capitals between Moscow and Saint-Petersburg occurred within a fairly centralized state that certainly qualifies as a country. Similar changes are known in recent history of various countries, such as Brazil, which has had three capitals in the last three centuries: Salvador (1549–1763), Rio de Janeiro (1763–1960) and Brazilia (1960–). Thus, the defining role of capitals in geopolitical units appears to be weaker than that of types in RN. This is hardly surprising because countries, at least in their modern conception (this was not always thus), derive their identity from cultural, linguistic, religious, ethnic and historical (factual or mythical) factors, in addition to a capital city, which is not necessarily of paramount importance.

### 5.3.2.3 Delimitation of Geopolitical Units

Geopolitical units differ from taxa under RN in having long been explicitly delimited. The sharpness of this delimitation has increased over time. Thus, the antique Roman Empire focused on controlling people and cities rather than defining fixed, sharp borders (Haselsberger 2014: 507). Similarly, for much of the Middle Ages, Europe was composed of dynastic realms that were neither firmly united nor sharply delimited. The Peace of Westphalia (1648) both coincides with the end of the Holy (Germanic) "Roman Empire" and a marking point in the geopolitical history of Europe that allowed the emergence of the current concept of "nation-state." While the events altered people's understanding of state borders, "exclusive border lines" (rather than permeable ones) were established only around the early 19th century (Haselsberger 2014: 507).

The great importance of delimitation of geopolitical units was explained eloquently by Haselsberger (2014: 510):

> A world without borders and boundaries is a **utopia**. Unbounded functional activities, be they social, cultural or economic, would be formless. And no territory could be administered unless it was clearly demarcated. In short, we live in a world full of compartments that enable us to manage our lives collectively. How could we acquire a national identity

if we were not able to distinguish between "us" and "them"? How could we draw up a plan or "develop" a given territory unless the appropriate region was defined beforehand? Thus, borders are essential to our everyday life and attempting to eradicate them is a meaningless project. But not all borders are perfect fits.

Since the 1930s, four kinds of processes of border demarcation have been recognized: antecedent, subsequent, superimposed and natural (Newman 2006: 174). Antecedent borders are delimited before human settlement and surround an area that was previously unsettled, often wild lands (for instance, primary forests or natural prairies). Subsequent borders delimit existing settlements according to ethnic or cultural patterns. Superimposed boundaries are imposed by an external (typically colonial) power in an area that it controls, often with little consideration of ethnic or tribal patterns. Many superimposed borders were drawn by European colonial powers in Africa and Asia and negatively impacted them (Newman 2006: 176), especially in the Middle East (Biger 2012). Natural borders follow physical features of the landscape, such as mountain ranges, rivers, deserts and oceans (e.g., Purcell 2013; Haselsberger 2014: 509).

It is perhaps no coincidence that superimposed borders are both the least beneficial (for the nearby populations) and the most artificial of the four kinds because they follow neither ethnic nor cultural patterns, and typically do not match landscape features either. Good examples of the impact of poor, artificial border choice on local populations are mentioned by Biger (2012: 62):

> National desires, colonial aspirations, imperial needs, all led to the unrealistic picture of the boundaries in the Middle East. Several new nations emerged as the outcome of these processes (Iraq, Syria, Trans Jordan, later Jordan), while old nations (The Kurd, the Armenians, the Druzs, the Alawis, etc.) never got a state of their own. The main players in the boundary game were the British and the French, who were the dominant forces in the Middle East between 1918 and 1950.

The extremely negative consequences that such artificial (superimposed) borders can have on peoples is illustrated by the history of the Middle East since the 19th century, as summarized by this quote (Biger 2012: 65–66):

> The Middle East was the last world region in which modern boundary lines were established. There are still some disputes concerning the delimitation of those lines. The most controversial are the boundaries of Israel. Only its boundaries with Egypt are settled, while nearly two segments of its boundary with Jordan are settled. The peace talks between Israel and Syria and Israel with the Palestinian authority were suspended because of the boundary problem. The Palestinian authority still waits for its first established international boundaries, which will **create a new state** in the Middle East.

This quote also illustrates the importance of international boundaries for the existence of states, given that such a boundary was deemed required to "create" a Palestinian state.

As we say, the rest is history; the Middle East is still suffering from these poorly selected borders today (e.g., Lacoste 2012). A vague parallel with RN can be made with the drawbacks of recognizing paraphyletic taxa, which creates problems in studies on the evolution of biodiversity (see Section 6.1), but fortunately does not result in blood baths.

Natural borders were no longer fashionable among geographers and political scientists after the 1940s because by then, borders were seen as social constructions and hence, were formed by people or, most accurately, by their governments (Newman 2006: 174). Thus, potential natural borders were convenient when they suited rulers, but were ignored when this was not judged advantageous. Nevertheless, such borders were undoubtedly established at least well into the 19th century, as reported by Biger (2012: 63):

> The outcome of the large number of international boundaries, which were established during the 19th century by the European powers all over the world was the acceptance of these lines, which are based on physical geographical features, as the best lines (Curzon 1907). Military requirements, philosophical ideas of the good in nature, the permanent existence of the river, mountain range, desert and other physical features and the simplicity in recognition, all these made those lines the best choice.

Among such natural borders, rivers figure prominently. As summarized by Popelka and Smith (2020: 294), the use of rivers as borders has several advantages, including "their inherent linearity, perceived impassibility, and the fact that they physically incise and divide the landscape." It is thus no surprise that rivers were used as important landmarks to delimit boundaries by many ancient empires, including the Neo-Assyrians and the Romans (Popelka and Smith 2020: 294), even though they also served more broadly as landmarks and even as means to enhance communication and commerce (Purcell 2013).

The use of natural borders remains pervasive. Popelka and Smith (2020: table 2) conservatively estimated that coastal borders (delimited by oceans and seas) represent 86% of the current international borders (in km), 57% of the "Level 1" borders (between states or provinces) and 30% of the Level 2 (county or local) borders. Rivers account for another 3% of the international borders, 7% of the Level 1 borders and 9% of the Level 2 borders. This leaves a measly 11% of the international borders that are not either coastal or defined by rivers.

The pervasive use of natural borders can be illustrated by several examples. Many geopolitical units correspond to islands or archipelagos, or even entire continents. These include countries (for instance, Australia, Cyprus, Japan, Madagascar and New Zealand), states or provinces (for instance, Newfoundland and Prince Edward Island in Canada; Crete and the Ionian islands in Greece; Sardinia and Sicily in Italy; the Balearic and Canary islands in Spain; Hawaii in the USA), and even cities (such as Laval, in the province of Quebec in Canada). Examples of rivers and lakes that have historically delimited territories include the Amur River, between Russia and China; the African Great Lakes, two of which (Tanganyika and Malawi) mark part of the western border of Tanzania; the Congo River, which constitutes part of the border between the Congo-Brazzaville and the Democratic Republic of Congo; the North American Great Lakes, which delimit part of the border between Canada and the USA; the Rhine, which delimited France from Germany for many centuries; the Rio Grande (called Rio Bravo del Norte by the Mexicans) between Texas (USA) and Mexico; and the Rio Uruguay, which delimits parts of the borders between Argentina, Brazil and Uruguay. Mountain ranges have also been important geopolitical landmarks: the Alps between Italy (to the South) and France, Switzerland, Germany and Austria (to the North); the Pyrenees between France and Spain; the Andes between Argentina and Chile; and the Himalaya, which define parts of the norther border of India, among others. Less often, more diffuse natural barriers have delimited states. For instance, antique Egypt was isolated from other countries to the west by the Sahara.

In some cases, when natural landmarks were unavailable to delimit countries, signaling posts (sometimes called "boundary stones") were designed. This presumably fits into the subsequent type of border in Newman's (2006) classification, and it represents a "man-made feature" in Biger's (2012) classification. The custom of using manmade territorial boundary markers harks back at least to 3000 BCE (the dawn of history, as the first written records known from both Egypt and Mesopotamia are from that time), as documented for the Sumerians in Mesopotamia (Winter 1985: fig. 17).

More recently, technological advances have allowed delimiting large territories without requiring physical posts. For instance, most of the border between Canada and the USA between the Great Lakes and the Pacific Ocean follows the 49th parallel (degrees North of the equator), and within Canada, several provinces extend north up to the 60th parallel, beyond which are located the three territories (Yukon, Northwest Territories and Nunavut).

Explicit delimitation constitutes a major difference between geopolitical units and taxa under RN. The borders of most states changed over time, but at any given time, delimitation has been fairly precise since a fairly remote antiquity, at least where it mattered (where cities, cultivated fields, water sources and mines occurred, for instance). An analogy can be drawn with clades, which diversify or dwindle (or even become extinct) over time, and the limits of these clades are precisely known under PN, but not under RN. In this respect, geopolitical territorial delimitation is more similar to PN than to RN.

Many wars were triggered by border disputes, with the leader (and possibly people generally) wishing to secure more territory at the expense of another country, kingdom or city-state. More globally, Haselsberger (2014: 518) indicated that most borders "are imposed on the world's surface through violence, force and intimidation in the course of wars, conquests and state formation." This phenomenon is ancient, as shown by the Stele of the Vultures (**Figure 5.1**), now displayed in the Louvre Museum (Paris), which relates a territorial war that occurred around about 2460 BCE between the Sumerian city-states of Umma and Lagash (Winter 1985).

**FIGURE 5.1** Detail from the Stele of the Vultures (about 2460 BCE), showing soldiers led by Eannatum, the ruler of Lagash, marching against the forces of the neighboring city-state of Umma. Picture by the French diplomat and archaeologist Ernest de Sarzec (1832–1901); public domain in France, in the United States and in most countries.

*Source:* Downloaded from Wikimedia Commons at https://en.wikipedia.org/wiki/Stele_of_the_Vultures#/media/File:Stele_of_the_vultures_(war).jpg.

Thus, such territory-motivated wars hark back at least back to the middle of the 3rd millennium BCE (they may even be much more ancient because we simply lack records for earlier periods), and they continue today, which shows that great importance has been given to precise delimitation of geopolitical units for more than 4,000 years. This is unsurprising because territory is precious. Other, less dramatic territorial disputes and even "negative reactions to immigration, are directly related to the demarcation and enactment of borders" (Popelka and Smith 2020: 294).

Conversely, the benefits of more objective, fairly stable borders (as natural borders arguably are) have been documented. For instance, Rutherford et al. (2014) were able to model accurately where violence occurred in Switzerland through the distribution of linguistic and religious groups, of administrative and natural (such as high mountains and lakes) borders. This model, which had previously been applied to India with good results (Lim et al. 2007) suggests that the administrative and natural borders explain well why Switzerland experiences little violence. Conversely, the model also explains why the

former Yugoslavia experienced far more violence (Rutherford et al. 2014); this results both from a poor choice of administrative boundaries, which do not match boundaries between ethnic, linguistic or religious groups, and because topographical features are not steep enough and do not delimit these groups well enough. An important conclusion of Rutherford et al. (2014: 2) is that "Violence arises due to the structure of boundaries between groups rather than as a result of inherent conflicts between the groups themselves." Thus, clearly, in geopolitics, the existence and location of boundaries is of great importance.

The precision of geopolitical borders and the energy spent on making these respected reflect the value that we attach to territory. But this begs the question of why proponents of RN claim that nomenclature should be applied to undelimited taxa (whose limits are reassessed in each study, even in the absence of changes in the phylogeny). Should we understand that these systematists consider that the main product of their work (nomenclatures and taxonomies) is worthless or trivial?

## 5.4 Chemistry and Physics

Comparisons between biological nomenclature and chemistry and physics are more difficult because the atoms that we will use as examples for these fields form classes rather than individuals (although taxa have been viewed as classes too; see Section 6.2.1). Current atom classification is usually summarized in the periodic table of the elements. Development of this table is often credited to the Russian chemist Dmitri Ivanovich Mendeleev (1834–1907), who published an influential paper (Mendeleev 1871) that proposed such a table (fairly different from the current version). However, as for most major scientific achievements, several other scientists made major contributions toward this goal, both before and after Mendeleev's main contributions to this problem. As summarized by Gordin (2019), these include the symbols used to designate atoms, first proposed by the Swedish chemist Jöns Jacob Berzelius (1779–1844), the work on the determination of atomic weights by the Italian chemist Amedeo Carlo Avogadro (1776–1856), the discovery of electrons by the British physicist Joseph John Thomson (1856–1940) and the work on atomic numbers by the English physicist Henry Moseley (1887–1915).

The current classification and nomenclature of chemical elements thus matured over many decades and is useful to facilitate communication in chemistry and physics. In this system, atoms are classified by their atomic number, which is the number of protons contained in their nucleus. These elements are also sorted into groups defined mostly by their chemical properties. The number of electrons in their highest orbital is an important factor to determine their chemical properties, but the relationship between electronic configuration in the highest orbital and chemical properties (and hence, attribution to chemical groups) is not nearly as straightforward as commonly portrayed (Scerri 1997: 553). Nevertheless, lithium (Li) has three protons, sodium (Na) has 11 and potassium (K) has 19, and these three elements belong to the same chemical group (alkali metals, which also includes heavier elements, such as Rubidium) partly because they possess a single valence electron, which makes them highly reactive.

Atoms of a given number can be subdivided into isotopes, which differ only by the number of neutrons in their nucleus, and isotopes are designated by the chemical symbol of their atom preceded by a superscript number that designates the number of nucleon (protons and neutrons). For instance, lithium exists in two naturally occurring isotopes, $^7$Li (with four neutrons, which represents 95% of the naturally occurring lithium) and $^6$Li (with only three neutrons), along with much less stable, very ephemeral radioisotopes, such as $^4$Li, $^8$Li and $^9$Li. The classification of atoms can thus be viewed as having three levels, which are, from the lowest to the highest, the isotopes, elements and groups of elements.

Contrary to RN, the three levels of this classification of elements are defined by objective criteria: number of nucleons for the isotopes, number of protons for the atoms and chemical properties partly reflecting the number of valence electrons for the groups of atoms. Thus, there is no ambiguity in how to "rank" a newly discovered isotope or element, contrary to a newly discovered taxon, which could be considered a new species, genus or family, depending on the systematist.

The meaning of names under this nomenclature is clear because isotopes and elements are clearly delimited, and groups (families) of elements are perhaps slightly less clearly delimited, but delimitation of these groups does not seem to be controversial, contrary to taxon delimitation under RN. Perhaps proponents of RN should try to convince chemists and physicists of the benefits of a system in which Li sometimes designates only lithium (or possibly, only $^7$Li), sometimes lithium plus sodium, and, sometimes, lithium, sodium and potassium, and in which all these meanings are simultaneously and indefinitely valid! A consensus on what Li actually means could emerge spontaneously (or not) and perhaps temporarily. This would be approximately equivalent to the fuzzy meaning of most taxon names under RN.

## 5.5 Folk Taxonomy

As we saw in Chapter 1, folk taxonomies may be ranked. Such rank can be represented by size of circles in Venn diagrams (Berlin 2014: 43, figure 1.7), which have also been used in biological taxonomy, although typically without a simple relationship between diagram size and rank. However, there are obvious differences between the putative folk taxonomic ranks and principles of RN as implemented in the RN codes.

First and most importantly, folk taxonomic levels (as opposed to the taxa that they include) are **never named** (except by the anthropologists who study them); this is perhaps the main reason why some anthropologists (Ellen 1998; Hunn 1998), cognitive scientists (such as Hatano 1998) and some systematists (Mishler and Wilkins 2018) doubt their existence. At best, these ranks may be present in the subconscious of the indigenous peoples, who would then use them without realizing it.

Second, there are few, if any, mandatory categories. Arguably, the level named "generic" (Berlin et al. 1973) or generic-species (Atran 1998) is the only such mandatory rank. Taxa of lower ranks are not ubiquitous. Thus, many generics are monotypic; among in Tzeltal Maya, Berlin et al. (1973: 221) report that 398 out of 471 generic taxa are monotypic (they do not include distinct named species or varieties). Furthermore, generics are not always included in a taxon of the next higher rank (either an unnamed intermediate category or a life form); these are the unaffiliated taxa, which belong to no life form, or may alternatively be viewed as monotypic life forms (Atran 1998: 593). This is as if we had a single name to refer to a genus and species, and occasionally (for unaffiliated generics), that name also applied to a class or a phylum.

Folk taxonomies differ in other fundamental ways from RN. Thus, ethnobiological taxa show basic (or core) and extended (peripheral) ranges (Berlin 2014: 41). The core taxa are somewhat analogous to type-taxa of RN, and the extended range is composed of taxa that resemble the core taxa. Taxa in the peripheral range are often named "like core taxon X," a practice that is not formalized in RN, except for provisionally unnamed taxa that may contain the abbreviation "aff." to indicate affinities with a given taxon (typically a species). However, core taxa in folk taxonomies are thought (by the natives) to be more central (or typical) of a taxon, which is more reminiscent of the ideal types (harking back to Plato) than to nomenclatural types used under RN. Furthermore, taxa may be in the peripheral range of two core taxa simultaneously. Under RN, the closest equivalent is the treatment of hybrids (under PN hybrids may belong to two otherwise non-overlapping clades). The boundaries of the peripheral ranges of folk taxonomies are fuzzy (Berlin 2014: 42), just like the boundaries of taxa under RN.

Beyond the question of ranks, folk taxonomies show that well before the advent of cladistics, people did not delimit taxa purely by appearance, contrary to pheneticists (Sneath and Sokal 1962). Thus, Wierzbicka (1992: 4) wrote:

> it is becoming increasingly clear that the notion of category (discrete category) [in folk taxonomy, meaning "taxon"] plays a fundamental role in human thinking and **cannot be reduced to the notion of resemblance**. To put it in terms of the theory of universal semantic primitives, the notion of kind ("X is a kind of Y") cannot be reduced to the notion of like ("X is like Y").[1] Similarity is not necessarily an adequate indicator of conceptual closeness (e.g., an Alsatian may be more "similar" to a wolf than to a Pekinese, but the folk category "dog" disregards this).

For systematists, dogs are domesticated descendants of wolves, which is a historical, phylogenetic concept. For pre-literate people (and probably also a good proportion of literate laymen), dogs are presumably defined partly by functional, utilitarian criteria: dogs are domesticated canids that can help hunting game, that can guard a tent (or house), that you feed, keep for company and so on.

Folk taxa show evidence of delimitation. Thus, Atran (1998: 548) concluded that folk taxonomies are comprised of "fairly well delimited groups of plants and animals." However, this delimitation is often fuzzy because not all people of a given speech community delimit a given taxon in exactly the same way (Pawley 2011: 426), and among several language communities, this variability in delimitation is even greater. Pawley (2011) documented this in various reflexes (versions) of the taxon *ikan, which can be translated loosely as "fish." Folk taxonomies also show evidence of polysemy (a single name being applied to two nested taxa). This typically occurs (within a given language community) when a given low-level taxon appears to form the core of the higher-level taxon, whereas the latter includes a variable number of peripheral taxa (Pawley 2011), as explained in Section 1.1.1.6. I am unaware of a study that has compared the precision of delimitation of taxa in folk taxonomies and under RN, and this would be a challenging task, but anecdotal evidence suggests that in at least some cases, folk taxa have had a more stable delimitation that taxa under RN. Section 3.4.1 provided such an example for the English folk taxon "ape" compared to *Hominidae*, but it would be interesting to make such comparisons more systematically for several taxa and among several linguistic communities.

Thus, while the widespread but increasingly controversial idea that folk taxa are ranked lends some historical legitimacy to RN, the lack of explicit reference to ranks by pre-literate peoples indicates profound differences between RN and folk taxonomies. More importantly, folk taxa are apparently not as precisely delimited as taxa under PN (not all natives place the boundary at the same place), but they do suggest a reasonably good delimitation, and plausibly a stronger relationship between a name and content (in terms of included taxa and phenotypic range) than under RN. This last point deserves further investigation.

## 5.6 Conclusions from Comparisons between the Nomenclatures of Various Fields

The previous review allows us to draw some conclusions about the use of absolute ranks (categories), the use of types and delimitation in various fields, and comparisons allow to determine to what extent the implementation of RN in the current codes (especially in botany and zoology) is representative of academic culture in general.

Ranking is used in some fields, often informally, as seen in geography and possibly in folk taxonomies, but the ranks appear to play a less important role in these nomenclatures than under RN. In the nomenclature used in folk taxonomies, these ranks, if they exist at all, are not named by the users (mostly native peoples) and may thus play a minor role in this kind of nomenclature. In other fields, the rank of a given entity may vary in space (stratigraphic units) or time (paleobiogeographical units). No field gives as much importance to ranks as RN.

The use of types varies among disciplines. In some fields such as folk taxonomies, some "core" taxa exist in the concept of a higher-ranking taxon, but this is necessarily an informal notion. In other fields, such as biogeography, the type of a high-ranking unit is not necessarily a lower-ranking unit, contrary to RN (for taxa above the species level). But this conclusion concerns the few proposals that have been made to regulate this nomenclature, and these proposed rules have not gained general acceptance. In yet other fields, such as geopolitics, true types may not exist, though it could be argued that capital cities and possibly wherever rulers reside may perhaps be conceptualized as a vague analogue of type localities (in RN) or internal specifiers (in PN).

This review of various fields of human knowledge shows that, in most cases, entities are fairly precisely delimited. The notable exceptions are folk taxonomy and biogeography. However, both exceptions are easily explained. Folk taxonomies rest on pre-scientific knowledge and methods. Thus, tradition may weight more in how folk taxa are conceptualized than the entities defined in the other fields discussed here, notably because folk taxa have not been reevaluated following scientific studies. Biogeography is still unregulated because the two main proposals reviewed here (*ICAN* and the proposals of Cecca and Westermann 2003) have yet to gain acceptance, a situation that may be partly explained by the lack of biogeographic societies (see Section 5.3.1.4), and these circumstances have not favored a thorough discussion on how biogeographic units could be defined and precisely delimited. Geopolitical units are fairly well delimited, notably for obvious economic and practical reasons, but this delimitation varies in time. Stratigraphic units and the chemical elements are precisely defined, and these limits appear to be fairly stable through time; this may reflect the greater maturity of these fields. Thus, RN appears to be fairly isolated in being designed to deliberately not delimit taxa. The emergence of PN appears to reflect the general trend observed in other fields of human knowledge (especially in stratigraphy and geopolitics) toward more precise delimitation over time.

# REFERENCES

American Commission on Stratigraphic Nomenclature. 1961. Code of stratigraphic nomenclature. American Association of Petroleum Geologists Bulletin 45:645–655.

Atran, S. 1998. Folk biology and the anthropology of science: Cognitive universals and cultural particulars. Behavioral and Brain Sciences 21:547–569.

Bardet, N. 2012. Maastrichtian marine reptiles of the Mediterranean Tethys: A palaeobiogeographical approach. Bulletin de la Société géologique de France 183:573–596.

Barker, N. P., P. H. Weston, F. Rutschmann, and H. Sauquet. 2007. Molecular dating of the 'Gondwanan' plant family Proteaceae is only partially congruent with the timing of the break-up of Gondwana. Journal of Biogeography 34:2012–2027.

Benison, K. C. 1997a. Field descriptions of sedimentary and diagenetic features in red beds and evaporites of the Nippewalla Group (Middle Permian), Kansas and Oklahoma. Kansas Geological Survey, Open-File Report 97–21:1–62.

Benison, K. C. 1997b. Acid lake and groundwater deposition and diagenesis in Permian red bed-hosted evaporites, midcontinent, U.S.A. PhD dissertation, University of Kansas, 480 pp.

Benison, K. C., J. J. Zambito IV, and J. Knapp. 2015. Contrasting siliciclastic-evaporite strata in subsurface and outcrop: An example from the Permian Nippewalla Group of Kansas, USA. Journal of Sedimentary Research 85:626–645.

Berlin, B. 2014. Ethnobiological classification: Principles of categorization of plants and animals in traditional societies. Princeton University Press, Princeton, NJ, 354 pp.

Berlin, B., D. E. Breedlove, and P. H. Raven. 1973. General principles of classification and nomenclature in folk biology. American Anthropologist 75:214–242.

Biger, G. 2012. The boundaries of the Middle East: Past, present and future. Studia z Geografii Politycznej i Historycznej 1:61–67.

Bjørlykke, H. 1961. Code of stratigraphic nomenclature for Norway. Norges geologiske undersøkelse 213:229–233.

Buchanan, R. and J. R. McCauley. 1987. Roadside Kansas: A traveler's guide to its geology and landmarks. University Press of Kansas, Lawrence, Kansas, 365 pp.

Candolle, A. P. De. 1820. Essai élémentaire de géographie botanique. F. G. Levrault, Strasbourg, 64 pp.

Cecca, F. and G. Westermann. 2003. Towards a guide to palaeobiogeographic classification. Palaeogeography, Palaeoclimatology, Palaeoecology 201:179–181.

Cecca, F. and R. Zaragüeta i Bagils. 2015. Paléobiogéographie. EDP Sciences, Les Ulis, 193 pp.

Cowie, J. W. 1986. Guidelines for boundary stratotypes. Episodes 9:78–82.

Cox, A., R. R. Doell, and G. B. Dalrymple. 1963. Geomagnetic polarity epochs and Pleistocene geochronometry. Nature 198:1049–1051.

Cragin, F. 1896. The Permian system in Kansas. Colorado College Studies 6:1–54.

Curzon, G. N. and L. Kedleston. 1907. Frontiers. Oxford University Press, Oxford, 59 pp.

Dowding, E. M., M. C. Ebach, and E. V. Mavrodiev. 2019. Temporal area approach for distributional data in biogeography. Cladistics 35:435–445.

Dubois, A., A. M. Bauer, L. M. Ceríaco, F. Dusoulier, T. Frétey, I. Löbl, O. Lorvelec, A. Ohler, R. Stopiglia, and E. Aescht. 2019. The Linz *Zoocode* project: A set of new proposals regarding the terminology, the Principles and rules of zoological nomenclature: First report of activities (2014–2019). Bionomina 17:1–111.

Dubois, A., A. Ohler, and R. Pyron. 2021. New concepts and methods for phylogenetic taxonomy and nomenclature in zoology, exemplified by a new ranked cladonomy of recent amphibians (Lissamphibia). Megataxa 5:1–738.

Ebach, M. C., J. J. Morrone, L. R. Parenti, and A. L. Viloria. 2008. International code of area nomenclature. Journal of Biogeography 35:1153–1157.

Ellen, R. 1998. Doubts about a unified cognitive theory of taxonomic knowledge and its memic status. Behavioral and Brain Sciences 21:572–573.

Fay, R. O. 1964. The Blaine and related formations of northwestern Oklahoma and Southern Kansas. Oklahoma Geological Survey Bulletin 98:1–238.

Foster, T. M., G. S. Soreghan, M. J. Soreghan, K. C. Benison, and R. D. Elmore. 2014. Climatic and paleogeographic significance of eolian sediment in the Middle Permian Dog Creek Shale (Midcontinent US). Palaeogeography, Palaeoclimatology, Palaeoecology 402:12–29.

Goloboff, P. A., C. I. Mattoni, and A. S. Quinteros. 2008. TNT, a free program for phylogenetic analysis. Cladistics 24:774–786.

Gordin, M. D. 2019. Ordering the elements. Science 363:471–473.

Haselsberger, B. 2014. Decoding borders: Appreciating border impacts on space and people. Planning Theory & Practice 15:505–526.

Hatano, G. 1998. Informal biology is a core domain, but its construction needs experience. Behavioral and Brain Sciences 21:575–575.

Hennig, W. 1966. Phylogenetic systematics. University of Illinois Press, Urbana, Chicago, London, 263 pp.

Holdoway, K. A. 1978. Deposition of evaporites and red beds of the Nippewalla Group, Permian, western Kansas. Kansas Geological Survey, Bulletin 215:1–43.

Hunn, E. S. 1998. Atran's biodiversity parser: Doubts about hierarchy and autonomy. Behavioral and Brain Sciences 21:576–577.

IUGS (International Subcommission on Stratigraphic Classification). 1979. Magnetostratigraphic polarity units: A supplementary chapter of the ISSC International Stratigraphic Guide. Geology 7:578–583.

Johnson, K. S. 1991. Guidebook for geologic field Trips in Oklahoma: Northwest Oklahoma. Oklahoma Geological Survey, Educational Publications 3:1–42.

Jordan, L. and D. L. Vosburg. 1963. Permian salt and associated evaporites in the Anadarko Basin of Western Oklahoma-Texas Panhandle Region. Oklahoma Geological Survey, Bulletin 102:1–76.

Lacoste, Y. 2012. La géographie, la géopolitique et le raisonnement géographique. Hérodote 2012:14–44.

Laurin, M. 2008. The splendid isolation of biological nomenclature. Zoologica Scripta 37:223–233.

Laurin, M. 2016. Revue critique de "Paléobiogéographie", par Fabrizio Cecca et René Zaragüeta i Bagils, 193 pages, 26.00€, EDP Sciences, 2015. Comptes Rendus Palevol 15:607–614.

Laurin, M. and R. Hook. 2022. The age of North America's youngest Paleozoic continental vertebrates: A review of data from the Middle Permian Pease River (Texas) and El Reno (Oklahoma) Groups. Bulletin de la Société Géologique de France 193:1–30.

Lehman, T. M. 1994. The saga of the Dockum Group and the case of the Texas/New Mexico boundary fault. New Mexico Bureau of Mines and Mineral Resources Bulletin 150:37–51.

Licht, A., G. Métais, P. Coster, D. Ibilioğlu, F. Ocakoğlu, J. Westerweel, M. Mueller, C. Campbell, S. Mattingly, and M. C. Wood. 2022. Balkanatolia: The insular mammalian biogeographic province that partly paved the way to the Grande Coupure. Earth-Science Reviews 226:103929.

Lim, M., R. Metzler, and Y. Bar-Yam. 2007. Global pattern formation and ethnic/cultural violence. Science 317:1540–1544.

Lucas, S. G. and S.-Z. Shen. 2018. The Permian chronostratigraphic scale: History, status and prospectus. Geological Society, London, Special Publications 450:21–50.

Mendeleev, D. 1871. Die periodische Gesetzmassigkeit der chemischen Elemente. Annalen der Chemie und Pharmacie Supplement 8:133–229.

Mishler, B. D. 1999. Getting rid of species?; pp. 307–315 in R. Wilson (ed.), Species: New interdisciplinary essays. MIT Press, Cambridge, MA.

Mishler, B. D. and J. S. Wilkins. 2018. The hunting of the SNaRC: A snarky solution to the species problem. Philosophy, Theory, and Practice in Biology 10:1–18.

Morrone, J. J., T. Escalante, and G. Rodriguez-Tapia. 2017. Mexican biogeographic provinces: Map and shapefiles. Zootaxa 4277:277–279.

Murphy, M. A. and A. Salvador. 1999. International stratigraphic guide: An abridged version. Episodes 22:255–271.

Murphy, S. J. 2021. Sampling units derived from geopolitical boundaries bias biodiversity analyses. Global Ecology and Biogeography 30:1876–1888.

Nelson, W. J. and R. W. Hook. 2005. Pease river group (Leonardian-Guadalupian) of Texas: an overview. New Mexico Museum of Natural History and Science Bulletin 30 (The Nonmarine Permian):243.

Nelson, W. J., R. W. Hook, and S. Elrick. 2013. Subsurface nomenclature in the Permian Basin (Texas-New Mexico): Lithostratigraphic chaos or fixable problem? New Mexico Museum of Natural History and Science Bulletin 60 (The Carboniferous-Permian Transition):312–313.

Nelson, G. J. and P. Y. Ladiges. 1996. Paralogy in cladistic biogeography and analysis of paralogy-free subtrees. American Museum Novitates 3167:1–58.

Newman, D. 2006. Borders and bordering: Towards an interdisciplinary dialogue. European Journal of Social Theory 9:171–186.

North American Commission on Stratigraphic Nomenclature. 2021. North American stratigraphic code. Stratigraphy 18:153–204.

Norton, G. H. 1939. Permian redbeds of Kansas. AAPG Bulletin 23:1751–1819.

Parenti, L. R., Á. L. Viloria, M. C. Ebach, and J. J. Morrone. 2009. On the International Code of Area Nomenclature (ICAN): A reply to Zaragüeta-Bagils et al. Journal of biogeography 36:1619–1621.

Pawley, A. 2011. Were turtles fish in Proto Oceanic? Semantic reconstruction and change in some terms for animal categories in Oceanic languages; pp. 421–452 in M. Ross, A. Pawley, and M. Osmond (eds.), The lexicon of Proto Oceanic: The culture and environment of ancestral Oceanic society. Research School of Pacific and Asian Studies, The Australian National University, Canberra.

Pendery, E. C. 1963. Stratigraphy of Blaine Formation (Permian), north-central Texas. AAPG Bulletin 47:1828–1839.

Philippe, M., G. Cuny, M. Bamford, E. Jaillard, G. Barale, B. Gomez, M. Ouaja, F. Thévenard, M. Thiébaut, and P. Von Sengbusch. 2003. The palaeoxylological record of *Metapodocarpoxylon libanoticum* (Edwards) Dupéron-Laudoueneix et Pons and the Gondwana Late Jurassic: Early Cretaceous continental biogeography. Journal of Biogeography 30:389–400.

Popelka, S. J. and L. C. Smith. 2020. Rivers as political borders: A new subnational geospatial dataset. Water Policy 22:293–312.

Purcell, N. 2013. Rivers and the geography of power. Pallas. Revue d'études antiques 90:373–387.

Pyron, R. A. 2011. Divergence-time estimation using fossils as terminal taxa and the origins of Lissamphibia. Systematic Biology 60:466–481.

Remane, J., M. G. Bassett, J. W. Cowie, K. H. Gohrbandt, H. R. Lane, O. Michelsen, and W. Naiwen. 1996. Revised guidelines for the establishment of global chronostratigraphic standards by the International Commission on Stratigraphy (ICS). Episodes Journal of International Geoscience 19:77–81.

Ronquist, F., S. Klopfstein, L. Vilhelmsen, S. Schulmeister, D. L. Murray, and A. Rasnitsyn. 2012b. A total-evidence approach to dating with fossils, applied to the early radiation of the Hymenoptera. Systematic Biology 61:973–999.

Ronquist, F., M. Teslenko, P. van der Mark, D. L. Ayres, A. Darling, S. Höhna, B. Larget, L. Liu, M. A. Suchard, and J. P. Huelsenbeck. 2012a. MrBayes 3.2: Efficient Bayesian phylogenetic inference and model choice across a large model space. Systematic Biology 61:539–542.

Rutherford, A., D. Harmon, J. Werfel, A. S. Gard-Murray, S. Bar-Yam, A. Gros, R. Xulvi-Brunet, and Y. Bar-Yam. 2014. Good fences: The importance of setting boundaries for peaceful coexistence. PloS One 9:e95660.

Sauquet, H. 2013. A practical guide to molecular dating. Comptes Rendus Palevol 12:355–367.

Scerri, E. R. 1997. The periodic table and the electron: Although electronic configurations are traditionally invoked to explain the periodic system, their explanatory power remains only approximate. American Scientist 85:546–553.

Servais, T., F. Cecca, D. A. Harper, Y. Isozaki, and C. Mac Niocaill. 2013. Palaeozoic palaeogeographical and palaeobiogeographical nomenclature; pp. 25–33 in D. A. T. Harper and T. Servais (eds.), Memoirs. Geological Society, London.

Servais, T. and M. Sintubin. 2009. Avalonia, Armorica, Perunica: Terranes, microcontinents, microplates or palaeobiogeographical provinces? Geological Society, London, Special Publications 325:103–115.

Sneath, P. H. and R. R. Sokal. 1962. Numerical taxonomy. Nature 193:855–860.

Swineford, A. 1955. Composition and texture of Upper Permian sediments in South-Central Kansas. Kansas Geological Society, Field Conference Guidebook 18:57–59.

Swofford, D. L. 2002. PAUP* phylogenetic analysis using parsimony (*and other methods). Version 4.0a, build 167 (updated on Feb. 1, 2020).

Viloria, A. L. 2004. Biogeografía: la dimensión espacial de la evolución. Interciencia 29:163–164.

Westermann, G. E. 2000a. Biochore classification and nomenclature in paleobiogeography: An attempt at order. Palaeogeography, Palaeoclimatology, Palaeoecology 158:1–13.

Westermann, G. E. 2000b. Marine faunal realms of the Mesozoic: Review and revision under the new guidelines for biogeographic classification and nomenclature. Palaeogeography, Palaeoclimatology, Palaeoecology 163:49–68.

Wierzbicka, A. 1992. What is a life form? Conceptual issues in ethnobiology. Journal of Linguistic Anthropology 2:3–29.

Winter, I. J. 1985. After the battle is over: The *Stele of the Vultures* and the beginning of historical narrative in the art of the Ancient Near East. Studies in the History of Art 16:11–32.

Zaragüeta-Bagils, R., E. Bourdon, V. Ung, R. Vignes-Lebbe, and V. Malécot. 2009. On the International Code of Area Nomenclature (ICAN). Journal of Biogeography 36:1617–1619.

Zaragüeta Bagils, R., H. Lelièvre, and P. Tassy. 2004. Temporal paralogy, cladograms, and the quality of the fossil record. Geodiversitas 26:381–389.

Zuckerkandl, E. and L. Pauling. 1965. Molecules as documents of evolutionary history. Journal of Theoretical Biology 8:357–366.

# 6

## Controversies

### 6.1 Controversies in Nomenclature

The preface mentions how controversial the introduction of phylogenetic nomenclature (PN from here on) has been. Now that we have summarized the development of systematics with emphasis on nomenclature, as well as the nomenclature of other fields, it is time to reexamine these controversies to determine which ones are based on misunderstandings, and which ones reflect unresolved, substantial issues. As Greuter et al. (2000: xvii–xviii) pointed out, nomenclatural discussion are often passionate, even when concerning seemingly trivial details. Thus, the harsh language used by some of the opponents of PN is only moderately surprising. This even transpires in the titles of many critiques of PN. A few "amusing" (and simultaneously deeply saddening) examples include "Dead on arrival: a postmortem assessment of 'phylogenetic nomenclature', 20+ years on" (Brower 2020), "The Phylocode: Beating a dead horse?" (Benton 2007) or "The PhyloCode is fatally flawed, and the 'Linnaean' System can easily be fixed" (Nixon et al. 2003). Perhaps most extreme in this vein is this quote from Brower (2020), with language that is more reminiscent of 16th century Inquisition than of a rational scientific debate:

> At last, the long-awaited PhyloCode has been published in a permanent and immutable form. Rather than haunting the ether as an unsubstantiated, protean sketch, now it exists as a physical entity which we may examine and **lay to rest**, or as the case may be, **commit to the flames.**

> **(Bold emphasis mine,
> unless stated otherwise.)**

In addition to an anachronistic criticism of electronic publications, which are becoming the norm, this passage seems to exclude any positive outcome of an examination of the *PhyloCode*. To add one positive touch to this bleak picture, I note that all such exaggerated (and mostly unsubstantiated) attacks have been one-sided; I am not aware of any proponent of PN having used such language in print, and I do not intend on starting a new trend on this front in this book.

The strong feelings that transpire from these few titles and extracts attest to the importance of this topic. A quote from Hennig (1981: xvii) can be used to explain this point, even though the context of Hennig's text was slightly different (its tone is ostensibly more positive, and it was actually written much earlier, in the original German text published in 1969, of which the 1981 book is an English translation): "I must particularly stress that critical debate with an author should not be seen as diminishing the value of his work. **The reverse is actually true**. It is not worth expending critical energy on unimportant work."

Acute, sometimes acrimonious, controversies are not unique to biological nomenclature; indeed, they have also plagued phylogenetics, as shown by the following amusing quote from Felsenstein (2001: 466–467), who is undoubtedly one of the greatest phylogeneticists and evolutionary biologists of the last century. It shows how polarized the systematic community became over the incorporation of statistics into phylogenetics:

> The fighting in systematics grew more intense than in any field I have known. I used to think that we fought a lot in when I worked in population genetics, but in that field we used to sit side by side at meetings without **growing red-faced, hissing at each other, or spreading scurrilous rumors**. In systematics, however, the controversy attracted extreme personalities, mostly to the other side of the issue (probably both sides felt this way; I still do).

Similar problems thus plague at least various fields of systematics and beyond. A second and last example comes from the debate about Null Hypothesis Statistical Testing (NHST) in the field of statistics. In the context of an extensive review of the benefits and pitfalls of NHST, Nickerson (2000: 289) noted:

> Although many of the participants have stated their positions objectively and gracefully, I have been struck with the **stridency of the attacks** by some on views that oppose their own. NHST has been described as "thoroughly discredited," a "perversion of the scientific method," a "bone-headedly misguided procedure," "grotesquely fallacious," a "disaster," and "mindless," among other things. Positions, pro or con, have been labeled "absurd," "senseless," "nonsensical," "ridiculous," and "silly." The surety of the pronouncements of some participants on both sides of the debate is remarkable.

Nickerson's (2000: 290) conclusion are noteworthy and seem to be applicable in most fields, including biological nomenclature: "However, there is little virtue in being confident of one's opinions when the confidence depends on not being aware of reasoned alternatives that exist."

The more civilized tone that has prevailed in the important debates on the ontology of taxa (see the following) gives us reasons to hope for more constructive exchanges in the future. The proponents of such debates would do well to read and reflect upon the following quote from Griffiths (1973: 342), written about various schools of systematics (cladistics, evolutionary systematics and phenetics) and which seems to apply well to the recent debates (in the last 30 years) in biological nomenclature:

> Many polemical articles have appeared in recent years, some unfortunately written by persons who **have not taken the trouble to understand properly** the views they are criticizing. I would commend to you the opinion, which I believe originates from Karl Popper, that, if criticism is to be useful, then the views criticized must be presented **in their strongest possible form**. We will not contribute to the advance of science if we try to discredit an opponent's views by **misrepresenting** them.

Most arguments against PN and the *PhyloCode* that have been invoked by opponents of PN (who are all proponents of RN [rank-based nomenclature], though not all proponents of RN are opponents of PN) have been shown to be flawed, either in dedicated, short response papers (de Queiroz 1995, 1997; Laurin et al. 2005) or in more substantial publications on PN (for instance, Cantino 2000; Bryant and Cantino 2002; Laurin 2005, 2008; Cantino and Queiroz 2020). These are discussed briefly in this section; more serious arguments are discussed in Sections 6.3 to 6.5. It is unnecessary to summarize all brief exchanges about weak or misguided criticism of PN here. Instead, the arguments presented in one of these studies (Platnick 2012) will be presented and analyzed to exemplify the weakness of the arguments used by opponents of PN.

Platnick's (2012) argument focuses on the information content that he claims is associated with family-group names regulated by the *International Code of Zoological Nomenclature*, or "*Zoological Code*," for short (ICZN 1999). Platnick (2012: 360) stated that:

> Thus, one needs to know only that an organism is a member of the spider family Oonopidae to know also that it is not a member of any other family of animals (or plants, for that matter), and that no member of any other family is also a member of the Oonopidae. Moreover, one can make all those inferences without having available a complete classification of the species included, either in the Oonopidae or in any other family.

This statement is almost correct, but the information alluded to here is slight, and much of it is obvious; would anybody wonder if a spider belongs to a family of vertebrates or plants? Under PN (or RN, for that matter), clade addresses (a list of nested taxa to which a lower-ranking taxon belongs) can provide far more detailed, useful information. Clade addresses were initially proposed in the context of a new species naming scheme (Dayrat et al. 2004), but could be used in other contexts when useful to briefly convey taxonomic information. Clade addresses can be adapted to the context to provide the desired level of information; thus, "*Seymouriamorpha Discosauriscus austriacus*" would be sufficient in the context of a vertebrate paleontology paper aimed at specialists of Paleozoic vertebrates, but "*Vertebrata Stegocephali Seymouriamorpha Discosauriscus austriacus*" might be more appropriate in the

context of a zoological paper aimed at a much broader audience. In any case, indicating that *Discosauriscus austriacus* is a member of the family Discosauriscidae (under RN) instead of the unranked taxon *Discosauriscidae* would arguably add little information.

Previously, I indicated that Platnick's (2012) argument is "almost" correct. The argument would be correct if monophyly were required by the *Zoological Code*; but it is not! In fact, monophyly is not even mentioned in that code (and neither is it in the *Botanical Code*). As a result, there are many paraphyletic families, and thus, some low-ranking taxa (considered genera and species under RN) are indeed nested within two families simultaneously, even though under RN, this situation remains hidden. For instance, some authors recognize a paraphyletic *Pongidae* (such as Ebua et al. 2018; Supriatna 2022), in which *Hominidae* is cladistically nested (though formally excluded under RN), even though a recent trend is to re-delimit *Pongidae* to make it monophyletic (Groves 2018). Given the lack of delimitation provided by the RN codes, these various delimitations of both families can persist indefinitely. As a result, under at least some schemes, all hominids are members of both nested taxa, or at least would be considered as such if monophyly were enforced, and should be considered as such in comparative studies in which phylogeny matters. No need to search hard for exotic examples to see where Platnick's (2012) arguments fail! This is but one example, but new phylogenies often reveal that various taxa, as formerly delimited, are paraphyletic, and given that the *Zoological Code* does not require monophyly, taxa ranked as families are not necessarily re-delimited to ensure monophyly in such situations (e.g., Boy and Martens 1991).

The persistence of paraphyletic taxa in published taxonomies and databases appears to be a pervasive problem, as documented by Smith and Patterson (1988: table 1), who found that out of 63 nominal families of echinoids and vertebrates that had been used by Van Valen (1973) to assess the presence of possible extinction peaks through time, only 21 (33%) were monophyletic and polytypic (included at least two nominal genera), whereas 29 (46%) were not monophyletic and 13 (21%) were monotypic (included a single nominal genus). This problem is not restricted to a single study by a single author because Smith and Patterson (1988: 145) found qualitatively similar problems in the data from Sepkoski and Raup (1986). Note that the effect of these taxonomic errors, as well as dating errors in the database of these studies, led to spurious results, so this debate is not of purely theoretical or nomenclatural interest. Thus, Sepkoski (1986) had argued that his data on 5,594 extinct mid-Permian to Pleistocene nominal genera suggested cyclical extinction peak with a periodicity of about 26 to 29 million years. However, Smith and Patterson (1988: 145) reported that when they separated the correct data (correctly dated extinctions of clades) from noise, which comprised both incorrectly dated clade extinctions and correctly dated pseudoextinctions (disappearances of paraphyletic groups, which is simply their continued existence under another name, typically because their phenotype changed over time), the cyclicity was visible only in the noise, not in the signal! This is hardly surprising because Smith and Patterson (1988: 156) concluded that in the data used by Sepkoski and

Raup (1986), "real extinctions generally contribute less to peaks of extinction [from the Carnian to the late Eocene] than do pseudoextinctions."

Platnick (2012) insisted on the information provided by the names, but under PN, names are considered only labels; they only need to be unique and stable. Information is included in the definition and diagnoses, and this is arguably better because much of what we think we know about a taxon may turn out to be incorrect or imprecise; diagnoses can easily be updated to reflect scientific progress, but names should not need to be updated in such situations. It is of course useful to have mnemonic names, such as *Tetrapoda* for four-limbed vertebrates, but there are obvious limits to such principles. Thus, snakes and gymnophionans are tetrapods even though they lack limbs (Laurin 2010). The reason why is clear under the phylogenetic definition (the crown-group of limbed vertebrates; Laurin 2020), but the name cannot, by itself, resolve this apparent paradox.

Thus, this chapter will try to expose the various arguments objectively (an admittedly difficult, though necessary task), tease apart legitimate concerns from exaggerated fears and present the numerous reasons that have pushed several systematists to develop or adopt a radically new paradigm (PN) and break away with two and a half centuries of rank-based (Linnaean) nomenclature (RN from here on).

Rather than waste time publishing poorly thought-out criticism of PN, it is more productive to concentrate on scientific questions that should be at the core of our choice of a nomenclatural system. These include the following (this list is not exhaustive):

Is there any biological justification (or objective basis) for Linnaean categories (the best-known kind of absolute ranks) and, if not, what are the consequences of continuing using a system based on these categories? These issues were discussed at length in Sections 3.3 and 3.4, which obviates the need to discuss them in detail later.

Should our biological nomenclatural system be theory-free, or should it be "custom-made" for the kind of entities that we think taxa are?

Does the Tree of Life metaphor adequately explain the biodiversity? This metaphor steadily grew in popularity and it has become better documented over time in eukaryotes, but many specialists of bacteria and archaea seriously question this, given the pervasive presence of horizontal transfer in these taxa (Mallet et al. 2015).

This chapter will try to address these central questions and explore their implications for biological nomenclature.

## 6.2 Ontology of Taxa Revisited

### 6.2.1 Taxa, Clades and Species: Classes, Homeostatic Property Clusters or Individuals?

In the preface, we briefly saw the various points of view about the ontology of taxa, as classes (the initially dominant point of view), individuals or homeostatic property clusters (HPCs). Now is time to revisit this issue to determine which perspective appears to be most correct or most productive.

Note that while PN has been developed based on the premises that taxa are individuals, PN is also compatible with the thesis that taxa are classes, as summarized by Pleijel and Härlin (2004). This is because classes can also be defined by extrinsic properties such as relationships. Hence, systematists who prefer to consider taxa as classes, those who think that this is a false dichotomy, and those who argue that the difference between classes and individuals is less clear-cut than previously argued (such as Wiley 1980; Platnick 1985: 91) can consider that phylogenetic definitions define classes. Under this ontology of taxa, the defining property is historical and consists in belonging to a given clade, or having descended from a given hypothetical ancestor. For instance, under this perspective, *Tetrapoda* (as established under the *PhyoCode*, as a crown-group) can be considered to be the class of organisms that derive from the last common ancestor of amniotes and lissamphibians, or the class of organisms that belongs to the smallest clade that includes amniotes and lissamphibians.

Some systematists and historians (such as Winsor 2003: 388) or philosophers of science still view (or recently viewed) taxa as homeostatic property clusters (HPCs). These are classes or universals characterized by a combination of features, most of which are present in each element of the HPC, but many or all of which may be missing in some elements. Boyd (1999) argued that this combination of characteristic elements remains relatively stable because of cohesion mechanisms.

For species, plausible cohesive mechanisms include gene flow and stabilizing selection. However, neither mechanism operates universally to the same extent. Gene flow is most intense in sexually reproducing species, but it is virtually absent (within nominal species) in asexual eukaryotes (such as parthenogenetic taxa). True sexuality is also inoperant outside eukaryotes, though bacteria and archaeans have another gene exchange mechanism that ensures transmission of some genetic material to even fairly distantly related taxa and, in some cases, even between archaeans, bacteria and eukaryotes. It is difficult to know the exact proportion of their genes that gets transferred that way, but it is clearly important; Koonin et al. (2001: 712) went so far as stating that the contribution of "horizontal gene transfer and lineage-specific gene loss to the gene repertoire of prokaryotes was comparable to that of vertical descent." But clearly, horizontal gene transfer in bacteria and archaeans cannot be viewed as a cohesive mechanism operating within species given that it operates even between the three domains of Life; about 3% of the genes of free-living bacteria and 4%–8% of archaean genes may result from inter-domain horizontal transfer (Koonin et al. 2001: 719). Stabilizing selection is likewise not universal; selection may also be directional and create diverging evolutionary pressures within a species (for instance, if it occupies a heterogeneous habitat) and diverging evolutionary trends among descendants of a single ancestral stock.

Anagenesis creates problems, as shown by chronospecies, which have to be cut at arbitrary points (Lhermin[i]er and Solignac 2000: 160). Ghiselin (2002: 158) presented a hypothetical solution (a lineage becomes a different species when it evolves a different mate recognition mechanism), but this would be unworkable in the only context in which the concept of chronospecies really matters, namely paleontology. In fact, delimiting species in time is problematic even if species are viewed as individuals under most species concepts (Rieppel 1986: 303).

Boyd (1999: 181) suggested that higher taxa also have cohesive mechanisms (otherwise, they could hardly be HPCs), and he provided the example of adaptive evolutionary innovations that constrain subsequent evolution and generate "tendencies toward stasis." However, this view had already been refuted by Ghiselin (1987) and it seems that only a small proportion of the Tree of Life is thus constrained to relative stasis. Take some taxa like *Mammalia* and *Dinosauria*, for instance. Are these defined by stasis? Body size has certainly evolved much within both (e.g., Benson et al. 2018; Didier et al. 2019). Metabolic rate is fairly high throughout Mammalia, but it has varied much in the more inclusive *Synapsida* and in *Sauropsida*. Boyd's (1999: 181) suggestion that taxa are "loci of evolutionary stasis" also implies that evolutionary rates should vary widely, with branches with high rates delimiting higher taxa, whereas branches with much lower rates would be located within less inclusive taxa. Given that all life forms belong to taxa, it is unclear under Boyd's hypothesis when evolution should occur. This is not entirely compatible with current views on taxonomy and evolutionary biology, and empirical support for this claim is scarce, among other reasons because it is difficult to get a global measure of evolutionary rate. For instance, molecular and morphological rates are not necessarily tightly correlated, and within these categories of characters, large variations can be found, for instance between genes or codon positions, between cranial and postcranial characters, and so on. Many higher taxa appear to be neatly delimited simply because extinction has created large gaps in the biodiversity, but once the fossil record is taken into consideration, these gaps disappear to the point that intense debates can result about the delimitation of taxa. Rowe and Gauthier (1992) provide a well-documented case of such a debate about the taxon *Mammalia*, which is extremely easy to delimit when the fossil record is ignored, but much harder when fossils are considered. To sum up, the view that taxa are HPCs is not entirely compatible with the state of the art in systematics and evolutionary biology. Ghiselin (2002: 153) seemed to summarize the established consensus much better when he stated that "Above the species level there is an important break: after speciation has taken place, cohesion is lost and we get purely historical entities, called clades, which do not participate as units in evolutionary processes."

Considering taxa as HPCs also has practical drawbacks. Given that none of the HPCs' properties are necessary or sufficient, their boundaries may well be fuzzy, whether these HPCs are considered species or clades (many of which are higher taxa). Indeed, the taxonomic history of *Mammalia* in the paleontological literature illustrates how imprecise definitions lead to fuzzy delimitation (Rowe and Gauthier 1992). However, if we view taxa as individuals, it is easy to pick events in the Tree of Life to delimit various clades. Indeed, this is the principle underlying PN.

If taxa are individuals, it should be clear that contrary to what has often been stated, higher taxa are no less real (or more arbitrary) than species. At least since Buffon's days, several systematists have viewed species as real, whereas higher

taxa were often considered artificial constructs (Buffon 1749; Prichard 1813: 7; Dubois 1982: 10; Mishler and Donoghue 1982: 491). There were exceptions, of course. For instance, the limited evolution within genera admitted in some of Linnaeus' and Buffon's writings imply that genera are more real than species because the limits of species can be crossed, but not those between genera (Lherminier and Solignac 2005: 34). Agassiz (1859, 1866) considered that higher taxa were no more and no less real than species, as noted by Wilkins (2018: 132), and Whewell (1840) had expressed similar views. Much more recently, Boyd (1999: 173), while considering taxa as HPCs, also concluded that higher taxa were no less real than species. Mishler (2010) considered that all taxa are clades, and that species are just the lowest level of clades that are formally recognized as taxa; hence, they are neither more real nor more comparable with each other than taxa of any other level in the Linnaean hierarchy. Ontologically, species as conceptualized by Mishler (2010) are similar to LITUs (Pleijel and Rouse 2000a), which are described in Section 6.2.2.6.

In fact, it could be argued that higher taxa are more real than species because whereas the ontology of species is extremely vague, as shown by the various competing species concepts, most systematists (Hennig 1981; Ax 1987; Brochu and Sumrall 2001) agree in seeing higher taxa as clades.

Another important point is that individuals change through time, but classes don't (Ghiselin 1987: 129). Individual members of a class can change, but not the class itself. Thus, the simple fact that taxa evolve indicates that they are individuals. Ghiselin (1987) focused his argument on species, but the same point could be made about higher taxa. Indeed, taxa change through time in many ways: in their geographic range, their phenotype and diversity, and at some point, most become extinct. For instance, when *Amniota* appeared about 330 Ma ago (Didier and Laurin 2020), it was composed of only two lineages of relatively small ectotherms (Didier et al. 2019), whereas systematists consider that it includes over 26,000 extant species (Burgin et al. 2018; Sues 2019), at least half of which are endotherms (and many of which must belong to independent evolutionary lineages). The individuality thesis requires that taxa be defined ostensively (by pointing at them, or at some of their members) rather than by listing some intrinsic properties (Ghiselin 2002: 155), and this applies to species and clades alike.

## 6.2.2 Species Names and the Nature of Species

### 6.2.2.1 Coexistence of Many Species Concepts

Over 140 species definitions corresponding to many concepts have been used in biology (Lherminier and Solignac 2005), and these can be grouped into several large categories emphasizing gene flow ("biological species"), resemblance ("phenetic species" and "chronospecies"), monophyly, lineages or even ecology (in which the niche plays a central role). However, not all such species concepts are equally valid, and other classifications of species concepts, or ways of identifying species, have been proposed. For instance, Wilkins (2011) proposed to recognize seven basic concepts: "agamospecies (asexuals), biospecies (reproductively isolated sexual species), ecospecies (ecological niche occupiers), evolutionary species (evolving lineages), genetic species (common gene pool), morphospecies (species defined by their form, or phenotypes), and taxonomic species (whatever a taxonomist calls a species)." However, Wilkins (2011) also noted that some of these concepts do not clarify what species are, but rather, how we identify them (for instance, by morphology or gene flux). In reality, he concluded that there is a single valid species concept, which he called the **generative** species concept, and which arguably harks back to John Ray (but see Section 1.2.2.2 for a different point of view). This concept, which arguably is close to the biospecies or "biological species" concept, includes various situations because reproductive isolation may result from pre- or post-zygotic mechanisms. Among the pre-zygotic isolation mechanisms, mate recognition (visual or otherwise) mechanisms play a prominent role (Paterson 1978, 1980).

According to Ghiselin (1974), the core of the so-called "species problem" is that biologists want a basic unit of biodiversity, akin to the role that the atom long played with matter (till we discovered that atoms are not the smallest, most basic particles, after all), and they want to call it "species." An additional problem, pointed out by Minelli (2022), is that the word "species" is applied to three distinct types of entities (named taxa, such as *Panthera tigris*; a nomenclatural rank; and a set of species concepts, such as the "biological species concept"), and this generates some confusion in discussions of species. Differences in personal preferences combine with differing traditions, preferred methods used and differences in the reproductive mode of organisms (sexual or clonal, chiefly) to exacerbate the difficulty in finding a universally acceptable solution to this "species problem." Yet, as pointed out by Minelli (2022),

> Like any other set of objects, animals and plants could be classified according to a number of different criteria, nevertheless it is often taken for granted that there should be only one classification of living species. A frequently advocated reason for preferring a general-purpose classification is the expected possibility to get a stable and universal nomenclature to be used as common currency in the different biological disciplines and also in nonscientific contexts, e.g., agriculture, trade, legislation.

This idea of species as a "common currency" of biodiversity does seem to be widespread among scientists. For instance, Bock (2004: 183) stated that "What must be kept in mind is that species taxa, as all other taxa in biological classification, serve as the **foundation** for all other biological analyses and hence **should be as similar to one another as possible**."

Some authors have thus argued that the diversity of entities that we insist in naming "species" is too heterogeneous. For instance, Ghiselin (1987: 138, 2002: 157) suggested restricting the use of the word "species" for sexual organisms and calling the others "pseudospecies." This is coherent with the "biological species concept," which is only applicable to sexual organisms; this excludes, on the one hand, agamospecies, which reproduce asexually without genetic exchange between organisms, through phenomena such as budding, self-fertilization (which takes place in some flowers) and parthenogenesis, and, on the other hand, organisms (mostly, prokaryotes and viruses, if the latter are considered organisms) that exchange genetic material with very distant relatives.

However, Ghiselin's (1987, 2002) narrow definition of species appears to be held by a small minority of systematists; after all, the authors of the *Prokaryotic Code* seem to consider that it deals with species, not with pseudospecies. Under such a narrow definition of species, hybrids (both natural and artificial) are problematic (Wagner 1983; Minelli 2022). However, hybrids are not rare, especially among embryophytes. Unsurprisingly, botanists felt a need for a term to designate all populations that can hybridize. Thus, Lotsy (1925) proposed the term **syngameon** for these. Seehausen (2004: 198) redefined the syngameon as "a complex of selection-maintained, genetically weakly but ecologically highly distinctive species capable of exchanging genetic material." Many systematists appear to find the syngameon concept useful because a Google Scholar search (carried out on May 8, 2022) yielded 692 papers that used that word. In zoology, naturally occurring hybrids appear to be less common, but Dubois (1982) suggested that all species that can give rise to hybrids when crossed be attributed to the same genus. When discussing the quote from Bock (2004: 183), Minelli (2022) concluded, on a pessimistic but probably realistic note, that "general agreement on this issue [the scientific community agreeing on a single definition of biological species] has emerged as highly controversial and possibly beyond hope of definitive solution." He also added that:

> the use of the same Linnaean binomials in taxonomy, evolutionary biology, ecology and so on is a consequence of the fact that we call species all the different biodiversity units worth recognition in all these disciplines, but does not attest that these units are, or can be, coextensive.

Similar statements have been made by other systematists. For instance, Richards (2010: 5) stated that "there are multiple, inconsistent ways to divide biodiversity into species (on the basis of multiple, conflicting species concepts), without any obvious way of resolving the conflict. No single species concept seems adequate."

If we consider species in the context of the Tree of Life as opposed to over a very short period, it is clear that community of reproduction in sexual organisms does not delimit them in time. Such reproductive communities form lineages, but many of these persist over extremely long times; indeed, the first lineage of sexual eukaryotic organisms gave rise to all extant eukaryotic lineages, but these belong to numerous nominal species. The evolution of these lineages is marked by various events, such as anagenesis (change), cladogenesis (splits) and extinction. The most natural and objective way to delimit species in this tree of lineages is using cladogeneses (Rieppel 1986: 302–303) and (trivially) extinctions. According to this proposal, a species is simply a lineage that exists between two consecutive cladogeneses (Hennig 1966; Griffiths 1974: 85) or between a cladogenesis and an extinction event, and this species concept has been used in recent evolutionary studies (such as Didier et al. 2017: 967). Note that under this species concept (which is endorsed in this book, if species are deemed necessary), species may be paraphyletic because if species A undergoes cladogenesis and gives rise to species B and C, species A is paraphyletic. Species may also be paraphyletic under other recently proposed species concepts (e.g., Freudenstein et al. 2017), and some other recent species concepts even allow for polyphyly, even if only temporarily (Crisp and Chandler 1996: 830).

This vague ontological status of the species has been recognized even by proponents of RN. For instance, Lücking (2019: 221) wrote:

> Given that species are delimited through their phylogeny but recognized by means of a combination of a multitude of parameters in an integrative taxonomic approach (Goulding and Dayrat 2016; Vinarski 2019), **it is not possible to make the species rank consistent across species for any one of these parameters**.

### 6.2.2.2 Species in Nomenclature

The rank-based codes of biological nomenclature recognize species as one of the lowest formal categories. Thus, under the *Zoological Code*, sub-species are the only lower, formal regulated category, whereas some other codes, like the *Botanical Code*, recognize a few lower categories (see Section 2.4.3). This situation, especially in zoology, creates a compelling practical reason to continue discussing species, and this is reinforced by the fact that conservation efforts target threatened species (e.g., Mace 2004; Recuero et al. 2010), rather than higher-ranking taxa. The fact that species play an important role in RN also prevents different kinds of species (for instance, phenetic clusters, reproductive communities and clades) from being simultaneously recognized for the same taxa in the context of a given publication (but such coexistence does exist for a number of nominal species in a set of publications), as Boyd (1999: 172–173) already recognized.

It has been argued that the binominal nomenclature minimizes the taxonomic information included in the names. This is a desirable attribute because taxonomy needs to be regularly updated, whereas the names need to remain stable. Nicolson (1991: 33) stated that "The Linnaean nomenclature (binomial) system has two facets: divorcing the name from the diagnosis and minimizing classification. Systems that totally eliminate classification, such as uninomial systems, are unwieldy. Systems that try to diagnose, such as phrase names, are unstable." This is only partly true. The genus names contain taxonomic information whenever the genus is considered to include more than one species, so binominal nomenclature, as currently implemented in the RN codes, does not really minimize taxonomic information included in the names. Some systematists realized long ago (well before the advent of PN) that this is a problem, and this is best demonstrated by this short quote from Cain (1959: 242): "But stability is still too difficult to achieve, and in particular is threatened by the now **wholly anachronistic** system of requiring a generic name before a species can be named at all." Cain (1959: 242) explained

further that "The necessity for putting a species into a genus before it can be named at all is responsible for the fact that a great deal of uncertainty is wholly cloaked and concealed in modern classifications." According to Cain (1959: 235), binominal nomenclature arose under its current form because for Linnaeus, genera rather than species formed the primary category of taxa. Thus, this form of binominal nomenclature was well-suited for Linnaeus' taxonomy, but it is arguably not ideal for modern taxonomy that is more species-centered. Similarly, Pleijel and Rouse (2000a: 627) stated that "In fact, this binominal system is one of the worst possible imaginable, given an ultimate aim of [nomenclatural] stability."

Solutions to completely exclude taxonomic information from species names exist, the most notable one being from Lanham (1965). Under this system, the species name would be composed of the specific epithet, the name of the author who erected the species, the year of publication of the work in which that species was erected, along with the page number where the species is erected. This solution has a few key advantages: the names are unique and completely independent of taxonomy, given that genus names would not be part of species names under that system. The information is already in the literature, so adopting this system would not require erecting any new name for the species that have already been named; we would only need to search for this basic bibliographic data to determine the valid name of any species, and such information is often included in taxonomic reviews. Thus, priority under this system would apply as under the rank-based codes, and could maintain the same starting year as currently (that is, 1758 for zoology).

No regulating body of any rank-based code has seriously considered implementing such a system as far as I know, but several scientists have subsequently supported the use of uninominal nomenclature for species (such as Cantino et al. 1999) or for ontology-free, lowest-ranking taxonomic units that are not necessarily called "species" (e.g., Pleijel and Rouse 2000b). The ISPN (International Society for Phylogenetic Nomenclature) did consider a uninominal species nomenclature when it was contemplating producing a species code to accompany the *PhyloCode* (Cantino and de Queiroz 2020). Benoit Dayrat proposed to use Lanham's epithet-based nomenclature for species names in a future code at the First international phylogenetic nomenclature meeting that convened in Paris in 2004, and a clear majority of the 70 participants (from 11 countries) preferred it over the current binominal nomenclature (Dayrat et al. 2004; Laurin and Cantino 2004: 478). That proposal also included suggestions by Julia Clarke that require the authors to specify which species concept they meant to apply to the species that they erect (Laurin and Cantino 2004: 478)—a critical piece of information that is rarely explicitly mentioned in publications where species are erected under most RN codes. However, this proposal has so far not been implemented because the ISPN discontinued its effort to develop a species code (Laurin and Cantino 2007: 113) for reasons that were discussed in Section 4.1. Thus, the prospects of seeing Lanham's (1965) proposal to use uninominal species names in the near future are bleak, but this does not mean that the system will not be implemented in a code in a more distant future, by the ISPN, by a regulating body of a rank-based code or by other organizations. After all, Cain (1959: 242) predicted long ago that such a system would be adopted in the future:

> It seems clear that with the great alteration in the status of the genus since the time of Linnaeus, an effectively uninominal system will come into use. This will then be free from that part of the present instability caused by generic changes, which are mostly matters of opinion.

This should logically happen, but when and how?

### 6.2.2.3 Species: Profusion of Concepts and Paucity of Data

The profusion of species concepts, along with the weight of tradition and other factors, may explain why very little progress has been made in over half a century on the "species problem." John Locke (1632–1704) already considered species (among other entities) as constructs (Lhermin[i]er and Solignac 2000: 159), or at least our ideas of species are such constructs (Sloan 1972). Darwin (1859: 485) wrote:

> In short, we shall have to treat species in the same manner as those naturalists treat genera, who admit that genera are merely **artificial** combinations made for convenience. This may not be a cheering prospect; but we shall at least be freed from the **vain search** for the undiscovered and **undiscoverable essence** of the term **species**.

Nearly a century later, to summarize a Systematics Association symposium volume entitled "The species concept in paleontology," to which leading figures (such as the geneticist J. B. S. Haldane and the paleontologists N. D. Newell and T. S. Westoll) participated, Nitecki (1957: 378) wrote that "It is now realized that the orderly, beautiful concept of species is **shattered** and what is left is a vague, **not quite easily definable group**," and Cain (1959: 240) concurred with this view. These statements by Darwin and Nitecki (among many others) are still true! There is no universally accepted species concept in biology, and this problem is compounded by the variety of criteria used to try to delimit species, even though Ghiselin (1966: 127) correctly pointed out that the practical problem of how to delimit taxa is distinct from how to define and conceptualize them.

Studying the history of how species were conceptualized is made difficult by the fact that many systematists did not clearly express which species concept they adhered to (Winsor 2003: 389). As stated by Pleijel and Rouse (2000a: 629): "In virtually all cases, the connection between the named species and the empirical evidence that justifies its status is weak or non-existent." Indeed, the fact that a species has been erected to include a group of organisms does not indicate that they form a reproductive community or a clade (for instance) simply because most studies do not refer explicitly to a given species concept. This was more recently confirmed by Minelli (2022), who stated:

species concepts do not necessarily translate into operational guides allowing the assignment of specimens to species. For example, despite the popularity of the so-called biological species concept, the instances in which biologists actually check if living specimens x and y **actually interbreed**, to decide whether they belong to the same species, are an **extremely minor exception**. Most identifications are made on preserved museum specimens; additionally, the name-bearing type specimens are nearly always preserved specimens and the exceptions to this rule are the living types of bacteria to which the biological species criterion would hardly apply. A great many researchers would probably accept the biological species concept in the abstract but, in their daily practice, to establish conspecificity of specimens they will use proxies, such as morphological or genetic similarity.

In some cases, there are good reasons for this lack of data. For instance, several authors advocate the existence of gene flow rather than the potential for hybridization in constrained situations (such as in captivity) as the best criterion to delimit species under the "biological" concept (e.g., Lherminier and Solignac 2005: 6), and recent research suggests that "species can differentiate despite ongoing interbreeding" and that adjustments need to be made to the biological species concept (Hausdorf 2011). As stated by Minelli (2022):

> The real problem, however, is delimiting species in conditions of **allopatry**, i.e., when comparing similar but distinguishable populations living in geographically isolated area[s], e.g., on different islands or mountains. In this case, some degree of morphological divergence is often accepted as a **proxy** for a real proof of **reproductive isolation**. However, what "some degree" may actually mean, remains undecidable.

Similarly, I suspect that many systematists adhere to a given concept, but in practice, they apply criteria that are more relevant to another concept; this is certainly my experience as a paleontologist. Pleijel and Rouse (2000a: 629) expressed a similar idea:

> Species are currently designated as such by taxonomists who are forced by the existing codes of nomenclature to describe organisms as species when in fact they generally have no idea of what is going on in nature. . . . In other words, making taxonomists decide that a few dead specimens represent a species is an **extravagant extrapolation** that has no place in science.

For instance, the "biological species concept" (in the sense of a reproductive community; abbreviated **BSC** from now on) is probably the most popular among practicing systematists, but most species were erected and delimited using morphological criteria, which arguably matches a purely phenetic species concept (also see Section 3.3.2 for the limits of this concept). This situation prevails despite the fact that the BSC is an old concept that harks back at least to Buffon (see Chapter 1). Raven et al. (1971: 1210) noted that "For more than 99 percent of the described species, we know nothing more than a few morphological facts and one to several localities where they occur." In the five subsequent decades, the situation probably improved a bit, especially for molecular data, but this statement appears to still be true of a majority of nominal species.

In practice, most systematists probably use the phenetic species concept when erecting new species. These phenetic species may reflect resemblance based on morphology, genetics or other criteria. However, such clusters should not be recognized as species (or taxa in general) if these are viewed as individuals rather than classes, because individuals lack defining intrinsic properties and can be defined by their history (Ghiselin 1974: 539). Thus, the most abundant data that we have on nominal species suggests that these form phenetic clusters, and this is inconsistent with both the species concept (BSC) and taxon ontology most widely accepted by systematists.

Is this profusion of species concepts, combined with the fact that the concept to which systematists adhere are rarely mentioned in the relevant publications, problematic? Certainly, to the extent that many biological studies in various fields focus on species or species counts (for instance, conservation biology to assess biodiversity; see Section 3.4.2). If species continue being recognized (as ranked taxa), scientists must strive to find a consensus about a species concept (Minelli 2022). As Cracraft (1983: 162) wrote, to continue using simultaneously several species concepts would amount to abandoning "the search for general patterns of biotic diversification."

### 6.2.2.4 Cryptic Species

When other (non-morphological) aspects of various taxa were examined, many nominal species were split further, as happened with birds when their songs were studied (Gwee et al. 2019), or when barcoding (Saitoh et al. 2015) and other genetic methods (Pulido-Santacruz et al. 2018) were used to assess their biodiversity. This is likely to be a typical situation for many, perhaps most, metazoan taxa, and most of the cryptic species (along with many non-cryptic ones) that currently exist (according to the BSC) probably still await discovery. Minelli (2017: 658) argued that "Modern revisitation based on molecular evidence and sophisticated methods of species delimitation often reveal an unsuspected cryptic diversity even within what were hitherto regarded as taxonomically unproblematic, and even popular, species." Even more telling, he mentioned that in annelids, cryptic species are recognized in almost every thorough study, and he provided several examples among annelids in which five or more cryptic species were recently recognized within a single previously recognized species. Among the most spectacular examples of previously unsuspected, cryptic biodiversity is the recognition of between 62 and 78 "putative species" within a single nominal species (*Gyratrix hermaphroditus*) of cosmopolitan flatworm (*Platyhelminthes, Rhabdocoela*). This biodiversity had gone unnoticed by morphologists, but this may have been from lack of study because Tessens et al. (2021), who discovered this cryptic biodiversity,

recognized 14 morphotypes based on the male copulatory organ, and most of these correspond to clades found in their molecular phylogeny.

This phenomenon is so common that Blaxter et al. (2005) proposed the concept of MOTUs (molecular operational taxonomic units) for putative taxa identified by molecular sequences, many of which may be new (and potentially cryptic) taxa. However, these taxa do not necessarily correspond to species as they have been conceptualized so far (Minelli 2022). Ironically, the expression "cryptic species" was introduced by a botanist (Darlington 1940), but cryptic species may be more common in animals than in embryophytes, according to an extensive literature search (Bickford et al. 2007; Minelli 2022). This conclusion remains preliminary because botanists use the expression "cryptic species" less frequently than zoologists (preferring, instead, to simply report new species), and this may have biased the literature search performed by Bickford et al. (2007). Furthermore, the search for adequate molecular markers of cryptic species has proven more difficult for embryophytes than for metazoans, and this may further bias our perspective on this issue (Shneyer and Kotseruba 2015).

Given the pervasive nature of cryptic biodiversity, the task awaiting systematists to properly describe this biodiversity and name the new taxa is daunting. So far, most studies did not go far beyond identifying the putative new taxa based on a few molecular markers. Thus, in their survey of 606 publications that had described cryptic biodiversity, Struck et al. (2018) reported that only 14% referred to a definition of the term "cryptic species," 47% of them presented no phenotypic data, only 25% reported differences in at least one phenotypic character between putative cryptic species and only 19% provided a formal description. Cryptic biodiversity studies are obviously fertile grounds for grey nomenclature (see Section 2.5.1)!

### 6.2.2.5 Limits of the Biological Species Concept

We can look at the problems raised by the most popular species concept, the "biological species concept" (BSC). The frequently quoted BSC based on gene flow between individuals and population, for instance, is applicable only to sexually reproducing organisms, and only over a short period of time. In paleontology, gene flow is generally unobservable (exceptions concern Pleistocene fossils that can sometimes be sequenced), so this criterion cannot be used. In the early days of paleontology, small variations that could plausibly be intraspecific justified the erection of new taxa, and this led to a proliferation of species names that subsequently created much nomenclatural confusion. A spectacular example is provided by the work of Robert Broom, who erected 369 therapsid (stem-mammal) species (Angielczyk and Kammerer 2018: 152), nearly half of which are no longer considered valid (Wyllie 2003).

Some widely used systematic methods analyze biodiversity in a way that is agnostic about species concepts. Thus, Hull (1979: 428) explained that:

> Thus, those cladists who continue to maintain that biological species are the ultimate units in their investigations are forced to admit that their

cladograms and classifications might be mistaken. They might have lumped several species into one or divided a single species into two or more species. For extinct forms, these mistakes are difficult, if not impossible, to remedy. Cladists who have abandoned the biological species concept are spared this problem. They are not classifying species but are grouping organisms according to the possession of particular traits.

Other authors, such as Cracraft (1983: 160), pointed out that the history of reproductive isolation (hence, of "biological species") may not reflect the history of taxonomic diversification in which many, perhaps most, systematists are interested. Thus, Cracraft's (1983: 170) phylogenetic species concept consists of a population or set of populations characterized by a unique combination of primitive and derived characters and corresponds to many taxa currently recognized as subspecies.

Even if we had access to genetic information in fossils, it would not always be clear how to delimit species in time, especially when important anagenetic change has led paleontologists to recognize "chronospecies," a concept that was proposed in the 1950s (Nitecki 1957). Such species succeed each other within a lineage in which no cladogenesis needs to occur. Ghiselin (1974: 539) used an analogy with economy to illustrate why this concept is flawed. Would we erect "chronofirms" to describe the same firm at different times?

Mishler and Donoghue (1982) raised three important problems with the BSC. Perhaps the most serious one is that it rests on the assumption that there is a discrete reproductive boundary between species, within which gene flow is abundant and beyond which it is nil. Empirical studies do not really support this claim (see Section 3.3.2). Mishler (1999) thus concluded that "the probability of intercrossability decreases gradually as you compare more and more inclusive groups," which makes this criterion difficult to use, even in the most favorable case, namely of organisms with sexual reproduction!

This quote raises the second problem raised by Mishler and Donoghue (1982: 494): the discontinuities in gene flow may not (and frequently do not) delimit the same entities as the discontinuities in morphology or ecology, which reduces the usefulness of the reproductive criterion in systematics because all these discontinuities are interesting. This mismatch between various kinds of discontinuities evokes cryptic species, in which one morphotype may hide two or more reproductive units, but the reverse also occurs; in many plant taxa, one interbreeding unit may encompass two or more morphotypes (Burger 1975; Mishler and Donoghue 1982). Because of this, the relationship between biological and phenetic species is complex.

This same lack of congruence between discontinuities observed in reproductive patterns and morphology also raise doubts about the ontological status of the so-called "fossil species" because this expression implies a coincidence between reproductive and morphological discontinuities, which many empirical examples disprove. On a nomenclatural note, the expression "fossil species," often used to discuss low-level taxa documented in the fossil record, is a misnomer because organisms fossilize and taxa can eventually become extinct,

but both phenomena are uncoupled, as shown by the fact that many extant taxa (*Homo sapiens*, *Primates*, *Mammalia* and *Vertebrata*, for instance) have a rich fossil record. Who would call *Mammalia* a "fossil taxon"? It would be better to qualify taxa of extant or extinct, and use the qualifiers "fossil" (or "living," "fresh," "dried" or "preserved in alcohol") for specimens (organisms).

These problems were not discovered recently. Camp and Gilly (1943) had already clearly exposed the complexity of the "species problem" in spermatophytes. They proposed to recognize no less than 12 kinds of species differing from each other in the nature of genetic (especially karyotypic) and morphological variability, as well as by the reproductive isolation (or lack thereof) of included subspecies, and proposed to call "biosystematy" the field that "seeks to explain the causes of these differences in the structure of species and so permit them to be arranged in a functional nomenclatural system" (Camp and Gilly 1943: 324).

Given these objections, it is not surprising that acceptance of the BSC appears to be uneven in various biological fields. Mishler and Donoghue (1982: 493) thus reported that the BSC is better accepted by zoologists than by botanists, a point also made by Minelli (2022), based partly on the opinion of Raven (1980).

Mishler and Donoghue (1982: 493) also refuted the main arguments that are often considered to justify a preference of the BSC over other species concepts. First, some systematists may think that reproductive units have more clear-cut boundaries than groups matching other species concepts. This assumption was refuted, at least for some angiosperms, long ago (Ornduff 1969). Second, the assumption that gene flow prevents divergence while lack of gene flow allows it might not be fully justified. Cohesion may be provided by stabilizing selection, even in the absence of gene flow, whereas disrupting selection may cause divergence even in the presence of gene flow (Jain and Bradshaw 1966). Furthermore, Ehrlich and Raven (1969) showed that gene flow is limited in many organisms and that this phenomenon may not account for the morphological integrity displayed by many nominal species. Endler (1973: 249) went further and stated that "gene flow may be unimportant in the differentiation of populations along environmental gradients." Along similar lines, Lande (1980: 467) argued that "of the major forces conserving phenotypic uniformity in time and space, stabilizing selection is by far the most powerful," and Grant (1980: 167), quoted by Mishler and Donoghue (1982), suggested that "the homogeneity of species is due more to descent from a common ancestor than to gene exchange across significant parts of the species area." Thus, phenotypic divergence and reproductive isolation need not coincide, nor be causally related. Mishler and Donoghue (1982: 495) concluded that:

> Because of the complex nature of variation in each of these factors [reproductive barriers, ecological role and homeostatic inertia], and because different factors may be "most important" in the evolution of different groups, a universal criterion for delimiting fundamental, cohesive evolutionary entities does not exist.

Systematic practice seems to validate this rather pessimistic outlook on the perspective of objectively delimiting
species, a procedure that should be possible if systematists
used the BSC and if biodiversity was actually structured into
discrete reproductive communities. Let's take the example
of great apes, taken here in the sense of a clade *Hominidae*
as delimited by Groves (2018) to encompass the orangutans,
gorillas, chimpanzees and humans. This small clade (four
genera, seven species and 11 subspecies, according to the
Mammal Species of the World website, accessed April 5,
2022) must be the most intensively studied of all, especially
if considering the resources allocated to research per species (e.g., Prado-Martinez et al. 2013) because of our obvious
anthropocentric bias. Yet, Groves (2018) depicted a situation
of persistent debate about the number of species and subspecies in orangutan (two or three species) and in chimpanzees
(two or three species). The number of subspecies is equally
uncertain; some authors recognize three subspecies in *Pan
troglodytes*, but others do not recognize any subspecies in this
taxon (Groves 2018: 21). In gorillas, a single species divided
into two subspecies were recognized until the 1990s, but now,
two species, one of which is divided into two subspecies, are
recognized by Groves (2018: 19), but the Mammal Species of
the World website (accessed April 5, 2022) recognizes two
pairs of subspecies (one in each species of *Gorilla*). Groves
(2018: 21) summarized the situation thus:

> There is a tendency to speak of the "orangutan,
> gorilla, chimpanzee and bonobo" as if they were
> four equal species [or genera]. In fact, Sumatran and
> Bornean orangutans are in most accounts at least
> as distinct as are chimpanzees and Bonobos, and
> Western and Eastern gorillas may be almost as dis-
> tinct (although see Prado-Martinez et al. 2013). It is
> difficult to escape the conclusion that it is customary
> to treat the Bonobo separately simply because it has
> its own common name!

### 6.2.2.6 Do We Need Species? Alternatives

To conclude, it seems best to consider a lineage between two
cladogenetic events as belonging to a single species, **if** species
must be referred to at all. This would reconcile (better than
other options) the species concept with the individual ontological status of taxa and with the Tree of Life metaphor. But
could systematics exist without referring to species? Mishler
(1999) argued for this and considered "the species problem as
a special case of the taxon problem," meaning that attempts at
defining the species level objectively have failed and that this
level is not more objective than any other (genus, family and
so on). He further suggested that getting rid of the species rank
would even be beneficial for users of biodiversity data outside
systematics. For instance, conservation biology, instead of
relying on the unsatisfactory method of taxon counts at a given
level (families, genera or species) could use more rigorous
methods to quantify biodiversity, such as Faith's (1992) phylogenetic diversity index (see Section 3.1.5.5), which would yield
better science. Dropping species would also bypass the old
dilemma faced by naturalists at least since the 18th century,

but even known among ethnobiologists (such as Ludwig 2018:
417): either the species is real, but then, there can be no evolution (under some species concepts), or evolution is real, and
species (under some concepts) are artificial constructs. This
dilemma vanishes if we accept that there are no species.

Solutions to drop the species rank were proposed long ago,
like LITUs (Pleijel and Rouse 2000a). LITUs (Least Inclusive
Taxonomic Units) are ontologically agnostic; they are simply
the smallest diagnosable taxa. In this, they resemble Cracraft's
(1983) phylogenetic species concept, except that the LITU
designation does not entail ranking (hence, specific status).
LITUs may or may not be reproductive units. Future inves-
tigations could reveal this and, in so doing, might also reveal
internal structure. What was one LITU would then become
a clade (if applicable) and smaller LITUs would be erected
therein. This could be seen, to some extent, as further development of Rosen's (1979: 277) suggestion that species are "the
smallest natural aggregation of individuals with a specifiable
geographic integrity that can be defined by any current set of
analytical techniques," or to similar species concepts formulated by Cronquist (1978: 15) and Nelson and Platnick (1981:
12), who stated that "In this book, then, species are simply the
smallest detected samples of self-perpetuating organisms that
have unique sets of characters." However, with LITUs, we dis-
pense with species altogether. Recently, Mishler and Wilkins
(2018) proposed a similar concept, the SNaRC, for "Smallest
Named and Registered Clade." Mishler and Wilkins (2018: 7)
argue that the SNaRC concept differs from the LITU in only
one important way: the SNaRC is necessarily monophyletic,
whereas they claim that this condition is not expressly mentioned for LITUs. However, this is dubious because Pleijel and
Rouse (2000a) insisted that taxa must be monophyletic, so we
could interpret this as implying monophyly of LITUs as well,
even though the authors did not state so explicitly. This point
was clarified by Pleijel and Rouse (2000b: 157), who stated:

> Nevertheless, we acknowledge the usefulness in
> specifying whenever a name refers to **a smallest
> known clade** which at present cannot be further
> subdivided; in accordance with Pleijel and Rouse
> (2000a) such taxa are termed LITUs (Least Inclusive
> Taxonomic Units) and are distinguished from other
> taxon names by an initial lower case letter.

Thus, both expressions refer to the same concept and, in this
book, I use the principle of chronological priority to select
LITU as the name for this concept.

Molecular systematists discover so many potentially new
taxa (many of which would be species according to several
concepts) that they often do not even take the time to properly describe and name them, at least in the study that evi-
dences them. Instead of erecting proper binominal names in
conformity with the rank-based codes, a variety of informal
kinds of names (grey nomenclature) are used, sometimes
in a single study. Minelli (2017: 659) thus cited a study that
gave the following names to new hydrozoan (cnidarian)
taxa: "*Aglaophenia* sp. 1, *Aglaophenia* sp. 2, *Lytocarpia* sp.
1; (b) *Macrorhynchia* nov. sp.; (c) *Macrorhynchia phoenicea*
morpho-type A; *Macrorhynchia phoenicea* morpho-type B."

Many such names have been entered into major molecular databases such as *Genbank* (www.ncbi.nlm.nih.gov/genbank/) and *BOLD* (Barcode of Life Data Systems; www.boldsystems.org/). As Minelli (2017) pointed out, beyond the lack of equivalence between groups defined solely on molecular data and species defined based on morphology, reproductive isolation and a variety of other criteria, it is not always clear if these names refer to an organism or group of organisms identified only at the genus or species group level that has probably already been properly named and described, or a new species that is identified at the genus or species group level. This practice thus creates substantial nomenclatural problems and hampers biodiversity assessment. The extent of the problem is shown by some basic stats. Minelli (2017: 662) reported that:

> Of the 306 entries in the taxonomic list for Lumbricidae (earthworms) in GenBank [accessed 28 09 2016], only 115 (37.6%) are standard binomens; another 15 entries are binomens followed by letters and/or numbers, e.g. *Allolobophora chlorotica* L5 (in two cases, the species epithet is preceded by „aff." of „cf."); two entries are for species complexes, e.g. *Lumbricus rubellus* complex; 59 entries are for a genus name followed by a formula, e.g. *Lumbricus* sp. SL-2003; another 6 for genus name only; and 109 entries are for samples identified to family level only, and just distinguished by a formula that in most cases does not correspond to any recognizable standard, e.g. Lumbricidae sp. Esik50. A small number of formulae (e.g. *Eophila* sp. BOLD:ACJ0004 and Lumbricidae sp. BOLD:AAV0357) point to corresponding entries in the BOLD database [described below].

Thus, for *Lumbricidae* at least, most of the information in *Genbank* cannot be unambiguously attributed to proper species. This chaotic nomenclatural situation yielded by some recent molecular research (see also the discussion of MOTUs in Section 6.2.2.4) is not a fatality. The nomenclature of these groups recognized by molecular techniques could be standardized.

The BIN (Barcode Index Number) system, which was elaborated using the *BOLD* database (Ratnasingham and Hebert 2007), makes substantial progress on this front (Ratnasingham and Hebert 2013). The BIN exploits the fact that when it was developed, more than two million records were already available for a 648 base pair region of the cytochrome *c* oxidase I (COI) gene, and that this gene seemed to be well-suited to assign specimens to species as they were then delimited. Thus, Ratnasingham and Hebert (2013) reported that "More than 95% of animal species examined possess a diagnostic COI sequence array," and that "COI divergences rarely exceed 2% within a named species, while members of different species typically show higher divergence." A sophisticated and computationally efficient method was developed to sort sequences into BINs that match previously recognized taxa reasonably well. Ratnasingham and Hebert (2013: 7) reported that, on average (for eight empirical test datasets), 89.2% of the BINs that they recovered matched described species; the other 10.8% represent BINs that either included members of more

than one described species, BINs that represented only part of the specimens currently assigned to a nominal species (some of these could represent cryptic species) or BINs that included only parts of specimens of more than one species. The fact that these fairly good results were obtained from a single gene region is encouraging. When a BIN matches a described species, this information is entered into the database. It could even be argued that, ontologically, the status of BINs is clearer than that of many nominal species, given that systematists are not always explicit about the species concept that they use when erecting species. The names of BINs are unique and stable, even though some BINs may subsequently be split or (more rarely) lumped. BINs have two identifiers, a *BOLD*-generated URI (Uniform Resource Identifier) and a DOI (Document Object Identifier). The form of the *BOLD*-generated URI consists in a code starting with "BOLD," followed by a colon, three letters and four numbers (such as BOLD:AAA9566), which cannot be confused with names of rank-based taxa of any level (species, genera and so on). When an existing BIN needs to be split (as might happen when we realize that a BIN that coincides with a nominal species actually represents two species under a given concept), a decimal is added to identify the new BINs. Thus, the BIN initially called BOLD:AAB2314 would be split into BOLD:AAB2314.1 and BOLD:AAB2314.2. The reverse situation, when two BINs need to be merged, as happens when subsequent research shows that two or more BINs should be considered to belong to a single species, is handled by synonymizing the most recent BINs and keeping only the first (oldest) one. This is reminiscent of how synonymy is handled under RN. In these processes, the *DOI* is amended to resolve lookups to the new or merged BINs. This system also allows for community annotation and validation. The high level of automation of these processes accelerates the rate at which molecular biodiversity can be described, although this is no substitute to sampling additional genes, studying other aspects (morphology, behavior and so on) of organisms and properly naming them, which are much more time-consuming tasks. Unsurprisingly, this approach has proven popular, judging by the 1,746 Google Scholar citations (search performed on November 13, 2022) of Ratnasingham and Hebert (2013) and the 22,400 results returned by a Google Scholar search (performed on the same date) for the expression "Barcode Index Number (BIN) system." The BIN database can be consulted at "www.boldsystems/org/bin."

All the solutions evoked in this section (except the BIN) may seem radical, but the challenge facing systematists in the description of the quickly vanishing extant biodiversity calls for bold suggestions and decisions. After all, Ratnasingham and Hebert (2013: 1) reported that, "It has been estimated that the cost of describing all animal species will exceed US$270 billion and require centuries." Bold decisions have long been beneficial for scientific progress, but are arguably too rare, and this can be illustrated by the following quote from a letter that Darwin sent to Bates (about Bates' descriptions of insects from Amazonia and what is now called Bayesian mimicry): "You have spoken out boldly on Species; & boldness on this subject seems to get rarer & rarer" (Egerton 2012: 49). Rank-based nomenclature has long been thought to be less than ideal by some systematists, as shown by this quote from Michener

(1963: 163): "Nomenclature should provide a means of communication but **should not straightjacket systematic thought** as it does." Perhaps Minelli (2000: 342–343), a former president of the International Commission of Zoological Nomenclature, was correct in summarizing the situation:

> With the time passing, the simple and convenient practice of naming species and genera seems to have gained such an importance, that concepts have to be looked for, in order to justify the nomenclatural practice. Literally, we witness sometimes, rather than an effort to establish a convenient nomenclature to convey theoretically sound concepts, a deliberate effort to establish concepts to justify continuing use of a traditional nomenclature.

## 6.3 Should the Codes Be Theory-Free?

At least some rank-based codes, such as the *Zoological Code* and *Botanical Code*, have been argued to be theory-free because they do not rely on any assumptions about the ontology of taxa, not even monophyly. On the contrary, the *PhyloCode* (Cantino and de Queiroz 2020) and PN in general rest on the assumption that taxa result from divergent evolution and that they should be monophyletic (with the possible exception of species, which are not currently regulated by the *PhyloCode*). Determining which system is preferable depends largely on what we know of how biodiversity arose (see Section 6.4).

Information content of names also differs under both systems. Names formulated under relatively theory-free nomenclature as enforced in the *Zoological Code* and *Botanical Code* convey little information, other than reference to a type and a rank. This problem was succinctly stated by Mishler (2010: 116):

> Only a search into the literature can uncover the basis [monophyly, overall resemblance, etc.] for a particular taxon name under the current codes, while under the *PhyloCode* one knows that the author of the name hypothesized it to be a monophyletic group. A name that can be used to convey anything [as under RN] really conveys nothing.

In addition to the problem of the low information content of named taxa under RN, Pavlinov (2021: 12) has argued that the common belief that biological nomenclature is, or should be, theory-neutral is "wishful thinking."

Hayden (2020) recently demonstrated that clades maximize information content, compared to para- or polyphyletic taxa, if their diagnoses are used to try to reconstruct character distribution data. Indeed, the advantages of recognizing only monophyletic taxa (with the possible exception of species) was a major impetus in the development of numericlature (even though not all such schemes were developed to identify clades), which used numerical codes to convey information about taxa. These numerical codes were not necessarily aimed at replacing taxon names, although they would have become part of these names under some proposals (e.g., Little 1964). Under some proposals, these codes provided a unique identifier

(analogous to a DOI) and information about the hierarchy of (monophyletic) taxa (e.g., Hull 1966, 1968), very much like clade addresses under the *PhyloCode*. Other proposed numericlatures were aimed at either replacing binomial nomenclature (reviewed in Griffiths 1981) or simply at codifying phenotypic characters (for instance, Gilmour 1973; Griffiths 1984); the latter were aimed at comparative analyses rather than at reflecting the taxonomic hierarchy. Numericlatures never gained much popularity (a Google Scholar search for "numericlature" carried out on November 24, 2022, yielded only 79 papers); they were discussed in the 1960s and used until the 1980s. However, they were never seriously considered by governing bodies of codes of biological nomenclature.

The nature of some numericlature proposals starting in the 1960s (e.g., Hull 1966) shows that systematists realized well before the advent of PN that it would be advantageous to require monophyly of taxa and thus represent phylogenetic relationships, which is not coherent with theory-free taxonomy. If numericlature fell out of fashion, it is obviously not because monophyly lost its importance in systematics, as shown by the spectacular development in phylogenetics since the 1960s (see Section 3.1). Rather, the development of other approaches may explain the decline of numericlature. Thus, comparative methods using dated trees and data matrices allow finer analyses and easier data retrieval than numericlatures combined with classical statistical analyses. Indeed, modern comparative methods account for the phylogenetic signal (e.g., Felsenstein 1985), and the combined use of phylogenetic trees and PN arguably solves nomenclatural problems better than numericlature, notably by not requiring a definitive taxonomy.

If the TOL is an appropriate evolutionary model (at least for eukaryotes), using PN rather than RN enhances coherence between nomenclatural practice and ontology of taxa. As de Queiroz and Gauthier (1992: 465) stated:

> the very acts of lumping and splitting—which are intimately tied to changes in the meanings of taxon names under the current system—are difficult to interpret in phylogenetic terms. Taxonomists can neither lump nor split taxa as named clades, for **clades are not things that taxonomists form, erect, unite, or divide,** but rather things to which they **give names**. Outside the context of Linnaean categories, the notions of lumping and splitting make little sense, and they are irreconcilable with the phylogenetic meanings of taxon names.

Even a former president of the ICZN stated that RN as implemented in the *Zoological Code* was "too permissive, in so far as it can be equally applied lo paraphyletic as to monophyletic groups" (Minelli 2003: 656). Indeed, Platnick (1985: 88) argued that "All modern systematists agree that natural groups exist." This begs the question of what natural groups are (for Platnick, these appear to be clades), but this suggests that most "modern systematists" do not favor nomenclature for theory-free taxa, at least if Platnick's (1985: 88) statement is accurate. It could even be argued that there are no theory-free taxa (except if these are conceptualized as purely arbitrary sets of organisms) because Platnick (1985: 88) argued that "phenetics is no more theory-free than is cladistics—it's just based on a different theory" and that "Phenetics is the theory that clustering by 'overall' or raw similarity, that is by both the presence and absence of characters, will resolve natural taxa."

## 6.4 The Web of Life: The Greatest Challenge for Phylogenetic Nomenclature

If most biodiversity arose through divergent evolution, the Tree of Life (TOL) metaphor is useful, and a code built on principles of PN will be preferable over a theory-free code that can adapt to a variety of ontologies of taxa. This is because a custom-made solution is typically better-suited for the case it was designed to handle than a general solution. An analogy can be made with parametric statistics, which are usually more powerful than their non-parametric equivalents (Mumby 2002: 85). Similarly, a suit made to match someone's measurements will typically fit better than a ready-to-wear suit designed for a population of men of a given height or mass. Conversely, opponents of PN have argued that the mere existence of reticulation implies that the TOL metaphor is inadequate (Dubois 2005: 374; Dubois et al. 2021: 18), and that this undermines the foundations of PN.

Thus, the key question that needs to be answered to determine if PN is advantageous is whether or not divergent evolution is prevalent (Dubois 2005: 374). For eukaryotes, this appears to be the case for most taxa. However, even among eukaryotes, hybridization, which often produces allopolyploidization, can create reticulation between closely related taxa, especially among embryophytes, in which the prevalence of this phenomenon is still being assessed (e.g., Debray et al. 2022). For bacteria and archaeans, evolution may be even less predominantly divergent than in embryophytes. Koonin et al. (2001: 717) estimated that 15%–20% of the prokaryotic genes were recently acquired through horizontal transfer. These approximate values hide much heterogeneity, from about 1.6% of the genes of *Mycoplasma genitalium* to 32.6% in *Treponema pallidum*. Koonin et al. (2001: 712) even stated:

> Thus, a true tree of life, a species tree, could not be constructed, not because of the complexity of the problem and erosion of the phylogenetic signal from ancient divergence events, but perhaps in principle (22, 23). The best one could hope for was a consensus tree that would reflect the history of a gene core conserved in all or the great majority of species and not subject to horizontal gene transfer. But the very existence of such a stable core and more so its actual delineation remain questionable.

The Network of Life preceded the Tree of Life model historically, as Ragan's (2009) historical survey showed. Ragan (2009) argued persuasively that Vitaliano Donati's *Della storia naturale marina dell'Adriatico*, published in 1750, is actually an early Network of Life. He also argued that the Tree of Life model was being replaced, for many prokaryote taxa, by a Network of Life, but that some taxa (especially among eukaryotes) diversified in a tree-like fashion.

This idea of a Network of Life for many prokaryotic taxa remains controversial, as shown by the various comments by reviewers (especially Eric Bapteste and Patrick Forterre) at the end of Ragan's (2009) review, but it is supported by several bacteriologists. Thus, Doolittle (2009: 2225) argued that probably less than 10% of the genes of these taxa have not been subject to lateral gene transfer (LGT). In fact, a study that attempted to produce a tree of major prokaryotic taxa used a procedure to filter gene sequences to exclude genes that had undergone LGT and ended up with 31 genes (Ciccarelli et al. 2006). Dagan and Martin (2006) pointed out that the average prokaryotic genome includes about 3,000 protein-coding genes and that 31 genes out of about 3,000 is a mere one percent; they thus nicknamed this tree the "*Tree of One Percent*"! However, there are good reasons why Ciccarelli et al. (2006) used only 31 genes; namely, they excluded not only genes that had undergone lateral transfer, but also paralogues, pseudogenes and retained genes present in bacteria, archaea and eukaryotes. Nevertheless, some systematists have argued that the search for the Tree of Life (TOL), like that of the last universal common ancestor (often called LUCA), is doomed to failure if the actual pattern is a Web of Life as argued by some prokaryote specialists (such as Doolittle 2009). However, the extent of horizontal gene transfer is "still extremely controversial" (Ciccarelli et al. 2006: 1284), and even the bacteriologists who argue for a Web of Life admit that horizontal gene transfer is much less pervasive among eukaryotes, in which the TOL still explains much of their biodiversity. Thus, the relative importance of LGT vs divergent evolution may vary between distantly related life forms, as summarized by this quote from Doolittle (2009: 2227):

> One can agree that LGT will **likely not interfere** with a **robust phylogeny for primates** that indeed recreates ancient splittings of their populations, without buying in to the notion that the supposed sisterhood of Thermotogales and Aquificales revealed by a small subset of their shared ribosomal protein genes reflects an even remotely similar evolutionary process. We know the mechanisms by which prokaryotes evolve, and like the driving forces listed by Gould, they are **various in their frequency, intensity and consequences** for the evolution of genomes. Why should we expect them together to produce a **single pattern**, a fractal one at that, good over all time and for **organisms of all types**?

Evolutionary mechanisms are much better understood among eukaryotes than among archaeans and bacteria. To emphasize this fact, Doolittle (2009: 2227) quoted a passage from Maynard Smith (1982: 5) that he indicated was still highly relevant:

> The modern synthesis of the 1940s was concerned with eukaryotes . . . . Its essential achievement was to bring together two previously separate disciplines—the chromosome theory of heredity and the study of natural populations. The same synthesis is now required for the prokaryotes. There is an abundant knowledge of their genetics, but as yet no

adequate synthesis of that knowledge with a study of the natural history of bacteria. For example, we have little idea of the significance of conjugation for bacterial populations; it is as if we had no idea of the significance of sexual reproduction for populations of birds and insects. Population thinking has been well developed for fully half a century, but has yet to be adopted by microbiology.

Recent research on bacteria and archaeans does support the conclusion that reticulation is widespread in these taxa (Mallet et al. 2015). In the case of eukaryotes, the predominantly divergent nature of evolution seems well-established, and it has been supported by the congruence between trees based on molecular data and on other types of evidence, chiefly morphology (Doolittle 2009: 2224; Lecointre and Le Guyader 2016: 6). However, even among eukaryotes, reticulation is probably more prevalent than previously thought (Mallet et al. 2015), but not sufficient to hamper productive application of PN to these taxa.

In addition to horizontal gene transfer, the phenomenon of endosymbiosis departs from a pure Tree of Life evolutionary model. Thus, it is commonly acknowledged that eukaryotes arose through endosymbiosis of alphaproteobacteria (which gave rise to the mitochondria present in most eukaryotes) and an archaean, which provided the DNA that is now in the nucleus. Subsequently, chloroplasts (present in photosynthetic eukaryotes) were probably acquired multiple times through endosymbiosis of cyanobacteria in various eukaryotes (Mallet et al. 2015: 140). A related phenomenon, endosymbiotic gene transfer, allows genes to be transmitted from organelles to the nucleus, which complicates the pattern further.

If the Web of Life model is correct for most life forms, the *PhyloCode* (Cantino and de Queiroz 2020) should not be expected to yield better results than RN. But is this likely? The proliferation of trees of prokaryotic taxa (e.g., Wolf et al. 2018) militates against Koonin et al.'s (2001) pessimistic prediction, although Doolittle (2009: 2223) is correct in stating that tree-building algorithms necessarily produce trees and hence, mere publication of trees is weak evidence that the TOL is the right evolutionary model. In his comments at the end of Ragan's (2009) review, Patrick Forterre summarized another point of view that supports the Tree of Life metaphor, even among prokaryotes:

> The hereditary history of living organisms can be depicted with a **tree-like** structure as long as new organisms originate by cell division. This seems to be the rule for most living organisms, since examples of fusion that prevent us from identifying the continuity of cellular lineages are rare. For instance plants are clearly a eukaryotic lineage that can be inserted into a eukaryotic tree, and not a peculiar lineage of cyanobacteria. Similarly, mammals are clearly a branch in the tree of animals, and not a peculiar form of retrovirus despite the fact that retroviruses and derived elements comprise up to 80% of their genomes. It was one of the great successes of science in the XIX century to realize that organisms are related to each other via a tree and not a network.

This was achieved thanks to progress in evolutionary theory and it's not a coincidence that the only illustration of the *Origin of species* is precisely such a tree.

In his comments, Forterre thus concluded that networks are a good metaphor to describe the movement of genes between genomes, but this occurs in a tree of organisms. Where much horizontal transfer has occurred, this tree is difficult to study, but it can theoretically be inferred.

Another recent development suggests that LGT is not so pervasive as to prevent the reconstruction of a biologically meaningful Tree of Life. Davín et al. (2018) devised a method to obtain relative node ages using LGT and successfully applied them to three portions of the Tree of Life: *Cyanobacteria, Archaea* and *Fungi*. A series of tests suggests that the method works well and yields plausible results. The fact that this method worked even for *Cyanobacteria* and *Archaea* suggests that the Tree of Life metaphor is useful even among prokaryotic organisms. All this bodes rather well for PN, but it may still be a little more difficult to apply PN among prokaryotic taxa than among eukaryotes.

## 6.5 Should the Codes Provide Rules to Delimit Taxa?

### 6.5.1 Deliberate Absence of Delimitation

A major source of nomenclatural instability in RN is the lack of delimitation of taxa provided by the rank-based codes. This is no accident; the regulating bodies of at least some of the rank-based codes aim at not regulating the delimitation of taxa. Principle 2 of the introduction of the *Zoological Code,* already quoted in Section 2.4.1 but highly relevant here, states:

> Nomenclature does not determine the inclusiveness or exclusiveness of any taxon, nor the rank to be accorded to any assemblage of animals, but, rather, provides the name that is to be used for a taxon whatever taxonomic limits and rank are given to it.
>
> **(ICZN 1999)**

This is actually inaccurate because rank affects names under RN, notably because of mandatory endings, and names differ more strongly between nominal series (for instance, between species, genus, family and class series). A similar statement regarding lack of delimitation occurs in the prokaryote code ("General Consideration 4"; Parker et al. 2019): "Rules of nomenclature do not govern the delimitation of taxa nor determine their relations." This feature has been claimed both to enhance nomenclatural flexibility by its proponents, and to generate chaos by its detractors (Laurin 2008). However, among fields of human knowledge, only RN seems to seek not to delimit the entities whose names it regulates (see Chapter 5). Even diagnoses characterize, rather than delimit, taxa (see Section 3.7). Contrast the previous quotes with this statement in the preamble of the *North American stratigraphic code:*

Any system of nomenclature must be sufficiently explicit to enable users to distinguish objects that are embraced in a class from those that are not. . . . Sufficient care is required in **defining the boundaries** of a unit to enable others to distinguish the material body from those adjoining it.

**(North American Commission on Stratigraphic Nomenclature 2005: 1556)**

Strangely, most proponents of RN do not seem to have realized that by not regulating the delimitation of taxa (indirectly, through definitions proposed by systematists), the goal of nomenclatural stability cannot be reached. Thus, Löbl (2015) admitted that the *Zoological Code* has had limited impact on nomenclatural stability, and discussed six sources of nomenclatural instability (under that code): changes of nomina due to changed taxonomic rank (1) or due to transfers of taxa (2); replacements of nomina due to homonymy (3), to synonymy (4), or to discovery of misidentified type species of genera (5) and changes in spelling of nomina (6). In his discussion of synonymy (4), Löbl (2015: 37) mentioned "increasing bureaucratic requirements coupled with insufficient resources allocated to curators of museums" and "unrecognized variations, overestimated value of characters, or erroneous choice of characters used to define taxa." One of these sources of instability (1, change in taxonomic rank) is illustrated by a large group of beetles. Löbl (2015: 36) reported that "The Palaearctic species have been placed in 91 subgenera . . . actually used as such by about half of the taxonomists, while the second half of workers holds these subgenera for genera." Strangely, the whole paper does not even mention that the main source of nomenclatural instability in two of these situations (1 and 4) simply reflects differences in preferences of systematists in the rank attributed to taxa and conflicts between splitters and lumpers, rather than genuine errors or limitations in human resources. Löbl (2015: 39) concluded that "As any taxon can be split and any related taxa might be lumped, **stability in nomenclature is utopian** and the devaluation of the principle of priority may easily turn into the opposite of needs." There are two problems with this statement. First, taxa can be discovered, not split (de Queiroz and Gauthier 1992: 465); Löbl (2015) confused choosing to apply a name to a more or less inclusive clade with splitting the taxa (clades) themselves; the latter is not up to taxonomists. Second, obviously, Löbl (2015) did not consider the simple solution to this central problem in nomenclature: provide a code that allows nomenclature to delimit taxa, and the main source of nomenclatural instability (that which depends on subjective decisions and preferences of systematists) will disappear.

However, some proponents of RN advocated delimiting taxa precisely, and this is not (always) a reaction to the development of phylogenetic nomenclature. Thus, Camp and Gilly (1943: 327) stated:

As here defined, Biosystematy seeks (1) to **delimit** the natural biotic units and (2) to apply to these units a system of **nomenclature** adequate to the task of conveying **precise information** regarding

their defined **limits**, relationships, variability, and dynamic structure.

This quote clearly shows that Camp and Gilly (1943) thought that biological nomenclature should allow systematists specify taxon delimitation precisely, and that this delimitation should be as objective and as permanent as possible (rather than being arbitrarily changed in every systematic revision of a taxon) decades before the foundations of PN were laid down.

## 6.5.2 DONS: Nearly Objective Name Allocation

### 6.5.2.1 Introduction to the DONS

Some of the staunchest proponents of RN who have criticized PN have nevertheless made proposals to allocate the names on a tree in a more objective manner than has been done so far in RN. Thus, Dubois (2015) proposed the *Duplostensional Nomenclatural System (DONS)*, which was further elaborated by Dubois et al. (2021). It is designed to apply to names of taxa of the "class-series," which covers all nomenclatural levels above the family-series. Under the *DONS*, taxon names of the class series that have been used in at least 100 titles of scientific publications after 1899 are attributed to the most inclusive taxon that includes all nucleogenera (genera initially included in the taxon) and excludes all alienogenera (those that were initially and explicitly excluded from it). By contrast, "class-series" names that have been more rarely used (in less than 100 titles of scientific publications after 1899) are attributed to the least inclusive taxon that includes all its conucleogenera, the set of genera initially included in the taxon (Dubois et al. 2021: 69).

This proposal (*DONS*) exhibits some similarities and differences compared to the *PhyloCode* and the *Zoological Code*. Like the *PhyloCode*, it uses objective criteria to allocate names to taxa (although it uses nucleogenera and alienogenera included in the original study, rather than phylogenetic definitions), and this system is meant to name clades (Dubois et al. 2021: 10). The alienogenera (Dubois et al. 2021: 367) are vaguely analogous to external specifiers because they are explicitly excluded from the taxon, with the substantial difference that the alienogenera are determined from the initial publication (or an early revision if this cannot be determined from the publication in which the name was erected). Likewise, the nucleogenera are vaguely analogous to internal specifiers, but like the alienogenera, they are determined from the publication in which the name was erected (or the first revision that mentioned nucleogenera, if the first publication failed to do that).

Like the *Zoological Code*, the *DONS* is a rank-based system; it uses information already in the literature and does not require formulating new definitions. However, its aim at allocating the names on a tree in a fairly objective manner bears vague resemblances with PN's aim of delimiting taxa, though name allocations under the *DONS* are less automatic and less permanent than delimitation under PN because the same clade could bear different names under different trees, even if these are mutually compatible. This proposal arguably goes against Principle 2 of the introduction of the *Zoological Code* (ICZN

1999), which states that "Nomenclature does not determine the inclusiveness or exclusiveness of any taxon." Unlike both the *Zoological Code* and the *PhyloCode*, this system in theory avoids having to make new subjective choices about a name/delimitation combination (but see the following for caveats).

### 6.5.2.2  Problems Raised by the DONS

The *DONS* thus has an interesting mix of properties that could appeal to several systematists, but it suffers from a major problem: its extreme complexity, which can be demonstrated in various ways. For instance, the section (pp. 6–112) of Dubois et al. (2021) that explains nomenclatural availability of taxon names and the *DONS* includes 75,344 words. However, it is not intelligible without the 40-page glossary, which includes the definitions of many new words invented by Dubois over the years, as well as new definitions (which may differ from those of the *Zoological Code*) of some older terms; this glossary includes 35,295 words. Thus, together, these two sections of Dubois et al. (2021) include 110,639 words, which is 2.58 times more than the *PhyloCode* (excluding the index), which includes 42,839 words. But length isn't everything. Writing style and vocabulary use can make a text easy or very difficult to read. In this respect, the presentation of *DONS* in Dubois et al. (2021) is no doubt one of the most difficult texts that I have read, as can be easily shown by a few quotes. Let's start with two paragraphs that appear near the end of the section that explains *DONS* (Dubois et al. 2021: 69):

> Allocation of CS [class series] nomina to taxa under the DONS Rules is simple and straightforward. It depends however on the ***category of usage*** to which the nomen is referred. According to this category of usage, a different nomenclatural subsystem of DONS will be used for the taxonomic allocation of the CS nomen.
>
> {b1} ***Metrostensional Nomenclatural Subsystem (MONS)***. If the CS nomen is a *distagmonym*, its taxonomic allocation relies solely on its ***onomatophore***, for example on the list of its ***conucleogenera*** (or on its single ***uninucleogenus***), for example all the ***available*** nominal genus-series nomina originally and unambiguously referred as ***valid*** to the taxon for which the CS nomen was proposed. This list is an ***indissoluble*** set of available nomina which act **altogether** as the onomatophore of the CS nomen at stake. Then, within any ergotaxonomy adopted as valid, this nomen is a ***nesonym***, which applies to the ***metronym***, for example the **least inclusive** CS taxon which contains **all** these nucleogenera (the ***metrotaxon*** of the nomen in this ergotaxonomy). This provides an unambiguous allocation of the nomen to a single CS taxon in the ergotaxonomy adopted.

> **(Emphasis in original; bold italics designates terms found in the glossary of Dubois et al. [2021]; bold [without italics] was used by Dubois et al. [2021] for emphasis.)**

To understand these two paragraphs, even a highly trained systematist will need to consult the glossary to learn the meaning of several terms. While some terms are defined using classical vocabulary, many others are not. Consider, for instance, the definition of nesonym (Dubois et al. 2021: 386):

> **Nesonym**, n. ● AL. ● **Ety**: G: νῆσος (nesos), 'island'; ὄνομα (onoma), 'name'. ● Class-series ***distagmonym***, taxonomically allocated within the frame of a given ***ergotaxonomy*** under DONS Criteria through its ***metrotaxon***, without reference to its ***orotaxon*** if present, and being therefore its ***metronym***. ● Dubois 2006a: 188. One of the two meanings of the term ***nesonym*** as defined by Dubois (2015c: 65), hereby distinguished from the term ***ellitonym*** and used in this restricted meaning. ● Code: no term.

To understand this glossary entry, the reader may want to read several more glossary entries, namely ***distagmonym***, ***ergotaxonomy***, ***metrotaxon***, ***orotaxon***, ***metronym*** and ***ellitonym***. Fortunately, most of these are explained using classical vocabulary, except for the last one (***ellitonym***), which reads (Dubois et al. 2021: 376):

> **Ellitonym**, n. ● AV. ● Ety: G: ἐλλιτής (ellites), 'lacking, defective'; ὄνομα (onoma), 'name'. ● Class-series nomen that misses an ***onomatostasis*** (***alienogenera***) and that therefore can be validated only as a ***metronym*** under the ***Ostensional Nomenclatural Systems***. ● One of the two meanings of the term ***nesonym*** as defined by Dubois (2015c: 65), hereby distinguished from the latter. ● Code: no term.

> **(Typography as in the original.)**

Again, to understand this term, the reader will need to consult additional glossary entries, namely ***onomatostasis***, ***alienogenera*** and ***Ostensional Nomenclatural Systems***. Some of these entries will require reading additional entries to be intelligible. Clearly, readers with an eidetic memory should read the entire glossary before reading this paper! And for the vast majority of readers not thus endowed, understanding this paper will require much determination and patience. Arguably, the *DONS* is more complex than need be, as can be illustrated by the justification provided by Dubois et al. (2021: 17–18) for not using the term "clade," which many authors consider to be unproblematic and easy to define:

> In the present work, we refrained from using the term ***clade***, as it is highly confusing (see Glossary below). It has been used in the literature in at least four distinct meanings, in zoological taxonomy and nomenclature to designate a nomenclatural CS rank and more recently as a CS and FS [family series] preudo-rank, and in evolutionary biology as a homophyletic or holophyletic group of organisms. In many recent publications it is used simultaneously in both the second and fourth of the meanings above.

> **(Typography as in the original.)**

Indeed, Haeckel (1866) first coined the word "cladus" (which became "clade" in other languages), along with a few other terms that acquired great importance in systematics (Tassy and Fischer 2021), such as "monophyletisch" ("monophyletic" in English) and "Phylogenie" ("phylogeny"). Haeckel (1866) proposed this word for a formal rank between Phylum and Class, as stated by Dubois et al. (2021: 373). A few other authors, such as Hatschek (1888) and Cuénot (1940a, 1940b), also used the word "clade" for a rank (equivalent to Phylum), but after these early studies, as far as I know, the word has been consistently used to designate monophyletic taxa in all studies that I know of, which are presumably representative of the relevant publications. As far as I know, Dubois (2006: 218) has been the only author to recently list "Cladus" and "Cladoma" as a rank (although Tassy and Fischer 2021 discussed this in their historical review, but argued against its use). Does the use of the word to designate a rank in a few, mostly very ancient studies, and its bare mention in a few more recent studies, render the meaning of the word "clade" so vague as to require its replacement? The extensive glossary of Dubois et al. (2021) and the highly technical language used in that paper seems to reflect the point of view that most of the vocabulary used by the vast majority of systematists requires replacement by a high number of new terms. The result, of course, is a language that is foreign to most systematists.

There are other problems with the *DONS*. Dubois et al. (2021: 61) state that the DONS:

> allows to find the valid and correct nomen of **any** taxon of a given group of organisms under **any** taxonomic arrangement, in **all** situations and in an **unambiguous, automatic, repeatable** and **universal** manner. This means that such a system does not leave room for interpretations, discussions and debates.

**(Typography as in the original.)**

However, the first of the ten criteria that Dubois et al. (2021) use to name suprageneric taxa, the "Consistent Naming Criterion" (CNC for short), requires an arbitrary choice. This criterion consists in recognizing a minimal node support value (in their case, 90%) to recognize these nodes as taxa. Obviously, some authors might think that a different threshold is more appropriate, and depending on the data and analytical method, these nodal support values would not be the same, and this would most likely result in a different nomenclature, even if the topology recovered were the same. For instance, parsimony bootstrap, maximum likelihood bootstrap or jacknife, and Bayesian posterior probabilities yield notoriously different support values (and often different topologies), although they tend to be congruent in identifying the best-supported clades (e.g., O'Reilly et al. 2018: fig. 6; Sanders et al. 2021: fig. 2). Thus, the *DONS* can neither ensure to recover the same nomenclature from two datasets that overlap extensively (or even from a single dataset analyzed with different methods to assess clade support), nor does it do so automatically. In this respect, it is much less automatic and unambiguous than the *PhyloCode*, and is much less efficient at stabilizing nomenclature given that with the *DONS*, the nomenclature needs to be reestablished for each tree (or even for a given tree, when support values change).

Using the *DONS* is also very time-consuming. For the very first step of the analysis, you must (Dubois et al. 2021: 103–104) "Build up a database of all available genus-series (GS) nomina in the group studied, with their nucleospecies ('type species') and a database of all available family-series (FS) nomina in the group, with their nucleogenera." The second step is not faster (Dubois et al. 2021: 104):

> Build up a database of all the FS nomina of the group ever used as valid at the rank family in at least one of 100 published comprehensive classifications or more since 1758, count their respective numbers of usages, sort them into four quarters and list those belonging in the upper quarter (UQ).

Other, later steps of the procedure are also time-consuming. It is difficult to envision that anybody but the most passionate about biological nomenclature would ever try applying this method. The *DONS* does not appear to meet one of the 11 desirable features of nomenclatural systems proposed by Dubois (2005: 375), namely, simplicity.

An additional problematic aspect of the *DONS* is that rather than minimizing the use of artificial ranks in nomenclature (something that will arguably need to be done in the future if we are ever to get a nomenclature that reflects the natural hierarchy of taxa), it increases dependency of zoological nomenclature on these nomenclatural ranks. This proposal to develop the *Zoological Code* is thus, in this respect, diametrically opposed to PN. The *DONS* can also destabilize nomenclature in at least some cases. Thus, in their discussion of *Gymnophiona* (which they consider to be an order), for which applying the *DONS* would lead to recognizing only three families, Dubois et al. (2021: 109) stated:

> This would not bother us much, but we are aware that most taxonomists have an immoderate fondness for "taxonomic stability", a non-scientific concept, and would probably be very "shocked" by a move from 10 families of caecilians as advocated by San Mauro et al. (2014) to three families! For this reason, **we decided to derogate**, at least provisionally, from our general Criteria in this case, and to recognise for the time being **five families** within this order. For this to be possible, it is necessary to add one **"superfluous" rank** to the taxonominal hierarchy in this order, and, by symmetry with the other two orders, we recognised two suborders in the latter. Then, we have one suborder with a single family and a second suborder with two superfamilies including two families each.

Clearly, applying the *DONS* to *Gymnophiona* would destabilize nomenclature of this taxon, given that San Mauro et al. (2014) largely followed an established trend (documented in Wilkinson et al. 2011) in recognizing an increasing number of families within *Gymnophiona*. It would also require recognition of redundant taxa, and all this certainly does not militate for integration of such rules into the *Zoological Code*.

These problems, especially the extreme (and arguably unnecessary) complexity, may explain why the *DONS* has not been adopted by many other systematists. Dubois et al. (2021) is well-cited, with 81 Google Scholar citations as of October 29, 2022, but apparently mostly for other aspects, such as the detailed phylogeny, taxonomy, and general points about amphibian conservation and the lack of herpetologists, according to my survey of the citing references. Perhaps more telling, Dubois (2015), who introduced the *DONS*, is cited by 29 references, but these are mostly papers authored (or co-authored) by Dubois (26 out of the 29 citing papers). Nevertheless, numerous concepts and most of the vocabulary used in the Linz *Zoocode* project (Dubois et al. 2019; see Section 6.9) are derived from the *DONS*. To conclude, the *DONS* is far too complex and does not provide enough nomenclatural stability to be a viable alternative to PN, and it seems that systematic community has realized this. However, as an attempt at providing objective, automatic name allocation to (ranked) taxa in the context of a tree, it exhibits some interesting parallels with PN and shows that some of the most enthusiastic proponents of RN realise the need for more objective allocation of names to taxa. In this perspective, it deserved to be presented in this chapter.

## 6.6 Should Phylogenetic Definitions Change Along with Phylogenies?

Härlin (2005) made the point that as phylogenies change, the content of clades whose names were defined according to the *PhyloCode* also changes. Hence, under PN as implemented under the *PhyloCode*, a given name applies to a different set of more or less inclusive taxa under various phylogenies. Härlin (2005) viewed this as problematic because these clades are not really the same, and he proposed that "Stability in either/or both names and content is a utopia that does not fit an evolutionary and scientific world-view. Both **names and reference should be allowed to change**. That is, taxon names need not be unique—only traceable."

But is such nomenclatural flexibility desirable? RN has always provided flexibility to re-delimit taxa, given that ranks do not exist in nature, but the result has often been nomenclatural chaos, as shown by Rowe and Gauthier (1992) for *Mammlia* and by Keesey (cited in Laurin and Bryant 2009: 336) for *Hominidae*. Härlin's (2005) proposal could minimize changes in contents of a given named clade, but this would come at the price of a steady flow of new taxon names and definitions for a given set of terminal taxa, which could generate confusion, with potentially several parallel proposals being developed for intensively studied, speciose clades. The fact that it has not been widely accepted (Härlin's 2005 paper had nine Google Scholar citations on August 27, 2022) suggests that some of these concerns are shared by other systematists.

## 6.7 Should the *PhyloCode* Regulate Name Selection?

As we saw in Sections 4.2.3 (about crown clades) and 4.4.3 (about total clades), the *PhyloCode* includes rules and recommendations about naming clades that differ rather strongly from those of the rank-based codes. In the latter, rules are technical requirements on proper name formation (notably regarding Latin grammar) but do little to constrain the selected name. In the *PhyloCode*, many of the best-known taxon names, such as *Biota, Tetrapoda, Amniota* and *Mammalia*, are tied to crown clades, and many total groups have received panclade names, such as *Pan-Biota, Pan-Amniota* and *Pan-Mammalia*, and the current version of that code ensures that such practices will continue. This approach has advantages and drawbacks. On the plus side, the meaning of many popular names will be obvious because it will refer to a crown clade, which will obviate the need to wonder where on the stem the limit of the taxon is, and the meaning of panclade names is also intuitive. On the downside, systematists are attached to their academic freedom, and in nomenclature, name formation is a key step. As McShea (2000: 330) commented, "It has been said that most scientists would rather use another scientist's toothbrush than his terminology."

Of course, if some articles of the *PhyloCode* prove unpopular, the CPN (Committee on Phylogenetic Nomenclature) has the authority to revise or remove them, and it has proven efficient in this task in the past, as shown by the history of the *PhyloCode*, which has changed substantially since the first version was posted on the Internet in 2000. Indeed, the articles and recommendations about crown clade and panclade names first appeared in version 3 of the *PhyloCode*, which was posted on June 16, 2006. These articles were adopted only after extensive discussion, to which I participated as CPN member, and while they represent to some extent a compromise, not all participants in these discussions were satisfied with the end result. These articles may well be revised in the future.

Note that this is not a debate about PN vs RN. Rather, this is a debate within the PN community, and PN does not require such naming conventions. Rather, PN facilitates adoption of such nomenclatural rules because crown and total clades are phylogenetic concepts; as such, their names can be regulated by PN, but not by RN, given that there is no relationship between rank and kind of clade (crown, total or neither) or even monophyly, and that some RN codes (*Zoological* and *Botanical*) do not even require that taxa be monophyletic (see Chapter 2).

Trying to thus constrain name choice has rarely (perhaps never) been attempted in the history of systematics. Thus, in this respect too, the *PhyloCode* departs from nomenclatural tradition, and it is still too soon to tell how the systematic community will react to these aspects of the *PhyloCode*. Will these constraints in name formation repel more systematists than the intuitive aspect will attract?

## 6.8 Confusion, Instability and Progress

### 6.8.1 How Much Instability Is the *PhyloCode* Causing?

Some proponents of RN have argued that the use of PN is generating too much nomenclatural instability. This is inaccurate because PN provides a way to fix delimitation of taxa to one of the most widely accepted usages in a given phylogenetic context. Dubois (2005: 393), who has criticized PN along these lines, nevertheless has long been advocating delimiting genera using hybridization potential (see Chapter 3), even though he admitted that this would strongly alter the nomenclature of many taxa (Dubois 1982: 30). As such, introducing this hybridization-based concept of genera would yield much more nomenclatural instability than PN and the *PhyloCode* (Cantino and de Queiroz 2020).

Confusion was also introduced by some recent changes in a rank-based code. Indeed, Dubois et al. (2013) argued that the new articles in the *Zoological Code* that allow nomenclatural acts to be validly published in electronic publications are confusing, a claim that has been disputed (ICZN 2014). Yet, on this point, zoological nomenclature has moved slowly, at least compared with stratigraphic nomenclature, which allowed electronic publication of "new and revised names" decades earlier, in 1983 (North American Commission on Stratigraphic Nomenclature 2005: 1555). It seems that even in RN, progress is closely linked with confusion, at least as a temporary condition until systematists become familiar with new rules. This was noted long ago in the American Ornithologists' Union (1886: 12) code:

> The case of an unstable and far from uniform system of nomenclature no more shows the need of improvement, than admits of those changes which are necessary; and though the **evils inseparable from all states of transition** may be obvious, they are themselves no less **transitory**, while the **good results** of the strict and consistent application of sound principles of nomenclature are **likely long to endure**.

Thus, nomenclatural instability has long accompanied development of RN, and it does not appear to be the real reason why some proponents of RN are hostile to PN and the *PhyloCode*. Furthermore, even the status quo entails much confusion within RN. As stated by de Queiroz and Gauthier (1992: 465):

> Under the current system, different authors use the same name for different clades and different names for the same clade, and this can happen as the result of **subjective differences** concerning assignments to Linnaean categories **even when the authors are in full agreement about what organisms and species make up the taxa**.

## 6.8.2 Minimal and Maximal Nomenclatural Stability

The controversy about which system (PN or RN) best promotes nomenclatural stability has persisted to an extent because previous discussions of this topic failed to distinguish between the **minimal** and the **maximal nomenclatural stability** ensured by both systems (Laurin 2008). **Minimal nomenclatural stability** (MiNS from here on) is provided automatically by a nomenclatural system (such as PN or RN) if systematists abide by the relevant codes, whereas **maximal nomenclatural stability** (MaNS from here on) is what can be achieved

if systematists agree on nomenclature, even though codes do not constrain them to do so. MiNS will thus be greater under nomenclatural rules that tightly link taxonomic content and names, whereas MaNS should be maximized by flexible nomenclatural rules that allow systematists to spontaneously agree on a set of names for a corresponding set of taxa, even if their contents change. Which of these two kinds of stability (minimal or maximal) should be optimized in nomenclature depends on how optimistic we are about systematists spontaneously agreeing about taxon names and delimitation.

The history of systematics in the last three centuries at least suggests that optimizing MiNS should be our goal because systematists do not generally agree spontaneously. This can be assessed by examining the **realized stability**, which is the stability or lack thereof that has prevailed, and which has arguably been rather poor. This is shown by various empirical examples, such as *Microsauria* (Carroll and Gaskill 1978), which lost all its original taxonomic content over time, *Rana* (Dubois 2007), which once included all anurans, and with which systematists are still struggling to name the hierarchy of its hundreds of nominal species (see Section 3.5), *Mammalia* (Rowe and Gauthier 1992) or *Hominidae* (Laurin and Bryant 2009; Groves 2018), which have designated a variety of nested clades (see Sections 4.1 and 3.4.1, respectively). Indeed, this was the main reason why codes of nomenclature were drafted in the first place (see Chapter 2), and the numerous nomenclatural debates that have continued to occur in the last decades (see Section 6.1) demonstrate that such codes are more needed than ever.

Which nomenclatural system (RN or PN) best promotes **MiNS**? Laurin (2008: fig. 1) already demonstrated, using an example with four terminal taxa, that RN yields very little MiNS (see Section 3.2.3 and **Figure 3.10**). This is unsurprising because it ties a name to a type, which exists in nature, and a rank, which is an artificial construct. By contrast, PN performs better because it links a name to a clade through a phylogenetic definition. The content of that clade (under PN) can change only if additional taxa are discovered or if the phylogeny changes, and such scientific progress may also require change in the contents associated with names under RN, at least if monophyly is to be maintained. To conclude, the greater relevance of minimal (rather than maximal) nomenclatural stability and the better performance of PN in this respect indicate that PN (and hence, the *PhyloCode*) enhance MiNS and arguably, realized nomenclatural stability.

## 6.8.3 How Much Confusion Do the Current Codes Generate?

Confusion is sometimes invoked as an argument for change. This was the very point made by Strickland et al. (1842: 2) to argue that a code of nomenclature was needed (quoted in Section 4.1). However, mere willingness to adhere to codes of nomenclature is not sufficient to stabilize nomenclature, as shown by the development of RN, in which ranks have not contributed as much to nomenclatural stability as systematists had hoped (see Chapter 3).

Another source of nomenclatural confusion and instability under RN is the complexity of its rules regarding name availability. As mentioned already in Section 2.4.3 on the *Botanical Code*, by the 1920s, the nomenclatural rules had become so complex that even experts in the field disagreed on their interpretation, and nomenclature has not improved in this respect since then, at least in zoology and botany. For instance, in the *Zoological Code*, two chapters (4, 7) and parts of others (8–10) are devoted to determining which names are available (this is called "validly published" under the *Botanical Code* and *Prokaryotic Code*, or "established" under the *BioCode* and *PhyloCode*). Rules to determine which names are valid (called "correct" under the *Botanical Code* and *Prokaryotic Code*, or "accepted" under the *BioCode* and *PhyloCode*) are also complex and are covered by chapters 8–10 and 12–16. The *Botanical Code* has a similar complexity in these respects. By contrast, the *PhyloCode* covers establishment ("availability" for the *Zoological Code*) of names in chapters 3–4, and acceptance ("validity" for the *Zoological Code*) of names in chapter 5. Much of this streamlining results from the absence of ranks, given that under the RN codes, name formation and validity are rank-dependent, so several sets of rules are required.

An additional, arguably unnecessary complexity in the RN codes stems from insistence on etymology and Latin grammar, especially for binominal names. This problem had already been noted long ago (e.g., Griffiths 1981: 547), and it may contribute to nomenclatural instability, notably through debates about name availability and correct spelling. Thus, under the *Zoological Code*, a family-name must "be a noun in the **nominative plural** formed from the stem of an available generic name" (Article 11.7.1.1), genus names must be nouns in the nominative singular (Article 11.8) and specific epithets can be "an adjective or participle in the nominative singular," "a noun in the nominative singular standing in apposition to the generic name," "a noun in the genitive case" or "an adjective used as a substantive in the genitive case and derived from the specific name of an organism with which the animal in question is associated" (Article 11.9.1). Latin was the language of the intellectual elite in Europe in the Middle Ages, but its prevalence has since dwindled, a trend presumably accelerated by the growth of the scientific community in emerging countries where Latin was never spoken much. It may no longer be appropriate to give that much weight to Latin grammar in scientific nomenclature. In line with this argument, regarding name formation, the *PhyloCode* requires only that names be composed of at least two letters from the Latin alphabet as used in English (considered to consist of 26 letters) and begin with a capital letter (Article 17.1).

While the RN codes contain many rules about formation and emendation of names, they leave room for ambiguity. This point was made by Dubois et al. (2021), who extensively discussed nomenclatural availability (over 33 pages of highly technical text and tables). They listed 36 situations that make names or nomenclatural acts unavailable (Dubois et al. 2021: table 4). The extent to which the *Zoological Code* fails to capture the complexity of this topic is explained in the following quote (Dubois et al. 2021: 45):

> These situations are much more varied than many taxonomists believe. Many authors think that the formula ***nomen nudum*** applies to all ***anoplonyms***

(unavailable nomina), but this is incorrect. The Glossary of the Code clearly defines *nomen nudum* as referring to a nomen that, if published before 1931, fails to conform to Article 12, or, if published after 1930, fails to conform to Article 13. This applies to only three of the 36 situations described in Table T4.AVN (Av-16, Av-31, Av-32).

This demonstrates, *ad absurdum*, that to correctly apply the current nomenclatural principles that underlie name availability in the *Zoological Code* (and presumably in other RN codes as well) is extremely complicated, and certainly much more cumbersome than warranted by the purpose of biological nomenclature. How much stability can nomenclatural rules provide if they are so complex that few systematists understand them?

### 6.8.4 Confusion and Progress: A Computer Science Analogy

Operating systems of personal computers provide a very good example of how progress may require drastic changes that generate confusion during a transition period (Laurin 2008). Shortly after it was released in 1981, the Microsoft Disk Operating System (MS-DOS) dominated the personal computer market and retained this position until approximately the mid-1990s. In retrospect, this command-driven operating system (OS) does not seem particularly easy to use because software written for it typically required memorizing dozens of commands to perform everyday tasks. Nevertheless, years after using it, many users had grown comfortable using MS-DOS.

The world of personal computing changed drastically with the advent of operating systems with a graphical user interface (GUI). First, Apple launched a new GUI-based OS (later called Mac OS) in 1984. It was quickly followed by Microsoft Windows in 1985 and by OS/2, which IBM released in 1987.

The release of all these GUI-based operating systems created much confusion among the millions of computer users who had become familiar with MS-DOS over the years, because the familiar command line was no longer there. In addition, software that worked under MS-DOS was incompatible with some of these OSs (many programs for MS-DOS worked under Windows, but not in the other OSs, and this must have contributed to give Windows its market dominance). A given computer could have more than one OS installed to allow users to use software written for these OSs.

This GUI revolution in computer science generated more confusion to many more people and was far costlier than any conceivable revolution in biological nomenclature. Yet, it is now difficult to imagine a world in which most computers are still driven by command lines. Should we have refrained from developing new OSs with GUIs simply because some computer users were unwilling or unable to learn a new, simpler interface? Sometimes, some transitory confusion is the price to pay for progress.

The GUI revolution also shows another relevant point: only successful projects (operating systems or nomenclatural systems) can create significant confusion. This is shown by the limited impact that OS/2 has had on computer users worldwide (simply because few users were exposed to it). Thus, PN and the *PhyloCode* may cause a transitory confusion among practicing systematists in the future only if a large proportion of systematists adopt them, and in this case, they will be key catalysts for the progress of biological nomenclature.

## 6.9 Do We Need the *PhyloCode*?

After reviewing all these controversies, we need to address one that we could also have started with: Do we need a *PhyloCode*? What about reforming the rank-based codes instead? This question is easily answered by looking at the history of RN. The short-lived draft *BioCode* exemplifies the difficulties that would be encountered if we tried to reform the rank-based codes to incorporate concepts of PN therein. By comparison, the *BioCode* project was modest; it only sought to homogenize RN as implemented in several codes into a single code. But it failed, as explained in Section 2.4.7, for a few reasons, especially the "unwillingness of many botanists and zoologists alike to part with their traditional rules and to accept registration of new names" (Minelli 2001: 167).

What would be the chances of getting the regulating body of each rank-based code to replace ranking principles that are the foundations of these codes by principles of PN, which are foreign to them, and to do so in a fairly coherent way? Needless to say, nearly none! In any case, the International Committee on Bionomenclature, which convened at the IUBS General Assembly in Naples in November 2000, already answered this question back in 2001 when it declared, "that a formal 'Phylocode' was not necessary, and at least for lower-level taxa would cause confusion" (ICZN 2001: 7). The IUBS apparently changed its stance later because it accepted the ISPN's bid for membership in August 2008, but the first argument (about the *BioCode*'s fate) convincingly shows that PN could most likely not be implemented by revising the RN codes and that such an approach would not be a viable alternative to the *PhyloCode*.

Sometimes, progress is better achieved by drafting a new code than by revising an old one. The case of the *PhyloCode* is emblematic, but even much more modest (less innovative) projects may be better carried out through new codes rather than revising current codes. This is shown by the growing dissatisfaction among several zoologists at the ICZN's inability to reform the code to better handle the availability and validity of taxon names. Early in his career, Alain Dubois sought to improve the *Zoological Code*. Thus, Dubois (2005: 367) explained that it was difficult to publish papers that suggested ways to improve the *Zoological Code*, but this was clearly his primary goal, as indicated in the title of the paper: "Proposed Rules for the incorporation of nomina of higher-ranked zoological taxa in the International Code of Zoological Nomenclature." But the practical difficulties of convincing the Zoological Commission to amend the code led him to revise his approach and initiate efforts to draft the Linz *Zoocode* (LZC). The following quote expresses doubts about the future course of events (Dubois et al. 2019: 51–52):

At the present stage of the work of the LZC, we are facing several uncertainties, concerning the future dates of completion of the work of the LZC concerning the *Zoocode* and of that of the Commission concerning the "5th edition" of the [*Zoological*] *Code*. Depending on the answers to these questions, the LZC might follow different pathways and take different decisions, extending from the dissolution of the LZC if the latter agrees with the next version of the *Code*, to the decision **not to follow the new Code** and to implement and **follow the Zoocode as an independent text (like the *Phylocode*)** and encourage zoologists to do so.

Dubois et al. (2019: 52) also argued that "the Commission has taken several ill-guided decisions under the Plenary Power, thus questioning its ability to solve such cases" and were concerned about "The slowness of action of the Commission on many cases, and the fact that a portion of them is simply abandoned without vote and without explanation."

The difficulty in solving such minor problems within the context of RN, despite the fact that this would require only minor adjustments to the rank-based codes, strongly suggests that attempts at introducing notions of PN in these codes would be doomed to failure. Clearly, a brand-new code, drafted from the ground up on PN principles was a much better choice. This is hardly surprising because as Oren Harari (quoted in Whalley 2018) once said, "The electric light [bulb] did not come from continuous improvement of candles"!

## REFERENCES

Agassiz, L. 1859. An essay on classification. Longman, Brown, Green, Longmans, & Roberts, London, viii + 381 pp.

Agassiz, L. 1866. Methods of study in natural history. Ticknor and Fields, Boston, viii + 319 pp.

American Ornithologists' Union. 1886. The Code of Nomenclature and check-list of North American birds adopted by the American Ornithologists' Union. American Ornithologists' Union, New York, viii + 392 pp.

Angielczyk, K. D. and C. F. Kammerer. 2018. 5. Non-mammalian synapsids: The deep roots of the mammalian family tree; pp. 117–198 in F. Zachos and R. Asher (eds.), Mammalian evolution, diversity and systematics. De Gruyter, Berlin.

Ax, P. 1987. The phylogenetic system: The systematization of organisms on the basis of their phylogenesis. John Wiley & Sons, Toronto, 340 pp.

Benson, R. B., G. Hunt, M. T. Carrano, and N. Campione. 2018. Cope's rule and the adaptive landscape of dinosaur body size evolution. Palaeontology 61:3–48.

Benton, M. J. 2007. The Phylocode: Beating a dead horse? Acta Palaeontologica Polonica 52:651–655.

Bickford, D., D. J. Lohman, N. S. Sodhi, P. K. L. Ng, R. Meier, K. Winker, K. K. Ingram, and I. Das. 2007. Cryptic species as a window on diversity and conservation. Trends in Ecology and Evolution 22:148–155.

Blaxter, M., J. Mann, T. Chapman, F. Thomas, C. Whitton, R. FLoyd, and E. Abebe. 2005. Defining operational taxonomic units using DNA barcode data. Philosophical Transactions of the Royal Society of London, Series B 360:1935–1943.

Bock, W. J. 2004. Species: The concept, category and taxon. Journal of Zoological Systematics and Evolutionary Research 42:178–190.

Boy, J. A. and T. Martens. 1991. Ein neues captorhinomorphes Reptil aus dem thüringischen Rotliegend (Unter-Perm; Ost-Deutschland). Paläontologische Zeitschrift 65:363–389.

Boyd, R. 1999. Homeostasis, species, and higher taxa; pp. 141–185 in R. A. Wilson (ed.), Species: New interdisciplinary essays. MIT Press, Cambridge, MA.

Brochu, C. A. and C. D. Sumrall. 2001. Phylogenetic nomenclature and paleontology. Journal of Paleontology 75:754–757.

Brower, A. V. 2020. Dead on arrival: A postmortem assessment of "phylogenetic nomenclature", 20+ years on. Cladistics 36:627–637. DOI: 10.1111/cla.12432

Bryant, H. N. and P. D. Cantino. 2002. A review of criticisms of phylogenetic nomenclature: Is taxonomic freedom the fundamental issue? Biological Reviews of the Cambridge Philosophical Society 77:39–55.

Buffon, G. L. L. 1749. Histoire naturelle, générale et particulière, avec la description du Cabinet du Roy. L'Imprimerie royale, Paris, 612 pp.

Burger, W. C. 1975. The species concept in *Quercus*. Taxon 24:45–50.

Burgin, C. J., J. P. Colella, P. L. Kahn, and N. S. Upham. 2018. How many species of mammals are there? Journal of Mammalogy 99:1–14.

Cain, A. J. 1959. The post-Linnaean development of taxonomy. Proceedings of the Linnean Society of London 170:234–244.

Camp, W. H. and C. L. Gilly. 1943. The structure and origin of species with a discussion of intraspecific variability and related nomenclatural problems. Brittonia 4:323–385.

Cantino, P. D. 2000. Phylogenetic nomenclature: addressing some concerns. Taxon 49:85–93.

Cantino, P. D., H. N. Bryant, K. de Queiroz, M. J. Donoghue, T. Eriksson, D. M. Hillis, and M. S. Y. Lee. 1999. Species names in phylogenetic nomenclature. Systematic Biology 48:790–807.

Cantino, P. D. and K. de Queiroz. 2020. International Code of Phylogenetic Nomenclature (PhyloCode): A Phylogenetic Code of Biological Nomenclature. CRC Press, Boca Raton, FL, xl + 149 pp.

Carroll, R. L. and P. Gaskill. 1978. The order Microsauria. American Philosophical Society, Philadelphia, 211 pp.

Ciccarelli, F. D., T. Doerks, C. Von Mering, C. J. Creevey, B. Snel, and P. Bork. 2006. Toward automatic reconstruction of a highly resolved tree of life. Science 311:1283–1287.

Cracraft, J. 1983. Species concepts and speciation analysis; pp. 159–187 in R. F. Johnston (ed.), Current ornithology. Springer, New York, NY.

Crisp, M. D. and G. T. Chandler. 1996. Paraphyletic species. Telopea 6:813–844.

Cronquist, A. 1978. Once again, what is a species? Biosystematics in agriculture; pp. 3–20 in J. A. Romberger (ed.), Biosystematics in agriculture. Allanheld & Osmun, Montclair.

Cuénot, L. 1940a. Remarques sur un essai d'arbre généalogique du règne animal. Comptes Rendus de l'Académie des Sciences de Paris 210:23–27.

Cuénot, L. 1940b. Essai d'arbre généalogique du règne animal. Revue Scientifique 4:222–229.

Dagan, T. and W. Martin. 2006. The tree of one percent. Genome Biology 7:1–7.

Darlington, C. D. 1940. Taxonomic species and genetic systems; pp. 137–160 in J. S. Huxley (ed.), The new systematics. Oxford University Press, Oxford.

Darwin, C. 1859. On the origin of species by means of natural selection or the preservation of favoured races in the struggle for life. John Murray, London, 502 pp.

Davín, A. A., E. Tannier, T. A. Williams, B. Boussau, V. Daubin, and G. J. Szöllősi. 2018. Gene transfers can date the tree of life. Nature Ecology & Evolution 2:904–909.

Dayrat, B., C. Schander, and K. D. Angielczyk. 2004. Suggestions for a new species nomenclature. Taxon 53:485–491.

Debray, K., M.-C. Le Paslier, A. Bérard, T. Thouroude, G. Michel, J. Marie-Magdelaine, A. Bruneau, F. Foucher, and V. Malécot. 2022. Unveiling the patterns of reticulated evolutionary processes with phylogenomics: Hybridization and polyploidy in the genus *Rosa*. Systematic Biology 71: 547–569.

de Queiroz, K. 1995. The definitions of species and clade names: A reply to Ghiselin. Biology and Philosophy 10:223–228.

de Queiroz, K. 1997. Misunderstandings about the phylogenetic approach to biological nomenclature: A reply to Lidén and Oxelman. Zoologica Scripta 26:67–70.

de Queiroz, K. and J. Gauthier. 1992. Phylogenetic taxonomy. Annual Review of Ecology and Systematics 23:449–480.

Didier, G., O. Chabrol, and M. Laurin. 2019. Parsimony-based test for identifying changes in evolutionary trends for quantitative characters: Implications for the origin of the amniotic egg. Cladistics 35:576–599.

Didier, G., M. Fau, and M. Laurin. 2017. Likelihood of tree topologies with fossils and diversification rate estimation. Systematic Biology 66:964–987.

Didier, G. and M. Laurin. 2020. Exact distribution of divergence times from fossil ages and tree topologies. Systematic Biology 69:1068–1087.

Doolittle, W. F. 2009. The practice of classification and the theory of evolution, and what the demise of Charles Darwin's tree of life hypothesis means for both of them. Philosophical Transactions of the Royal Society B: Biological Sciences 364:2221–2228.

Dubois, A. 1982. Les notions de genre, sous-genre et groupe d'espèces en zoologie à la lumière de la systématique évolutive. Monitore Zoologico Italiano-Italian Journal of Zoology 16:9–65.

Dubois, A. 2005. Proposed rules for the incorporation of nomina of higher-ranked zoological taxa in the International Code of Zoological Nomenclature: 1. Some general questions, concepts and terms of biological nomenclature. Zoosystema 27:365–426.

Dubois, A. 2006. Proposed rules for the incorporation of nomina of higher-ranked zoological taxa in the International Code of Zoological Nomenclature: 2. The proposed rules and their rationale. Zoosystema 28:165–258.

Dubois, A. 2007. Naming taxa from cladograms: A cautionary tale. Molecular Phylogenetics and Evolution 42:317–330.

Dubois, A. 2015. The Duplostensional Nomenclatural System for higher zoological nomenclature. Dumerilia 5:1–108.

Dubois, A., A. M. Bauer, L. M. Ceríaco, F. Dusoulier, T. Frétey, I. Löbl, O. Lorvelec, A. Ohler, R. Stopiglia, and E. Aescht. 2019. The Linz *Zoocode* project: A set of new proposals regarding the terminology, the principles and rules of zoological nomenclature: First report of activities (2014–2019). Bionomina 17:1–111.

Dubois, A., P.-A. Crochet, E. C. Dickinson, A. Nemésio, E. Aescht, A. M. Bauer, V. Blagoderov, R. Bour, M. R. de Carvalho, L. Desutter-Grandcolas, T. Frétey, P. Jäger, V. Koyamba, E. O. Lavilla, I. Löbl, A. Louchart, V. Malécot, H. Schatz, and A. Ohler. 2013. Nomenclatural and taxonomic problems related to the electronic publication of new nomina and nomenclatural acts in zoology, with brief comments on optical discs and on the situation in botany. Zootaxa 3735:1–94.

Dubois, A., A. Ohler, and R. Pyron. 2021. New concepts and methods for phylogenetic taxonomy and nomenclature in zoology, exemplified by a new ranked cladonomy of recent amphibians (Lissamphibia). Megataxa 5:1–738.

Ebua, A. C., T. E. Angwafo, and M. D. Chuo. 2018. Status of blue duiker (*Cephalophus monticola*) and bushbuck (*Tragelaphus scriptus*) in Kom-Wum forest reserve, North West region, Cameroon. International Journal of Environment, Agriculture and Biotechnology 3:619–636.

Egerton, F. N. 2012. History of ecological sciences, part 41: Victorian naturalists in Amazonia-Wallace, Bates, Spruce. Bulletin of the Ecological Society of America 93:35–60.

Ehrlich, P. R. and P. H. Raven. 1969. Differentiation of populations. Science 165:1228–1232.

Endler, J. A. 1973. Gene flow and population differentiation: Studies of clines suggest that differentiation along environmental gradients may be independent of gene flow. Science 179:243–250.

Faith, D. P. 1992. Conservation evaluation and phylogenetic diversity. Biological Conservation 61:1–10.

Felsenstein, J. 1985. Phylogenies and the comparative method. The American Naturalist 125:1–15.

Felsenstein, J. 2001. The troubled growth of statistical phylogenetics. Systematic Biology 50:465–467.

Freudenstein, J. V., M. B. Broe, R. A. Folk, and B. T. Sinn. 2017. Biodiversity and the species concept—lineages are not enough. Systematic Biology 66:644–656.

Ghiselin, M. T. 1966. An application of the theory of definitions to systematic principles. Systematic Zoology 15:127–130.

Ghiselin, M. T. 1974. A radical solution to the species problem. Systematic Zoology 23:536–544.

Ghiselin, M. T. 1987. Species concepts, individuality, and objectivity. Biology and Philosophy 2:127–143.

Ghiselin, M. T. 2002. Species concepts: The basis for controversy and reconciliation. Fish and Fisheries 3:151–160.

Gilmour, J. 1973. Octal notation for designating physiologic races of plant pathogens. Nature 242:620.

Goulding, T. C. and B. Dayrat. 2016. Integrative taxonomy: Ten years of practice and looking into the future. Archives of the Zoology Museum of the Lomonosov Moscow State University 54:116–133.

Grant, V. 1980. Gene flow and the homogeneity of species populations. Biologisches Zentralblatt 99:157–169.

Greuter, W., J. McNeill, F. R. Barrie, H. M. Burdet, V. Demoulin, T. S. Filgueiras, D. H. Nicolson, P. C. Silva, J. E. Skog, P. Trehane, N. J. Turland, and D. L. Hawksworth. 2000. International Code of Botanical Nomenclature. Koeltz Scientific Books, Königstein, Germany, xviii + 474 pp.

Griffiths, A. 1981. A numericlature of the yeasts. Antonie van Leeuwenhoek 47:547–563.

Griffiths, A. 1984. A descriptive numericlature for isolates of cyanobacteria. British Phycological Journal 19:233–238.

Griffiths, G. C. D. 1973. Some fundamental problems in biological classification. Systematic Zoology 22:338–343.

Griffiths, G. C. D. 1974. On the foundations of biological systematics. Acta Biotheoretica 23:85–131.

Groves, C. 2018. The latest thinking about the taxonomy of great apes. International Zoo Yearbook 52:16–24.

Gwee, C. Y., J. A. Eaton, K. M. Garg, P. Alström, S. Van Balen, R. O. Hutchinson, D. M. Prawiradilaga, M. H. Le, and F. E. Rheindt. 2019. Cryptic diversity in *Cyornis* (Aves: Muscicapidae) jungle-flycatchers flagged by simple bioacoustic approaches. Zoological Journal of the Linnean Society 186:725–741.

Haeckel, E. 1866. Generelle Morphologie der Organismen. Reimer, Berlin, 1036 pp.

Härlin, M. 2005. Definitions and phylogenetic nomenclature. Proceedings of the California Academy of Sciences 56:216–224.

Hatschek, B. 1888. Lehrbuch der Zoologie, eine morphologische Ubersicht des Thierreiches zur Einfuhrung in das Studium dieser Wissenschaft. Gustav Fischer, Jena, iv + 432 pp.

Hausdorf, B. 2011. Progress toward a general species concept. Evolution 65:923–931.

Hayden, J. E. 2020. Monophyletic classification and information content. Cladistics 36:424–436.

Hennig, W. 1966. Phylogenetic systematics. University of Illinois Press, Urbana, Chicago, London, 263 pp.

Hennig, W. 1981. Insect phylogeny. John Wiley & Sons, Chichester, xi + 514 pp.

Hull, D. L. 1966. Phylogenetic numericlature. Systematic Zoology 15:14–17.

Hull, D. L. 1968. The syntax of numericlature. Systematic Zoology 17:472–474.

Hull, D. L. 1979. The limits of cladism. Systematic Zoology 28:416–440.

ICZN. 1999. International Code of Zoological Nomenclature. The International Trust for Zoological Nomenclature, London, 306 pp. www.iczn.org/the-code/the-international-code-of-zoological-nomenclature/the-code-online/

ICZN. 2001. International Committee on Bionomenclature. Bulletin of Zoological Nomenclature 58:6–7.

ICZN. 2014. Zoological Nomenclature and electronic publication: A reply to Dubois et al. (2013). Zootaxa 3779:3–5.

Jain, S. and A. D. Bradshaw. 1966. Evolutionary divergence among adjacent plant populations: I. The evidence and its theoretical analysis. Heredity 21:407–441.

Koonin, E. V., K. S. Makarova, and L. Aravind. 2001. Horizontal gene transfer in prokaryotes: Quantification and classification. Annual Reviews in Microbiology 55:709–742.

Lande, R. 1980. Genetic variation and phenotypic evolution during allopatric speciation. The American Naturalist 116:463–479.

Lanham, U. 1965. Uninomial nomenclature. Systematic Zoology 14:144.

Laurin, M. 2005. The advantages of phylogenetic nomenclature over Linnean nomenclature; pp. 67–97 in A. Minelli, G. Ortalli, and G. Sanga (eds.), Instituto Veneto di Scienze, Lettere ed Arti, Venice.

Laurin, M. 2008. The splendid isolation of biological nomenclature. Zoologica Scripta 37:223–233.

Laurin, M. 2010. How vertebrates left the water. University of California Press, Berkeley, xv + 199 pp.

Laurin, M. 2020. Tetrapoda; pp. 759–764 in K. de Queiroz, P. D. Cantino, and J. A. Gauthier (eds.), Phylonyms: A companion to the PhyloCode. CRC Press, Boca Raton, FL.

Laurin, M. and H. N. Bryant. 2009. Third meeting of the International Society for Phylogenetic Nomenclature: A report. Zoologica Scripta 38:333–337.

Laurin, M. and P. D. Cantino. 2004. First international phylogenetic nomenclature meeting: A report. Zoologica Scripta 33:475–479.

Laurin, M. and P. D. Cantino. 2007. Second meeting of the International Society for Phylogenetic Nomenclature: A report. Zoologica Scripta 36:109–117.

Laurin, M., K. de Queiroz, P. D. Cantino, N. Cellinese, and R. Olmstead. 2005. The PhyloCode, types, ranks, and monophyly: A response to Pickett. Cladistics 21:605–607.

Lecointre, G. and H. Le Guyader. 2016. Classification phylogénétique du vivant: tome 1. Belin, Paris, 4th ed., 584 pp.

Lhermin[i]er, P. and M. Solignac. 2000. L'espèce: définitions d'auteurs. Comptes Rendus de l'Académie des Sciences—Series III—Sciences de la Vie 323:153–165.

Lherminier, P. and M. Solignac. 2005. De l'espèce. Syllepse, Paris, XI + 694 pp.

Little, F. J., Jr. 1964. The need for a uniform system of biological numericlature. Systematic Zoology 13:191–194.

Löbl, I. 2015. Stability under the International Code of Zoological Nomenclature: A bag of problems affecting nomenclature and taxonomy. Bionomina 9:35–40.

Lotsy, J. 1925. Species or Linneon. Genetica 7:487–506.

Lücking, R. 2019. Stop the abuse of time! Strict temporal banding is not the future of rank-based classifications in fungi (including lichens) and other organisms. Critical Reviews in Plant Sciences 38:199–253.

Ludwig, D. 2018. Revamping the metaphysics of ethnobiological classification. Current Anthropology 59:415–438.

Mace, G. M. 2004. The role of taxonomy in species conservation. Philosophical Transactions of the Royal Society of London, Series B 359:711–719.

Mallet, J., N. Besansky, and M. W. Hahn. 2015. How reticulated are species? BioEssays 38:140–149.

Maynard Smith, J. 1982. Evolution now: A century after Darwin. W. H. Freeman and Co, San Francisco, CA, 239 pp.

McShea, D. W. 2000. Trends, tools and terminology. Paleobiology 26:330–333.

Michener, C. D. 1963. Some future developments in taxonomy. Systematic Zoology 12:151–172.

Minelli, A. 2000. The ranks and the names of species and higher taxa, or a dangerous inertia of the language of natural history; pp. 339–351 in M. T. Ghiselin and A. E. Leviton (eds.), Cultures and institutions of natural history: Essays in the history and philosophy of science. California Academy of Sciences, San Francisco.

Minelli, A. 2001. Zoological nomenclature: Reflections on the recent past and ideas for our future agenda. Bulletin of Zoological Nomenclature 58:164–169.

Minelli, A. 2003. Zoological nomenclature after the publication of the Fourth Edition of the Code; pp. 649–658 in A. Legakis, S. Sfenthourakis, R. Polymeni, and M. Thessalou-Legaki (eds.), The new panorama of animal evolution. Pensoft, Sofia.

Minelli, A. 2017. Grey nomenclature needs rules. Ecologica Montenegrina 7:654–666.

Minelli, A. 2022. Species; in B. Hjørland and C. Gnoli (eds.), ISKO Encyclopedia of Knowledge Organization (IEKO). www.isko.org/cyclo/species

Mishler, B. D. 1999. Getting rid of species?; pp. 307–315 in R. Wilson (ed.), Species: New interdisciplinary essays. MIT Press, Cambridge, MA.

Mishler, B. D. 2010. Species are not uniquely real biological entities; pp. 110–122 in F. J. Ayala and R. Arp (eds.), Contemporary debates in philosophy of biology. Wiley-Blackwell, New York.

Mishler, B. D. and M. J. Donoghue. 1982. Species concepts: A case for pluralism. Systematic Zoology 31:491–503.

Mishler, B. D. and J. S. Wilkins. 2018. The hunting of the SNaRC: A snarky solution to the species problem. Philosophy, Theory, and Practice in Biology 10:1–18.

Mumby, P. J. 2002. Statistical power of non-parametric tests: A quick guide for designing sampling strategies. Marine Pollution Bulletin 44:85–87.

Nelson, G. and N. I. Platnick. 1981. Systematics and biogeography: Cladistics and vicariance. Columbia University Press, New York, xi + 567 pp.

Nickerson, R. S. 2000. Null hypothesis significance testing: A review of an old and continuing controversy. Psychological Methods 5:241–301.

Nicolson, D. H. 1991. A history of botanical nomenclature. Annals of the Missouri Botanical Garden 78:33–56.

Nitecki. 1957. What is a paleontological species? Evolution 11:378–380.

Nixon, K. C., J. M. Carpenter, and D. W. Stevenson. 2003. The PhyloCode is fatally flawed, and the "Linnaean" System can easily be fixed. The Botanical Review 69:111–120.

North American Commission on Stratigraphic Nomenclature. 2005. North American stratigraphic code. AAPG Bulletin 89:1547–1591.

O'Reilly, J. E., M. N. Puttick, D. Pisani, and P. C. Donoghue. 2018. Probabilistic methods surpass parsimony when assessing clade support in phylogenetic analyses of discrete morphological data. Palaeontology 61:105–118.

Ornduff, R. 1969. Reproductive biology in relation to systematics. Taxon 18:121–133.

Parker, C. T., B. J. Tindall, and G. M. Garrity. 2019. International Code of Nomenclature of Prokaryotes: Prokaryotic Code (2008 revision). International Journal of Systematic and Evolutionary Microbiology 69:S1–S111. https://doi.org/10.1099/ijsem.0.000778

Paterson, H. E. 1978. More evidence against speciation by reinforcement. South African Journal of Science 74:369–371.

Paterson, H. E. 1980. A comment on "mate recognition systems". Evolution 34:330–331.

Pavlinov, I. Y. 2021. Taxonomic Nomenclature: What's in a name—theory and history. CRC Press, Boca Raton, 276 pp.

Platnick, N. I. 1985. Philosophy and the transformation of cladistics revisited. Cladistics 1:87–94.

Platnick, N. I. 2012. The poverty of the PhyloCode: A reply to de Queiroz and Donoghue. Systematic Biology 61:360–361.

Pleijel, F. and M. Härlin. 2004. Phylogenetic nomenclature is compatible with diverse philosophical perspectives. Zoological Scripta 33:587–591.

Pleijel, F. and G. W. Rouse. 2000a. Least-inclusive taxonomic unit: A new taxonomic concept for biology. Proceedings of the Royal Society of London, Series B 267:627–630.

Pleijel, F. and G. W. Rouse. 2000b. A new taxon, *capricornia (Hesionidae, Polychaeta)*, illustrating the LITU ("Least-Inclusive Taxonomic Unit") concept. Zoologica Scripta 29:157–168.

Prado-Martinez, J., P. H. Sudmant, J. M. Kidd, H. Li, J. L. Kelley, B. Lorente-Galdos, K. R. Veeramah, A. E. Woerner, T. D. O'Connor, G. Santpere, A. Cagan, C. Theunert, F. Casals, H. Laayouni, K. Munch, A. Hobolth, A. E. Halager, M. Malig, J. Hernandez-Rodriguez, I. Hernando-Herraez, K. Prüfer, M. Pybus, L. Johnstone, M. Lachmann, C. Alkan, D. Twigg, N. Petit, C. Baker, F. Hormozdiari, M. Fernandez-Callejo, M. Dabad, M. L. Wilson, L. Stevison, C. Camprubí, T. Carvalho, A. Ruiz-Herrera, L. Vives, M. Mele, T. Abello, I. Kondova, R. E. Bontrop, A. Pusey, F. Lankester, J. A. Kiyang, R. A. Bergl, E. Lonsdorf, S. Myers, M. Ventura, P. Gagneux, D. Comas, H. Siegismund, J. Blanc, L. Agueda-Calpena, M. Gut, L. Fulton, S. A. Tishkoff, J. C. Mullikin, R. K. Wilson, I. G. Gut, M. K. Gonder, O. A. Ryder, B. H. Hahn, A. Navarro, J. M. Akey, J. Bertranpetit, D. Reich, T. Mailund, M. H. Schierup, C. Hvilsom, A. M. Andrés, J. D. Wall, C. D. Bustamante, M. F. Hammer, E. E. Eichler, and T. Marques-Bonet. 2013. Great ape genetic diversity and population history. Nature 499:471–475.

Prichard, J. C. 1813. Researches into the physical history of man. John and Arthur Arch, London, viii + 558 pp.

Pulido-Santacruz, P., A. Aleixo, and J. T. Weir. 2018. Morphologically cryptic Amazonian bird species pairs exhibit strong postzygotic reproductive isolation. Proceedings of the Royal Society B: Biological Sciences 285:20172081.

Ragan, M. A. 2009. Trees and networks before and after Darwin. Biology Direct 4:43.

Ratnasingham, S. and P. D. Hebert. 2007. BOLD: The Barcode of Life Data System (www. barcodinglife. org). Molecular Ecology Notes 7:355–364.

Ratnasingham, S. and P. D. Hebert. 2013. A DNA-based registry for all animal species: The Barcode Index Number (BIN) system. PLoS One 8:e66213.

Raven, P. H. 1980. Hybridization and the nature of species in higher plants. Canadian Botanical Association Bulletin 13, supplement:3–10.

Raven, P. H., B. Berlin, and D. E. Breedlove. 1971. The origins of taxonomy. Science 174:1210–1213.

Recuero, E., J. Cruzado-Cortes, G. Parra-Olea, and K. R. Zamudio. 2010. Urban aquatic habitats and conservation of highly endangered species: The case of *Ambystoma mexicanum* (Caudata, Ambystomatidae). Acta Zoologica Fennici 47:223–238.

Richards, R. A. 2010. The species problem: A philosophical analysis. Cambridge University Press, Cambridge, X + 236 pp.

Rieppel, O. 1986. Species are individuals: A review and critique of the argument; pp. 283–317 in M. K. Hecht, B. Wallace, and G. T. Prance (eds.), Evolutionary biology, vol. 20. Plenum Publishing Corporation, New York.

Rosen, D. E. 1979. Fishes from the uplands and intermontane basins of Guatemala: Revisionary studies and comparative geography. Bulletin of the American Museum of Natural History 162:267–376.

Rowe, T. and J. Gauthier. 1992. Ancestry, paleontology, and definition of the name Mammalia. Systematic Biology 41:372–378.

Saitoh, T., N. Sugita, S. Someya, Y. Iwami, S. Kobayashi, H. Kamigaichi, A. Higuchi, S. Asai, Y. Yamamoto, and I. Nishiumi. 2015. DNA barcoding reveals 24 distinct lineages as cryptic bird species candidates in and around the Japanese Archipelago. Molecular Ecology Resources 15:177–186.

Sanders, M., D. Merle, M. Laurin, C. Bonillo, and N. Puillandre. 2021. Raising names from the dead: A time-calibrated phylogeny of frog shells (Bursidae, Tonnoidea, Gastropoda) using mitogenomic data. Molecular Phylogenetics and Evolution 156:107040.

San Mauro, D., D. J. Gower, H. Müller, S. P. Loader, R. Zardoya, R. A. Nussbaum, and M. Wilkinson. 2014. Life-history evolution and mitogenomic phylogeny of caecilian amphibians. Molecular Phylogenetics and Evolution 73:177–189.

Seehausen, O. 2004. Hybridization and adaptive radiation. Trends in Ecology & Evolution 19:198–207.

Sepkoski, J. J. 1986. Global bioevents and the question of periodicity; pp. 47–61 in O. Walliser (ed.), Global bio-events. Springer, Berlin.

Sepkoski, J. J., Jr. and D. M. Raup. 1986. Periodicity in marine extinction events; pp. 3–36 in D. K. Elliott (ed.), Dynamics of extinction. Wiley, New York.

Shneyer, V. and V. Kotseruba. 2015. Cryptic species in plants and their detection by genetic differentiation between populations. Russian Journal of Genetics: Applied Research 5:528–541.

Sloan, P. R. 1972. John Locke, John Ray, and the problem of the natural system. Journal of the History of Biology 5:1–53.

Smith, A. B. and C. Patterson. 1988. The influence of taxonomic method on the perception of patterns of evolution; pp. 127–216 in M. K. Hecht and B. Wallace (eds.), Evolutionary biology, vol. 23. Plenum Press, New York.

Strickland, H. E., J. S. Henslow, J. Phillips, W. E. Shuckard, J. B. Richardson, G. R. Waterhouse, R. Owen, W. Yarrell, L. Jenyns, C. Darwin, W. J. Broderip, and J. O. Westwood. 1842. Report of a committee appointed "to consider of the rules by which the Nomenclature of Zoology may be established on a uniform and permanent basis. Annals and Magazine of Natural History 11:1–17.

Struck, T. H., J. L. Feder, M. Bendiksby, S. Birkeland, J. Cerca, V. I. Gusarov, S. Kistenich, K.-H. Larsson, L. H. Liow, and M. D. Nowak. 2018. Finding evolutionary processes hidden in cryptic species. Trends in Ecology & Evolution 33:153–163.

Sues, H.-D. 2019. The rise of reptiles: 320 million years of evolution. John Hopkins University Press, Baltimore, xi + 385 pp.

Supriatna, J. 2022. Family Pongidae (Gray, 1870); pp. 185–201 in J. Supriatna (ed.), Field Guide to the Primates of Indonesia. Springer, Cham.

Tassy, P. and M. Fischer. 2021. "Cladus" and clade: A taxonomic odyssey. Theory in Biosciences 140:77–85.

Tessens, B., M. Monnens, T. Backeljau, K. Jordaens, N. Van Steenkiste, F. C. Breman, K. Smeets, and T. Artois. 2021. Is "everything everywhere"? Unprecedented cryptic diversity in the cosmopolitan flatworm *Gyratrix hermaphroditus*. Zoologica Scripta 50:837–851.

Van Valen, L. 1973. A new evolutionary law. Evolutionary Theory 1:1–30.

Vinarski, M. 2019. The roots of the taxonomic impediment: Is the "integrativeness" a remedy? Integrative Zoology 15:2–15.

Wagner, W. H. 1983. Reticulistics: The recognition of hybrids and their role in cladistics and classification; pp. 63–79 in

N. I. Platnick and V. A. Funk (eds.), Advances in cladistics. Columbia University Press, New York.

Whalley, G. A. 2018. Surrogate survival: Battle between left ventricular ejection fraction and global longitudinal strain. JAAC Journal 11:1580–1582.

Whewell, W. 1840. The philosophy of the inductive sciences: Founded upon their history. J. W. Parker, London, cxx + 523 pp.

Wiley, E. O. 1980. Is the evolutionary species fiction? A consideration of classes, individuals and historical entities. Systematic Zoology 29:76–80.

Wilkins, J. S. 2011. Philosophically speaking, how many species concepts are there? Zootaxa 2765:58–60.

Wilkins, J. S. 2018. Species: The evolution of the idea. CRC Press, Boca Raton, xxxviii + 389 pp.

Wilkinson, M., D. San Mauro, E. Sheratt, and D. J. Gower. 2011. A nine-family classification of caecilians (Amphibia: Gymnophiona). Zootaxa 2874:41–64.

Winsor, M. P. 2003. Non-essentialist methods in pre-Darwinian taxonomy. Biology and Philosophy 18:387–400.

Wolf, Y. I., D. Kazlauskas, J. Iranzo, A. Lucía-Sanz, J. H. Kuhn, M. Krupovic, V. V. Dolja, and E. V. Koonin. 2018. Origins and evolution of the global RNA virome. MBio 9:e02329–18.

Wyllie, A. 2003. A review of Robert Broom's therapsid holotypes: Have they survived the test of time? Palaeontologia Africana 39:1–19.

# 7

# Conclusion: The Future of Biological Nomenclature

## 7.1 Phylogenetic Nomenclature Is a Logical Outcome of Progress in Evolutionary Biology

Hopefully, this book has shown that progress in systematics harking back at least to Aristotle, and more importantly, the rise of evolutionary biology that started in the 18th century (see Section 1.4) and rapid progress in phylogenetics in the 20th century (see Chapter 3), created the conditions for the rise of phylogenetic nomenclature (PN). In fact, it could be argued that the development of PN was unavoidable, given this context. A point of view close to this was expressed more than two decades ago by Alessandro Minelli, when he was president of the International Commission on Zoological Nomenclature (ICZN). He stated that "In my 1995 paper, I wrote that 'We must expect that the development of cladistics will increasingly ask for a revised biological nomenclature', and this is exactly what is happening with the PhyloCode" (Minelli 2001: 167). Subsequent development of the *PhyloCode*, establishment of the *ISPN* in 2004 (Laurin and Cantino 2004), and the rising success of PN in the last three decades (**Figure 7.1**) suggest that PN fulfills a growing need in systematics. Brower (2020) argued that use of PN had declined in the last few years, but his argument was based only on the works cited by Cantino and de Queiroz (2020), and this may simply reflect the fact that the bulk of the *PhyloCode* was written well before 2020.

## 7.2 The Advent of Phylogenetic Nomenclature Will Be Beneficial for Evolutionary Biology

The rise of PN will facilitate some developments in evolutionary biology. Thus, after examining many problems created in previous paleontological studies on the evolution of biodiversity through time based on the number of taxa ranked as families and genera through time, Smith and Patterson (1988: 161) concluded that:

> In this chapter we have tried to demonstrate that traditional taxonomy, based on arbitrarily construed [non-monophyletic] taxa, cannot be used to identify meaningful paleobiological phenomena. No matter how sophisticated the statistical tests used on such data, **arbitrary taxa produce spurious results**. Such analyses generally tell us more about how taxonomists have worked in the past (the Red Queen hypothesis) than about patterns and processes of evolution. It is crucial to know whether the units

being analyzed reflect events in the real world or *ad hoc* taxonomic decisions. It is also vital to identify where sampling biases might be significant. **Only a hierarchically structured classification of monophyletic groups can provide the necessary taxonomic foundation for analyses of evolutionary pattern**.

Yet, some of the main rank-based codes, namely the *International Code of Zoological Nomenclature*, or *"Zoological Code"* for short (ICZN 1999), and the *International Code of Nomenclature for algae, fungi, and plants (Shenzhen Code)*, or *"Botanical Code"* for short (Turland et al. 2018), do not require taxa to be monophyletic (see Chapter 2). As of this writing, I am not aware of any serious discussions by the governing bodies of these codes to change the status quo. It appears that among the codes that regulate names of eukaryotic organisms, only the *PhyloCode* provides an adequate tool for 21st-century systematics and evolutionary biology.

## 7.3 Nomenclature Develops through Innovation

Some opponents of phylogenetic nomenclature (PN from here on) argue that it breaks with the past and that this will create confusion (see Chapter 6). This amounts to an appeal to tradition, which is unsurprising given that Papavero et al. (2001: 5) argued that "Now it is possibly **due to the power of tradition** that only few systematists dare to propose abandonment of the Linnaean categories, some of which are even obligatory according to the codes of nomenclature." In fact, even proponents of RN who try to improve the *Zoological Code* deplore that "The resistance, quite frequent in science, to novelty in a domain where a tradition has long been in force" prevents progress in this field (Dubois et al. 2019: 21). However, progress often requires such a break with the past, and not only in systematics. An analogy can be made with stratigraphic nomenclature through this quote extracted from the foreword to the 1983 edition of the *North American stratigraphic code*, reproduced in the 2005 edition (North American Commission on Stratigraphic Nomenclature 2005: 1547):

> The Commission considered whether to discard our codes, patch them over, or rewrite them fully, and chose the last. . . . Take this Code, use it, but **do not condemn it because it contains something new** or not of direct interest to you. Innovations that prove unacceptable to the profession will expire without damage to other concepts and procedures, just as did the geologic-climate units of the 1961 Code.

DOI: 10.1201/9781003092827-8

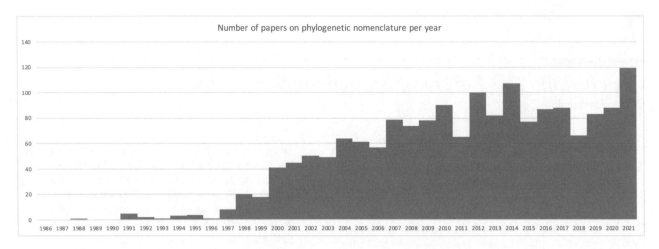

**FIGURE 7.1** Development of phylogenetic nomenclature as shown by the number of papers mentioning this method each year. The data for this figure are based on a Google Scholar search carried out on April 29, 2022, using the expressions "phylogenetic nomenclature."

Few systematists would disagree with the statement that important nomenclatural progress was made in the mid-18th century, when Linnaeus and his contemporaries published their landmark nomenclatural works. Yet, Nicolson (1991: 33) wrote:

> Priority had no part of the early schemes of nomenclature. Their authors, including Linnaeus, were focused on **replacing the past**. Linnaeus was the winner of this competition but contemporary and subsequent workers continued to devise new nomenclatural schemes and rules to overturn the past.

By comparison, the ISPN has paid much more attention to avoiding the disruption of previous usage (in terms of content of taxa and their names) than Linnaeus ever did. It is trying to build on the past, mostly by providing better ways of defining taxa to ensure more stability in their names and contents (through a delimitation that depends on phylogenetic context).

The pioneers in rank-based nomenclature long envisioned that this type of nomenclature would one day be replaced by something different. Thus, the introduction to Alphonse de Candolle's (1867) *Lois* includes this visionary prediction:

> Viendra pourtant une époque où les formes végétales actuelles ayant toutes été décrites, les herbiers en offrant des types certains, les botanistes ayant fait, défait, quelquefois refait, élevé ou abaissé, et surtout modifié plusieurs centaines de milliers de groupes, depuis les classes jusqu'aux simples variétés d'espèces, le nombre des synonymes étant devenu infiniment plus considérable que celui des groupes admis, la science aura besoin de quelque **grande rénovation** dans les formes. Cette nomenclature que nous nous efforçons d'améliorer, paraîtra alors comme un vieil échafaudage, formé de pièces renouvelées péniblement, une à une, et entouré de débris constitués par toutes les parties rejetées qui formeront un encombrement plus ou moins gênant.

L'édifice de la science aura été construit, mais il ne sera pas assez dégagé de tout ce qui a servi à l'élever. Alors, peut-être, **il surgira quelque chose de tout différent de la nomenclature linnéenne**, quelque chose qui sera imaginé pour donner **définitivement des noms à des groupes définitifs**.

Cela est le secret de l'avenir, et d'un avenir **encore bien éloigné**.

However, a time will come when the current plant forms having all been described, herbaria offering certain types, botanists having made, undone, sometimes remade, raised or lowered, and above all modified several hundred thousand groups, from classes to simple varieties of species, the number of synonyms having become infinitely more considerable than that of the valid taxa, science will need some **great renovation** in its principles. This nomenclature, which we are trying to improve, will then appear like an old scaffolding, formed of parts painfully renewed, one by one, and surrounded by debris made up of all the rejected parts which will form a more or less embarrassing clutter. The edifice of science will have been built, but it will not be sufficiently free from all that has served to raise it. Then, perhaps, **there will arise something completely different from Linnaean nomenclature**, something which will be imagined to give **stable names to definitive groups**.

This is the secret of the future, and of a future that is **still very distant**.

This quote could almost apply now, and a similar, though less explicit statement, is found in the code developed by the American Ornithologists' Union (1886: 2). We still lack "definitive groups" because the inventory of the extant biodiversity is far from exhaustive and because phylogenetics has not yet provided reliable trees for all taxa. These, however, are not nomenclatural problems.

Darwin (1859: 420) was more explicit in how we should conceptualize and delimit taxa, as exemplified by this quote:

> All the foregoing rules and aids and difficulties in classification are explained, if I do not greatly deceive myself, on the view that **the natural system is founded on descent with modification**; that the characters which naturalists consider as showing true affinity between any two or more species, are those which have been inherited from a common parent, and, in so far, **all true classification is genealogical**; that **community of descent is the hidden bond which naturalists have been unconsciously seeking**, and not some unknown plan of creation, or the enunciation of general propositions, and the mere putting together and separating objects more or less alike.

I can think of no better justification for phylogenetic nomenclature as it has been implemented in the *PhyloCode*.

Phylogenetic nomenclature is sufficiently different from rank-based nomenclature to be seen as the type of nomenclatural revolution envisioned by de Candolle more than one and a half century ago. If de Candolle could have witnessed the development of phylogenetic nomenclature, he would no doubt be pleased to see that his predicted distant revolution in biological nomenclature is now well under way. This is shown by the fact that PN has been used by a fair number of scientific papers, as shown by the extensive bibliography of the *PhyloCode* (Cantino and de Queiroz 2020), and some systematics textbooks (for instance, Laurin 2010; Sues 2019) also use this new, revolutionary nomenclature.

The previous quotes from the 19th century should not obscure the fact that calls for a drastic change (away from RN) were also made much more recently, but before the advent of PN. In this context, it is worth quoting the conclusion of Hull's (1964) essay on "Consistency and monophyly," with which most proponents of PN would probably agree:

> Perhaps it is an exaggeration to say that the purpose of phylogenetic taxonomy has been to make classification represent the form of the Linnaean hierarchy instead of phylogeny. It is no exaggeration to say that its purpose has been to represent phylogeny whenever the form of the Linnaean hierarchy permitted. **It has permitted very little**. A classification [compatible with RN] implies much in general about phylogeny but little that is specific enough to actually contradict phylogeny. What is worse, in the one area in which specific implications are possible, classification actually does contradict phylogeny. One solution to this predicament is a modification of the form of the Linnaean hierarchy. How extensive a modification is made depends on how extensively evolutionary taxonomists wish to represent phylogeny.

Hopefully, this book has shown that PN represents a rather extensive modification of previous nomenclatural practice and that it can faithfully reflect phylogenetic information.

What will determine its success is adoption by practicing systematists. What does the future hold for biological nomenclature? Härlin (2005: 221) wrote that "Phylogenetic nomenclature is likely here to stay. Exactly in what form is still to be settled. Any system intended for the future needs to get rid of [absolute] ranks." Since then, the hegemony of the *PhyloCode* over alternative proposed implementations of PN seems to have grown. These alternatives include Sereno's (2005) proposal of unregulated PN that entails development of an online database of phylogenetic definitions (see Section 4.3), and Härlin's (2005) "phylogenetic system of reference" that involves redefining names when the phylogeny changes (see Section 6.6).

So far, only a relatively small minority of systematists have adopted PN (hence, the *PhyloCode*). However, there seems to be increasing acceptance among systematists that taxa (with the possible exception of species) should be monophyletic and, to a lesser extent, we see an increasing awareness that Linnaean categories are artificial constructs that should be abandoned. The last point can be illustrated by some biology textbooks (not written by proponents of PN) that emphasized rankless nomenclatures (for instance, Ax 1987; Lecointre and Le Guyader 2001; Westheide and Rieger 2009), or applied it at least to high-ranking taxa above the family level (e.g., Pough et al. 2018: 23). One of these books was noticed beyond the field of biology, notably in anthropology (Lévi-Strauss 2002), and was so successful for teaching biology from the elementary school to the university level, and even among birdwatchers and other amateurs (entomologists, botanists and so on), that it was expanded into a two-book set and was translated into three other languages (Italian, German and English). This book is now in its fourth edition, and it explicitly emphasizes the subjective nature (and hence, the limited interest) of absolute ranks (Lecointre and Le Guyader 2016: 32).

Even a former president of the ICZN criticized the use of Linnaean categories in one of his papers and concluded that attempts to objectively rank higher taxa had failed (Minelli 2000). He also suggested (Minelli 1999) that "In the future, Linnaean and non-Linnaean classification might exist side-by-side." In fact, this had already happened by then, but PN has gained prominence since then, a trend that should accelerate following the publication of *Phylonyms* (de Queiroz et al. 2020). More recently, in his paper that announced the then-new fourth edition of the *Zoological Code*, Minelli (2003: 653) also wondered "are we sure that the latter [RN] can satisfy all conceptual and practical requirements of today's zoologists? Just browsing through the literature seems to suggest that it is really not so." He also evoked (p. 654) "an increasingly widespread dissatisfaction with the Linnaean hierarchy" to introduce a brief discussion of PN and the draft *PhyloCode*. These are presumably some of the main reasons why Minelli (2003: 649) indicated that "The fourth edition of the Code marks a few sensible changes in the way the scientific names of animals are produced and used, but **more momentous changes** may be waiting ahead." These shifts in consensus seem to pave the way for PN. Will PN prevail, and if so, how long will it take to become the mainstream biological nomenclature?

# REFERENCES

American Ornithologists' Union. 1886. The Code of Nomenclature and check-list of North American birds adopted by the American Ornithologists' Union. American Ornithologists' Union, New York, viii + 392 pp.

Ax, P. 1987. The phylogenetic system: The systematization of organisms on the basis of their phylogenesis. John Wiley & Sons, Toronto, xiii + 340 pp.

Brower, A. V. 2020. Dead on arrival: A postmortem assessment of "phylogenetic nomenclature", 20+ years on. Cladistics 36:627–637.

Cantino, P. D. and K. de Queiroz. 2020. International Code of Phylogenetic Nomenclature (PhyloCode): A Phylogenetic Code of Biological Nomenclature. CRC Press, Boca Raton, FL, xl + 149 pp.

Darwin, C. 1859. On the origin of species by means of natural selection or the preservation of favoured races in the struggle for life. John Murray, London, 502 pp.

de Candolle, A. 1867. Lois de la nomenclature botanique adoptées par le Congrès international de botanique tenu à Paris en août 1867: suivies d'une 2ᵉ édition de l'introduction historique et du commentaire qui accompagnaient la rédaction préparatoire présentée au Congrès. H. Georg, Genève, 64 pp.

de Queiroz, K., P. D. Cantino, and J. A. Gauthier, eds. 2020. Phylonyms: A companion to the PhyloCode. CRC Press, Boca Raton.

Dubois, A., A. M. Bauer, L. M. Ceríaco, F. Dusoulier, T. Frétey, I. Löbl, O. Lorvelec, A. Ohler, R. Stopiglia, and E. Aescht. 2019. The Linz *Zoocode* project: A set of new proposals regarding the terminology, the principles and rules of zoological nomenclature: First report of activities (2014–2019). Bionomina 17:1–111.

Härlin, M. 2005. Definitions and phylogenetic nomenclature. Proceedings of the California Academy of Sciences 56:216–224.

Hull, D. L. 1964. Consistency and monophyly. Systematic Zoology 13:1–11.

ICZN. 1999. International Code of Zoological Nomenclature. The International Trust for Zoological Nomenclature, London, 306 pp. www.iczn.org/the-code/the-international-code-of-zoological-nomenclature/the-code-online/

Laurin, M. 2010. How vertebrates left the water. University of California Press, Berkeley, xv + 199 pp.

Laurin, M. and P. D. Cantino. 2004. First international phylogenetic nomenclature meeting: A report. Zoologica Scripta 33:475–479.

Lecointre, G. and H. Le Guyader. 2001. Classification phylogénétique du vivant. Belin, Paris, 1st ed., 543 pp.

Lecointre, G. and H. Le Guyader. 2016. Classification phylogénétique du vivant: tome 1. Belin, Paris, 4th ed., 584 pp.

Lévi-Strauss, C. 2002. Guillaume Lecointe & Hervé Le Guyader, Classification phylogénétique du vivant. Illustrations de Dominique Visset. Publié avec le concours du Centre national du livre. Paris, Belin, 2001, 543 pp., annexes, bibl., index, tabl. L'Homme 162:309–312.

Minelli, A. 1999. The names of animals. Trends in Ecology and Evolution 14:462–463.

Minelli, A. 2000. The ranks and the names of species and higher taxa, or a dangerous inertia of the language of natural history; pp. 339–351 in M. T. Ghiselin and A. E. Leviton (eds.), Cultures and institutions of natural history: Essays in the history and philosophy of science. California Academy of Sciences, San Francisco.

Minelli, A. 2001. Zoological nomenclature: Reflections on the recent past and ideas for our future agenda. Bulletin of Zoological Nomenclature 58:164–169.

Minelli, A. 2003. Zoological nomenclature after the publication of the Fourth Edition of the Code; pp. 649–658 in A. Legakis, S. Sfenthourakis, R. Polymeni, and M. Thessalou-Legaki (eds.), The new panorama of animal evolution. Pensoft, Sofia.

Nicolson, D. H. 1991. A history of botanical nomenclature. Annals of the Missouri Botanical Garden 78:33–56.

North American Commission on Stratigraphic Nomenclature. 2005. North American stratigraphic code. AAPG Bulletin 89:1547–1591.

Papavero, N., J. Llorente-Bousquets, and J. M. Abe. 2001. Proposal of a new system of nomenclature for phylogenetic systematics. Arquivos de Zoologia Museu de Zoologia da Univeristade de São Paulo 36:1–145.

Pough, F. H., R. M. Andrews, M. L. Crump, A. H. Savitzky, K. D. Wells, and M. C. Brandley. 2018. Herpetology. Sinauer Associates, Sunderland, MA, xv + 591 pp.

Sereno, P. C. 2005. The logical basis of phylogenetic taxonomy. Systematic Biology 54:595–619.

Smith, A. B. and C. Patterson. 1988. The influence of taxonomic method on the perception of patterns of evolution; pp. 127–216 in M. K. Hecht and B. Wallace (eds.), Evolutionary Biology. Plenum Press, New York.

Sues, H.-D. 2019. The rise of reptiles: 320 million years of evolution. John Hopkins University Press, Baltimore, xi + 385 pp.

Turland, N. J., J. H. Wiersema, F. R. Barrie, W. Greuter, D. Hawksworth, P. S. Herendeen, S. Knapp, W.-H. Kusber, D.-Z. Li, K. Marhold, T. May, J. McNeill, A. Monro, J. Prado, M. Price, and G. Smith. 2018. International Code of Nomenclature for algae, fungi, and plants (Shenzhen Code) adopted by the Nineteenth International Botanical Congress Shenzhen, China, July 2017. Koeltz Botanical Books, Glashütten, xxxviii + 254 pp.

Westheide, W. and G. Rieger, eds. 2009. Spezielle Zoologie. Teil 2: Wirbel-oder Schädeltiere. Vol. 2. Springer-Verlag, Heidelberg.

# Glossary

**Absolute rank (noun):** See **rank**.

**Anagenesis (noun):** Evolutionary change in a **lineage** that does not result in splitting of the lineage. In this book, anagenesis is not considered to lead to speciation.

**Apomorphy (noun):** Derived (new) character-state. Antonym of **plesiomorphy**. For instance, for the character "vertebrate appendage," if considered as binary, the state "fin" is a **plesiomorphy** (primitive state) whereas the state "limb with digits" is an **apomorphy** (derived state) because the fin gave rise to the limb through evolution. A shared derived character is a **synapomorphy**, whereas a derived character unique to a taxon is an **autapomorphy**.

**Binominal (adjective):** Composed of two names. Rank-based nomenclature extensively uses binominal names (composed of genus name and a specific epithet), often called "binomial" names, an expression avoided in this book (except in quotes and titles of previous publications), partly because "binomial" has a different meaning in mathematics.

**Branch length (noun):** Evolutionary time or amount of evolutionary change represented by the branch. The latter is typically expressed in number of nucleotide substitutions, or number of changes in discrete phenotypic characters. An **evolutionary tree** consists of a **topology** and a set of **branch lengths**.

**Character (noun):** An attribute of biological organisms. In systematics, "taxonomic characters" have a distribution that suggests that they can be used to delimit taxa or to assess their affinities.

**Clade (noun):** A group of organisms that includes all the descendants of their last common ancestor. Each clade is composed of one or more **lineages**. Thus, the taxon *Mammalia* is a clade composed of thousands of **lineages**.

**Cladistics (noun):** Systematic method that classifies biological organisms on the basis of shared common ancestry. It recognizes only **monophyletic** taxa (**clades**), with a possible exception for species (according to some authors). The phylogenetic inferences used by cladists are typically based on **parsimony**, although a minority of authors also use a variant called three-taxon analysis, often abbreviated as 3ia.

**Cladogenesis (noun):** Splitting of a lineage into two. In this book, speciation is equivalent to cladogenesis.

**Cladogram (noun):** A simple diagram that represents nested patterns of relationships between taxa. The terminal taxa in cladograms are always represented as sister-groups of a terminal taxon or of a more inclusive taxon (composed of several terminal taxa), but some terminal taxa could actually be ancestors of other included taxa; neither this kind of information, nor branch lengths, are represented in cladograms, contrary to **evolutionary trees**.

**Class (noun):** a group of entities that can be unambiguously defined by essential and sufficient properties. Examples includes kinds of atoms, like H (hydrogen) or helium (He), which are defined by the number of protons in their nucleus. Elements of classes are often indistinguishable. Taxa were formerly considered to form classes (for instance, birds have been conceptualized as the class of vertebrates covered by feathers), but many systematists now consider them to be **individuals** instead.

**Classification (noun):** An arrangement of entities into sets. An indented list of **taxa** (where the indentation level reflects relative ranks) and a cladogram are two common forms of classifications used in systematics.

**Consensus tree (noun):** A tree that retains either only the clades found in all source trees (**strict consensus**), or only those found in at least a given frequency (typically 50% or more; **majority-rule consensus**) of these source trees. These are used when an analysis (typically, parsimony-based) yields more than one equally optimal trees.

**Crown clade (noun):** See Crown group.

**Crown group (noun):** A clade (also called **crown clade**) that stems from the last common ancestor of two extant taxa. Another way of stating this is that a crown-group is the smallest clade that includes two extant taxa. For instance, the taxon *Tetrapoda* is a crown-group if it is considered to be the smallest clade that includes *Lissamphibia* (the smallest clade that includes extant amphibians) and *Amniota* (the smallest clade that includes mammals and reptiles, the latter including birds).

**Delimited (adjective):** Applies to concepts (in this book, mostly about taxa) that have sharp boundaries. These include taxa defined through phylogenetic nomenclature, geographical entities such as countries or states, or geochronological units for which GSSPs have been set.

**Derived (adjective):** New, recent, antonym of **primitive**. Applies to character states. For instance, for the character "vertebrate appendage," if considered as binary, the state "fin" is **primitive** compared to the state "limb with digits," which is **derived** because the fin gave rise to the limb through evolution.

**Ectothermic (adjective):** Applies to organisms that cannot generate enough heat to maintain a stable body temperature over a wide range of environmental conditions. Such organisms largely thermoregulate (adjust their body temperature) through their behavior. For instance, snakes bask in the morning sun if they need to increase their body temperature and retire into burrows or into the shade if they get too hot. Most turtles, squamates and crocodilians similarly bask in the sun to heat up.

**Endothermic (adjective):** Applies to organisms that generate enough body heat to maintain a stable body temperature over a wide range of environmental conditions. Such taxa (represented today mostly by birds and mammals) have a faster metabolism than their ectothermic relatives, and often feature thermal insulation (fur or feathers, sometimes complemented by a subcutaneous layer of fat, as in whales).

**Evolutionary tree (noun):** Graphic depiction of evolutionary relationships that includes both a topology and a set of branch lengths. Such trees depict kinship relationships and the relative or absolute ages of last common ancestors represented by internal nodes.

**GSSP (noun):** Acronym for "Global Stratotype Section and Point." This is a golden spike that has been planted in a geological section to mark the bottom of a geochronological unit. GSSPs are selected after careful consideration by the geological community of various relevant criteria to serve as a global standard. The upper end of the geochronological unit is marked by the GSSP that defines the base of the overlying unit.

**Individual (noun):** A unique kind of entity that is typically spatio-temporally bounded. For instance, taxa can be considered individuals because each is unique, and each is bounded in time (with an origin and an extinction) and in space (each inhabits only parts of the Earth, sometimes even a very small part of our planet). Individuals are better defined by their history than by intrinsic properties. For instance, the taxon *Tetrapoda* can better be defined by its history (that is, the divergence between *Amniota* and *Lissamphibia*) than by the presence of four limbs, as shown by the fact that gymnophionans and snakes are universally considered to be tetrapods, even though they lack limbs.

**Kingdom (noun):** A high-ranking taxon that many systematists long viewed as belonging to the most inclusive unit of biodiversity. In biology, kingdoms long included only *Animalia* and *Plantae*. The kingdom is no longer the highest category because many biologists now recognize domains, which typically include *Archaea, Bacteria* and *Eukarya*. A second meaning is the nomenclatural level itself, between domain and phylum. All kingdoms in the first sense are considered to belong to this nomenclatural level.

**Lineage (noun):** A branch of the Tree of Life composed of a series of organisms that are ancestors or descendants of each other. In sexually reproducing organisms, cohesion of the lineage is ensured by gene flow within each lineage, but there is minimal gene flow between lineages. No internal branching structure should be present in a lineage at the taxonomic level, even though branching may be present at lower levels, such as cell or organelle lineages within organisms. **Lineages** should not be confused with **clades**, which may include one or more lineages. Thus, the taxon name *Homo sapiens* can designate both a

clade and a **lineage**, but the more inclusive taxon *Hominidae* (which also includes other nominal species of *Homo*, and other nominal genera of hominids, such as *Australopithecus*) is a **clade** composed of several **lineages**.

**Majority-rule consensus:** See **consensus tree**.

**Maximal nomenclatural stability (abbreviated MaNS):** nomenclatural stability (in the relationship between name and taxonomic content) that can be achieved if systematists agree spontaneously on nomenclatural issues without being thus constrained by nomenclatural rules and principles. This can be opposed to **minimal** and **realized nomenclatural stability**.

**Minimal nomenclatural stability (abbreviated MiNS):** nomenclatural stability (in the relationship between name and taxonomic content) automatically provided by a set of nomenclatural rules and principles if systematists abide by these rules and principles. This can be opposed to **maximal** and **realized nomenclatural stability**.

**Monophyletic (adjective):** Applied to groups of organisms (taxa), signifies that it includes all the descendants of their last common ancestor. A monophyletic group is a **clade**. Examples include birds, mammals and primates.

**Monophyly (noun):** See **monophyletic**.

**Morphocline (noun):** Variation in a character that suggests that it typically changes through small jumps (for discrete, meristic characters such as digit, vertebral or tooth counts) or gradually (for continuous characters, such as body size and shape). Not all discrete characters form clines. For instance, the number of axes of symmetry (often materialized by petals) in flowers may not form a cline because it may be easier to double the number of such axes (from three to six, or four to eight) than to add or subtract one (from three to four, for instance).

**Node (noun):** End point of branches of evolutionary trees. Nodes can be terminal, in which case they correspond to terminal taxa, or internal (where branches meet), in which case they represent hypothetical last common ancestors.

**Node height (noun):** Distance between a node and the root of an evolutionary tree. These can be measured in evolutionary time or in evolutionary change. The height can be obtained by adding the length of the branches that connect a given node to the root of the tree.

**Nomenclature (noun):** A set of terms, often technical. In biology, in addition to a large number of technical terms, including taxon names, "nomenclature" also designates the set of rules that dictate how names (especially taxon names) are formed and applied.

**Ontogenetic (adjective):** Which pertains to **ontogeny**. In systematics, characters are sometimes stated to reflect ontogeny if they change substantially through normal development.

**Ontogeny (noun):** The development of an organism from inception (zygote formation, for sexually reproducing organism) to its death.

**Panclade name (noun):** A name for a total group (that is, a **crown group** and all its stem group) formed by adding the prefix "*Pan-*" to the name of its **crown group**, which retains its capital first letter (see Article 10.3 of the *PhyloCode*).

**Paraphyletic (adjective):** Applied to groups of organisms (taxa), signifies that it includes some, but not all, of the descendants of the last common ancestor of group. Examples include fishes (which would need to include tetrapods to be monophyletic) and invertebrates (which would need to include vertebrates to be monophyletic).

**Parsimony (noun):** Principle used to infer evolutionary relationships (in the form of cladograms) based on minimizing the number of evolutionary steps (transformations from one state to another). Also used to infer character history, also by minimizing the number of steps.

**Phylogenetic nomenclature (noun):** A recent kind of biological nomenclature (abbreviated PN throughout the book) that defines taxon names using species (or specimens) and apomorphies (derived characters) and ignores absolute ranks (Linnaean categories), contrary to **rank-based nomenclature**. Phylogenetic nomenclature is currently regulated by the *PhyloCode*.

**Phylogenetic signal (noun):** Tendency for closely related taxa to resemble each other more closely than to more distantly related taxa. This can be evaluated on an individual character or on several characters at a time (a whole matrix or a partition thereof).

**Plesiomorphy (noun):** Primitive (old) character-state. Antonym of **apomorphy**. For instance, for the character "vertebrate appendage," if considered as binary, the state "fin" is a **plesiomorphy** (primitive) whereas the state "limb with digits" is an **apomorphy** (derived) because the fin gave rise to the limb through evolution.

**Polyphyletic (adjective):** Applied to groups of organisms (taxa), signifies that it includes two or more distantly related groups of organisms. Examples include worms, which include various animals without appendages, and pachyderms, which include thick-skinned mammals. Systematists have long refrained from using polyphyletic taxa.

**Primitive (adjective):** Ancient, antonym of **derived**. Applies to character states. For instance, for the character "vertebrate appendage," if considered as binary, the state "fin" is **primitive** compared to the state "limb with digits," which is **derived** because the fin gave rise to the limb through evolution.

**Rank (noun):** Hierarchical position in a taxonomy. All taxonomies have at least **relative ranks**, meaning that a given taxon includes other taxa of a lower rank, and is included in a more inclusive taxon of a higher rank. Under rank-based nomenclature, taxa are assigned an **absolute rank**, also called Linnaean category (such as species, genus, family, order, class, phylum, regnum and so on).

**Rank-based nomenclature (noun):** The nomenclatural system currently used by most systematists. Under that system (abbreviated RN throughout the book), taxon names are defined by a rank (often called "Linnaean category") and a type, which may be a specimen (for species) or a low-ranking taxon for a higher-ranking taxon. Thus, under the rank-based codes, a genus is typified by a species, and a family is typified by a genus.

**Relative rank (noun):** See **rank**.

**Realized nomenclatural stability:** nomenclatural stability (in the relationship between name and taxonomic content) that has been achieved among systematists for a set of names and taxa and over a given period. This can be opposed to **minimal** and **maximal nomenclatural stability**.

**Semaphoront (noun):** Term introduced by Hennig to refer to a specimen on which characters can be observed, and which is taken to represent a taxon and an ontogenetic stage. The ontogenetic dimension is important because characters displayed by an individual over the course of its life, from zygote to death, can vary tremendously.

**Skewness (of tree length distribution; noun):** Asymmetry on either side of the mode. A normal distribution (in the statistical sense) is symmetrical, with both tails of the distribution being equally far (at a given frequency) from the mode, which also coincides with the mean. In an asymmetrical distribution, one of the tails is farther from the mode (which does not coincide with the mean) than the other.

**Species (noun):** A low-ranking taxon that many systematists view as the basic unit of biodiversity. A second meaning is a nomenclatural level between genus and subspecies. All species in the first sense are considered to belong to this nomenclatural level. A third meaning is an ontological concept that systematists associate with species, such as the so-called "biological species concept" (abbreviated BSC in this book).

**Specifier (noun):** A specimen (normally a type-specimen), species or apomorphy that is used in the phylogenetic definition of a taxon name. Specifiers are thus vaguely analogous to **types** in RN, with several important differences: there are external specifiers, apomorphies can be specifiers (though apomorphy-based definitions also require having at least specimen or **species** as an additional specifier) and use of at least two specifiers (mandatory under PN) delimits **taxa**.

**Squamata (noun):** A clade that includes snakes, amphisbaenians and "lizards" (paraphyletic). Lizards (sometimes still formally recognized as the taxon *Lacertilia*) is a paraphyletic group because some lizards are more closely related to snakes than to other lizards.

**Stem group (noun):** Paraphyletic group that includes all extinct taxa and direct ancestors (the **stem lineage**, which is often undocumented in the fossil record) of a given **crown group**. Thus, for *Mammalia* (when considered as a **crown group**), the stem group includes all the taxa that are still sometimes called "pelycosaurs" (a paraphyletic group that includes caseids, ophiacodontids and sphenacodontids, among others) and therapsids (this taxon has been recently redefined as a clade that encompasses mammals, but initially excluded them and included Permian and Triassic taxa such as *Dinocephalia, Dicynodontia, Gorgonopsia* and *Therocephalia*), in addition to the unobserved **stem lineage**.

**Stem lineage (noun):** The lineage that includes direct ancestors of a taxon. It is often undocumented or difficult to recognize in the fossil record because identifying ancestors relies partly on negative evidence (the absence of autapomorphies), which is inherently weak, but it necessarily existed.

**Strict consensus (noun):** See **consensus tree**.

**Synapomorphy (noun):** See **apomorphy**.

**System (noun):** In systematics, this term designates relationships between taxa, but systems include more than the binary inclusion/exclusion information contained in **classifications**. Thus, maps and evolutionary trees are systems because they include information such as distances between cities and borders between states (for the former) or evolutionary distance (expressed in amount of time or quantity of change) between **taxa**. On the contrary, a cladogram, which displays no such information and can be represented by a nested set of taxa, is a classification.

**Taxon (plural "taxa"; noun):** In systematics, a group of biological organisms that is formally recognized by taxonomists. In the last decades, systematists have come to realize that taxa (species excepted) should be monophyletic, but this was not always the case. Thus, *Mammalia* and *Aves* are monophyletic, but other taxa that have since been dismantled, such as *Invertebrata* and *Pisces* (fishes), were paraphyletic.

**Topology (noun):** Information indicating evolutionary relationships between terminal taxa on a cladogram or an **evolutionary tree**. This information is only relative, and can be depicted in the form: ((A,B)C). This indicates that taxa A and B are more closely related to each other than either is to taxon C, but this does not indicate when the last common ancestors of these three taxa lived; such information can be conveyed by **branch lengths** if they reflect evolutionary time.

**Total group (noun):** Clade composed of a **crown group** and its **stem group**. For instance, the total group of *Mammalia* includes, in addition to mammals, all extinct taxa (such as ophiacodontids, edaphosaurids, sphenacodontids and dicynodonts) that are more closely related to mammals than to any crown group (in this case, *Reptilia*).

**Tree (noun):** See **evolutionary tree**.

**Type (noun):** In biology, this word has at least four meanings. **1)** Historically, this refers to an idealized concept of a taxon, which harks back at least to Antique Greek philosophers. Modern biology, including RN, is not typological in this sense. **2)** Sometimes, "type" is used in the sense of "typical," meaning normal or average. **3)** In RN, a type is either a specimen that is used to define a species name (and on which characters can be observed, sometimes distinguished as a "taxonomic type"), or **4)** a low-ranking taxon that is used to define the name of a higher-ranking taxon (such as a type-species is used to define a genus name and a type-genus is used to define a family name). These third and fourth meanings of "type" are used throughout the book, unless stated otherwise.

**Undelimited (adjective):** Applies to concepts (in this book, mostly about taxa) that have fuzzy boundaries. These include taxa defined through rank-based nomenclature or geochronological units for which GSSPs have not been selected.

**Weighting (of characters; noun):** Procedure that places greater importance on some characters than on others to establish phylogenetic trees. By default, in parsimony analyses, all characters are equally weighted.

# Index

Note: Page locators in **bold** indicate a table. Page locators in *italics* indicate a figure